# Studies in Fuzziness and Soft Computing

Volume 429

The series "Studies in Fuzziness and Soft Computing" contains publications on various topics in the area of soft computing, which include fuzzy sets, rough sets, neural networks, evolutionary computation, probabilistic and evidential reasoning, multi-valued logic, and related fields. The publications within "Studies in Fuzziness and Soft Computing" are primarily monographs and edited volumes. They cover significant recent developments in the field, both of a foundational and applicable character. An important feature of the series is its short publication time and world-wide distribution. This permits a rapid and broad dissemination of research results.

Indexed by SCOPUS, DBLP, WTI Frankfurt eG, zbMATH, SCImago.

All books published in the series are submitted for consideration in Web of Science.

Sunil Mathew · John N. Mordeson · M. Binu

# Weighted and Fuzzy Graph Theory

Springer

Sunil Mathew
Department of Mathematics
National Institute of Technology Calicut
Calicut, Kerala, India

John N. Mordeson
Department of Mathematics
Creighton University
Omaha, NE, USA

M. Binu
Department of Mathematics and Statistics
St. Albert's College (Autonomous)
Ernakulam, Kerala, India

ISSN 1434-9922 ISSN 1860-0808 (electronic)
Studies in Fuzziness and Soft Computing
ISBN 978-3-031-39755-4 ISBN 978-3-031-39756-1 (eBook)
https://doi.org/10.1007/978-3-031-39756-1



This Springer imprint is published by the registered company Springer Nature Switzerland AG
The registered company address is: Gewerbestrasse 11, 6330 Cham, Switzerland

*Sunil Mathew would like to dedicate the book to his wife Sonia.*

*John N. Mordeson would like to dedicate the book to his family.*

*Binu would like to dedicate the book to her husband Manesh Madhavan.*

# Preface I

The most serious threat the world is facing today may be climate change. All members of the United Nations adopted Agenda 2030. Climate change has a negative influence on the terrible crimes of human trafficking and modern slavery. There have been many strong papers written on these issues. One paper may use linguistics such as low, medium, and high to measure a county's achievement of various goals or targets. Another research article may use colors to measure a country's achievement. Yet another may use numbers to measure achievement. The issues of sustainability, climate change, human trafficking, and modern slavery are good candidates for the use of Mathematics of uncertainty due to the lack of accurate data available.

Climate change will mean one emergency after another, year after year, as heat waves, floods, fires, and storms create failures through our systems. Climate change is also going to damage the networks that make up our civilization. Its effects will be long-lasting. The networks involved are the transportation network, energy network, economics network, and healthcare network.

The US Centers for Disease Control and Prevention (CDCP) estimates that three out of four new infectious diseases come from human-animal contact. Certain outbreaks are triggered by a jump from animal to human in disturbed habitats. These disruptions are caused in part by global warming.

Human livelihoods, stable economies, good health, and high quality of life all hinge on a stable climate and Earth system, and on a diversity of species and ecosystems. Yet biodiversity is declining faster than any point in human history and time is running to limit global temperature rise. Climate change significantly impacts human trafficking and modern slavery. However, it is also the case that modern slavery has a negative impact on climate change. Accurate data concerning human trafficking and modern slavery is impossible to obtain.

Human trafficking is a multi-billion dollar criminal industry that denies freedom to nearly 25 million people around the world. Accurate data concerning the flow of trafficking in persons is impossible to obtain due to the very nature of the problem. The goal of the trafficker is to be undetected. The size of the problem also makes it very difficult to obtain accurate data. The victims' reluctance to report crimes or to testify for fear of reprisals. The notions of vulnerability of countries and government response data of these countries have been made available by various studies. However, none of these studies involved the amount of flow from country to country.

Different types of fuzzy sets have been used to examine the problems mentioned above, in particular fuzzy graphs. In an extremely important paper by Klement and Mesiar, it is shown that many of these families of fuzzy sets are lattice isomorphic. Consequently, theoretical results from one family can be immediately carried over to another. The question becomes, can results from applications be immediately carried over? We first consider an application from modern slavery involving government response and vulnerability. The initial data was taken from the Walk Free Foundation. It was normalized to determine a fuzzy measure for the government response of a country, $\mu_G$; the country's vulnerability, $\mu_V$; and the county's prevalence, $\mu_p$; for modern slavery. For the country Brazil, it was found that $\mu_G(\mathit{Brazil}) = 0.684$; $\mu_V(\mathit{Brazil}) = 0{:}441$; and $\mu_p$ $(\mathit{Brazil}) = 0{:}255$: It should be noted that as government response of a country increases its vulnerability and prevalence decreases. Hence, government response is in a sense the opposite of vulnerability and prevalence. We see that $\mu_G(\mathit{Brazil}) + \mu_V(\mathit{Brazil}) > 1$: Thus, we have an example of a neutrosophic set that occurred naturally from real-world data. Also $\mu_G(\mathit{Brazil})^2 + \mu_V(\mathit{Brazil})^2 < 1$ and so we also have a Pythagorean fuzzy set. Now, the lattice isomorphism between Pythagorean fuzzy sets and intuitionistic fuzzy sets is essentially $f(x_1, x_2) = (x_1^2, x_2^2)$ where $x_1, x_2 \in [0, 1]$. Now, $f$ preserves strict inequality and $f(0, 0) = (0, 0)$, $f(1, 1) = (1, 1)$ and so applied data could be easily converted from Pythagorean fuzzy data to intuitionistic fuzzy data. However, this does not seem to be the case for neutrosophic fuzzy sets, at least not at first glance. An interesting research project would be to consider this situation for other lattice isomorphic families of fuzzy sets.

A research group headed by Prof. Sunil Mathew has been using weighted and fuzzy graph theory to examine some of the above global problems. The research group consists of mostly Ph.D. candidates and Ph.D. graduates. This group has developed many interesting results concerning weighted and fuzzy graphs. It has been an honor to work with this group of talented researchers.

Omaha, USA John N. Mordeson

# Preface II

Three hundred years before Leonard Euler's work inspired many to think about connections between points, graphs were found to be a convenient way of expressing relations in a very systematic manner. Graphs stand at the top of mathematics and computer science as a very useful device today. As in the case of geometry, Euclidean Geometry portrays an idealization of objects seen in nature. Graphs do not really represent any object in this nature. Graph theory has grown into several different dimensions. The era of connections is gone and that of connectivity took its place. Large interconnection networks like the internet emerged. Capacities of vertices and edges of a graph became very important. The flow enhancement became essential. Even though computer scientists have come up with several problems and algorithms like bottleneck problems, quality of service problems, maximum bandwidth problems, etc., there is no strongly supporting mathematical theory even today. This book is a humble attempt to start from the scraps—from very old to very new. The authors do believe that this is the first book written solely on weighted graphs. Fuzzy graph theory also emerged in the recent times to deal with normalized weights. This motivated many authors to work in parallel with weighted versions. Some of the chapters in this book were written exclusively for weighted graphs and some are weighted adaptations of fuzzy graph findings. We are delighted to present this book to all graph theorists and computer scientists.

In Chap. 1, we present some of the classic works on weighted graphs. It mainly covers the works by Bondy, Dirac, Erdos, Fan, and Broersma. Chapter 1 also provides an introduction to graphs and weighted graphs. Heavy paths, heavy cycles, tritrees, and weighted digraphs are presented. Optimal cycles are characterized. Weighted Hamilton paths and cycles are studied extensively. The longest heaviest paths in blocks are identified. Section 1.4 provides a detailed discussion on weighted trees. Path extremal graphs are also discussed. Section 1.5 gives a discussion on weighted digraphs.

In Chap. 2, we present weighted graph versions of some of the results in fuzzy graphs. Introduction to partial cut vertices, bridges, partial trees, and partial blocks are made. A number of characterizations for these structures are also presented.

Chapter 3 continues further with connectivity concepts in weighted graphs. Precisely weighted graphs are studied and totally weighted networks are discussed in detail. Using partial cut sets, Menger's theorem is generalized. Vertex connectivity and edge connectivity of graphs are generalized in weighted graph point of view.

In Chap. 4, we discuss an unexplored concept in graph theory, namely the cycle connectivity. In an unweighted graph, cycle connectivity between two vertices is either 1 or 0, depending on whether a cycle passes through the pair or not. But it may be different in weighted graphs as cycles of different strengths can pass through such a pair. Cycle connectivity of graphs is examined in detail. Cyclic cutvertices and cyclic bridges are studied and a detailed study of cyclically balanced weighted graphs is provided.

In Chap. 5, different distances in weighted graphs are discussed. Self-centered weighted graphs are studied with respect to some new distances. Geodetic convex sets and geodetic blocks are presented in detail. Geodetic boundary and interior are discussed in Sect. 5.8 and monophonic convex sets in Sect. 5.9. Some characterizations of monophonic blocks are provided in Sect. 5.10. Results on monophonic boundary and interior are presented in the rest of the chapter.

In Chap. 6, we focus on two major themes, degree sequences and saturation of weighted graphs. Partial blocks are characterized in Sect. 6.2 and partial trees in Sect. 6.3. Vertex and edge saturation counts are calculated for different classes of weighted graphs and the new concept of saturation is briefly analyzed. Characterizations of partial blocks and trees are also made using the saturation concept.

Chapter 7, provides the concepts of intervals in weighted graphs. Intervals in graphs were studied by Mulder, Nebesky, etc. Some of their works are also presented. Different types of distances give rise to different types of intervals in weighted graphs. Interval operations are discussed in Sect. 7.3 and strong gate concept and gated subgraphs in Sect. 7.4.

In Chap. 8, we follow the lead of Klement and Mesiar to discuss the various lattice isomorphisms that allows for the immediate transfer of results from one family of fuzzy sets to another. These families of fuzzy sets include intuitionistic, Pythagorean, interval-valued, neutrosophic, and others. The transfer principle for nonstandard analysis is discussed as is the determination of fuzzy algebra results from crisp abstract algebra results. We concentrate on the connection between fuzzy graphs and weighted graphs. We present an isomorphism between the two that allow for the concepts of t-norms, t-conorms, and complements to be carried over from fuzzy set theory to weighted set theory.

In Chap. 9, we use the work of Head and Weinberger to demonstrate how results of fuzzy algebra can be determined from results of crisp algebra. We present some results on undirected power graphs which demonstrate how results from algebra can be used to form graphs. We then show how these results can be used to determine results for fuzzy graphs.

Calicut, India — Sunil Mathew
Omaha, USA — John N. Mordeson
Ernakulam, India — M. Binu

# Acknowledgments

The book is dependent on the journals Combinatorica, Discrete Mathematics, Discussions in Mathematical Graph Theory, Applied Mathematical Letters, Information Processing Letters, and Proyecciones Journal of Mathematics. We thank these journals for their support of our work involving partial blocks, partial trees, saturation, and sequences in weighted graphs. The authors are grateful to the editorial board and production staff of Springer International Publishing, especially to Janusz Kacprzyk. The authors are indebted to George and Sue Haddix for their support of Creighton University and mathematics of uncertainty. We are also indebted to Etienne Kerre, Rudolf Seising, Enric Trillas, Erich Klement, and Radko Mesiar for their inspiring work and support.

Calicut, India — Sunil Mathew
Omaha, USA — John N. Mordeson
Ernakulam, India — M. Binu

# Contents


**1 Graphs and Weighted Graphs** ..... 1
1.1 Graph Theory ..... 2
1.2 Weighted Graph Theory ..... 10
1.3 Heavy Paths and Optimal Cycles ..... 11
1.4 Tritrees and Cyclic Extremal Graphs ..... 37
1.5 Weighted Digraphs ..... 47
References ..... 50

**2 Connectivity** ..... 53
2.1 Strength of Connectedness ..... 53
2.2 Strong Edges and Strong Paths ..... 57
2.3 Partial Blocks ..... 61
2.4 Partial Trees ..... 66
References ..... 69

**3 More on Connectivity** ..... 71
3.1 Precisely Weighted Graphs ..... 71
3.2 Generalized Menger's Theorem ..... 80
3.3 Generalized Connectivity Parameters ..... 85
References ..... 88

**4 Cycle Connectivity** ..... 89
4.1 Strongest Strong Cycles ..... 89
4.2 Cycle Connectivity ..... 91
4.3 Cycle Connectivity of Weighted Graphs ..... 94
4.4 Cyclic Cut Vertices and Cyclic Bridges ..... 95
4.5 Cyclically Balanced Graphs ..... 98

**5 Distance and Convexity** ..... 105
5.1 Weighted Distance ..... 105
5.2 Self Centered Weighted Graphs ..... 111
5.3 The Distance Matrix ..... 113
5.4 Weighted Center of Trees ..... 115

5.5 Complement of a Weighted Graph ... 116
5.6 Geodetic Convex Sets ... 117
5.7 Geodetic Blocks ... 119
5.8 Geodetic Boundary and Interior ... 123
5.9 Monophonic Convex Sets ... 125
5.10 Monophonic Blocks ... 127
5.11 Monophonic Boundary and Interior ... 130
References ... 132

**6 Degree Sequences and Saturation** ... 133
6.1 Sequences in a Weighted Graph ... 133
6.2 Characterization of Partial Blocks ... 134
6.3 Characterization of Partial Trees ... 137
6.4 Vertex and Edge Saturation Counts ... 141
6.5 Saturation ... 146
References ... 154

**7 Intervals and Gates** ... 155
7.1 Distances in Weighted Graphs ... 155
7.2 Intervals in Weighted Graphs ... 161
7.3 Operations on Strong Intervals ... 167
7.4 Strong Gates in Weighted Graphs ... 168
References ... 177

**8 Weighted Graphs and Fuzzy Graphs** ... 179
8.1 Isomorphisms of Different Families of Fuzzy Sets ... 181
8.2 Weighted Graphs and Fuzzy Graphs ... 183
8.3 $t$-Conorm Fuzzy Graphs ... 187
8.4 Nonstandard Analysis ... 190
8.5 Nonstandard Weighted Graphs ... 194
References ... 195

**9 Fuzzy Results from Crisp Results** ... 197
9.1 Derivation of Fuzzy Results from Crisp Results ... 197
9.2 Fuzzy Results from Crisp Results Continued ... 200
9.3 Fuzzifiable Operations ... 201
9.4 Fuzzification of Fuzzifiable Operations ... 203
9.5 $\mathcal{FP}(A)$ and $\mathcal{P}(A)$ Generate the Same Variety of Algebras ... 204
9.6 Undirected Power Graphs of Semigroups ... 206
9.7 Fuzzy Subgraphs of Undirected Power Graphs ... 207
References ... 208

**Bibliography** ... 209

**Index** ... 213

# About the Authors

**Dr. Sunil Mathew** is an Associate Professor in the Department of Mathematics, NIT Calicut, India. He has acquired his masters from St. Joseph's College Devagiri, Calicut, and Ph.D. from the National Institute of Technology Calicut in the area of Fuzzy Graph Theory. He has published over 125 research papers and written 10 books. He is a member of several academic bodies and associations. He is editor and reviewer of several international journals. He has an experience of more than 20 years in teaching and research. His current research topics include fuzzy graph theory, bio-computational modeling, graph theory, fractal geometry, and chaos.

**Dr. John N. Mordeson** is a Professor Emeritus of Mathematics at Creighton University. He received his B.S., M.S., and Ph.D. from Iowa State University. He is a member of Phi Kappa Phi. He has published over 20 books and over 200 journal articles. He is on the editorial board of numerous journals. He has served as an external examiner of Ph.D. candidates from India, South Africa, Bulgaria, and Pakistan. He has refereed for numerous journals and granting agencies. He is particularly interested in applying mathematics of uncertainty to combat the problem of human trafficking.

**Dr. M. Binu** received her Ph.D. in 2019 from the Department of Mathematics, National Institute of Technology Calicut, India, in the area of Connectivity in Fuzzy Graph Theory. She is an Assistant Professor in Mathematics at the Department of Mathematics and Statistics, St. Albert's College (Autonomous), Ernakulam, Kerala, India. Her present research includes fuzzy logic, graph theory, and network science. She has published several research papers and co-authored two books.

# Chapter 1
# Graphs and Weighted Graphs

The theory of graphs has progressed a lot after Leonhard Euler used a graph-theoretic approach to solve the seven Konigsberg bridge problem in 1736. One of the most preeminent ways of applying mathematics in real-world scenario involves modeling using graph theory. It is because of its instinctual diagrammatic portrayal. Graphs have been found to be helpful in modeling real-world problems. Graphs are convenient to use in modeling of problems in numerous situations such as transportation problems, electrical circuits, networking, facility location problem and traffic flow problem. A natural way of modeling using graphs would be to consider the vertices of a graph as cities and edges as roads, telephone lines or pipelines between these cities. To be more precise, any situation involving a relation between entities can be represented by a graph. For example, graphs are helpful in representing chemical bonds in molecules, transfer of signals to brain cells, the descent of species, etc. Also, in the case of a large construction project, the vertices can be used to represent the different stages of construction. If one stage arises from another, then the vertices corresponding to these stages are connected by an edge.

A graph can be undirected or directed depending on whether the pairwise relationships among objects are symmetric or not. Nevertheless, in many real-world situations, representing a set of complex relational objects as directed or undirected graphs is not complete. Weighted graphs, a natural generalization of graphs, offer a framework that helps to overcome such conceptual limitations. As the name indicates, weighted graphs generalize graphs by allowing edges to have real numbers as weights. There are several discrete optimization related problems in a variety of disciplines such as electronics engineering, electrical engineering, statistics, operations research, computer science, and combinatorics, which can be modeled by weighted graph structures. Classic applications contain finding a minimum cost route or set of routes for delivery vehicles, designing of least cost telecommunication systems, maximizing throughput in a manufacturing system, and distributing electricity from a group of supply points to meet customer demands at minimum cost.



S. Mathew et al., *Weighted and Fuzzy Graph Theory*, Studies in Fuzziness and Soft Computing 429, https://doi.org/10.1007/978-3-031-39756-1_1

## 1.1 Graph Theory

Different from many branches of mathematics, the theory of graphs has an exact birthdate. The first paper on graph theory was published by Euler in 1736. The first book in the literature of graph theory was written by Konig[1]. Thereafter, Berge [2], Ore [3] and Harary [4] came up with the first set of books in this area. Later, in 1976, Roberts [5] has dealt with a variety of applications of graphs in different fields such as engineering, environmental sciences, ecology and biological sciences.

Graph theory, though a relative newcomer in the spectrum of mathematics, liberally extends its helping hand to many other areas of knowledge and chemistry is not an exception. Indeed Mathematical Chemistry uses the fruits of graph theory at its maximum. The concept of the energy of a graph was proposed quite some time ago where the first result was obtained as early as the 1940s [6]. The name was chosen because for some special graphs that represent certain types of molecules, graph energy is related to a part of their electron energy. After a long latent period, the concept of the energy of graphs attracted the attention of a large number of researchers and it became a popular topic of research, especially among mathematicians. One of the important classes of graphs is trees. Trees have been widely used in computer science to represent hierarchical data which makes storage and recollection easier. Tree structure was first studied by Cayley in 1857, who also termed it as kenograms [7]. The significance of trees is apparent from its applications in various fields, especially molecular evolution, analysis of algorithms, the compilation of algebraic expressions, theoretical models of computation, etc. Trees are very often used to represent hierarchical data and are the central structure for storing and organizing data in computer science. The dynamic expansion of graph theory has led to the development of many significant and applicable subareas with its own concepts and theorems such as metric graph theory [8], algebraic graph theory [9], algorithmic graph theory [10] and spectral graph theory [11].

A **graph** $G$ consists of a finite nonempty set $V$ of $p$ points (vertices) together with a prescribed set $X$ of $q$ unordered pairs of distinct points (vertices) of $V$. Each pair $x = \{u, v\}$ of points in $X$ is a line (edge) of $G$, and $x$ is said to join $u$ and $v$. We write $x = uv$ and say that $u$ and $v$ are **adjacent vertices** (sometimes denoted $u$ adj $v$); point $u$ and edge $x$ are incident with each other, as are $v$ and $x$. If two distinct edges $x$ and $y$ are incident with a common point, then they are **adjacent edges.** A graph with $p$ points and $q$ edges is called a $(p, q)$ graph. The $(1, 0)$ graph is **trivial**.

The cardinality of the vertex set of a graph $G$ is called the **order** of $G$ and is commonly denoted by $n(G)$ or more simply by $n$ if no confusion occurs, while the cardinality of its edge set is the **size** of $G$ and is often denoted by $m(G)$ or $m$. An $(n, m)$ graph has order $n$ and size $m$.

The **degree of a vertex** $v$ in a graph $G$ is the number of edges of $G$ incident with $v$, which is denoted by $deg_G\ v$ or simply $deg\ v$. A vertex is called even or odd according to whether its degree is even or odd. A vertex of degree 0 is called an **isolated vertex** and a vertex of degree 1 is an **end-vertex** (or **pendant vertex**).

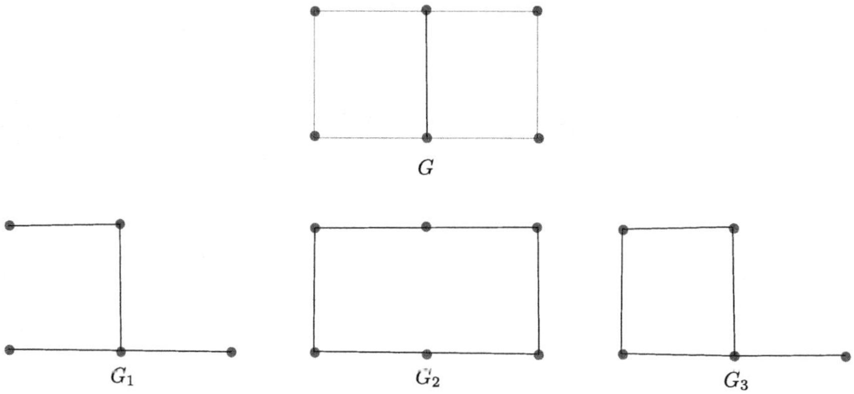


**Fig. 1.1** An illustration of different types of subgraphs

The **minimum degree of** $G$ is the minimum degree among the vertices of $G$ and is denoted by $\delta(G)$. The **maximum degree** is defined similarly and is denoted by $\Delta(G)$.

A **subgraph** of $G$ is a graph having all of its vertices and edges in $G$. It is a **spanning subgraph** if it contains all the vertices of $G$. If $H$ is a subgraph of $G$, then $G$ is a **supergraph** of $H$. For any set $S$ of vertices in $G$, the **induced subgraph** $\langle S \rangle$ is the maximal subgraph with vertex set $S$. Thus two vertices of $S$ are adjacent in $\langle S \rangle$ if and only if they are adjacent in $G$.

In Fig. 1.1, $G$ is a graph with 6 vertices and 7 edges. $G_1$, $G_2$ and $G_3$ are subgraphs of $G$. $G_1$ is neither a spanning subgraph of $G$, since it does not contain all vertices of $G$, nor an induced subgraph of $G$. $G_2$ is a spanning subgraph of $G$ as it contains all vertices of $G$. But it is not an induced subgraph of $G$. The induced subgraph of $G$ which contains all the vertices of $G$, is the graph $G$ itself. Here, $G_3$ is an induced subgraph of $G$ but not a spanning subgraph of $G$. $\delta(G) = 2$, $\Delta(G) = 3$. Also, $\delta(G_1) = 1$, $\Delta(G_1) = 3$, $\delta(G_2) = 2$, $\Delta(G_2) = 2$, $\delta(G_3) = 1$ and $\Delta(G_3) = 3$.

We have the first theorem of graph theory below. It is also known as 'the Handshaking lemma'

**Theorem 1.1.1** ([12]) *Let $G$ be an $(n, m)$ graph where $V(G) = \{v_1, v_2, v_3, \ldots, v_n\}$. Then*

$$\sum_{i=1}^{n} deg\ v_i = 2m. \tag{1.1}$$

Proof of this results is trivial from the fact that any edge in a graph $G$ contributes two towards the degree sum of the graph $G$. This result has interesting consequences. For example, it follows from the theorem that there exist an even number of odd vertices in any graph.

Edges that join the same pair of distinct vertices are called **parallel edges**. An edge represented by an unordered pair in which the two elements are not distinct is

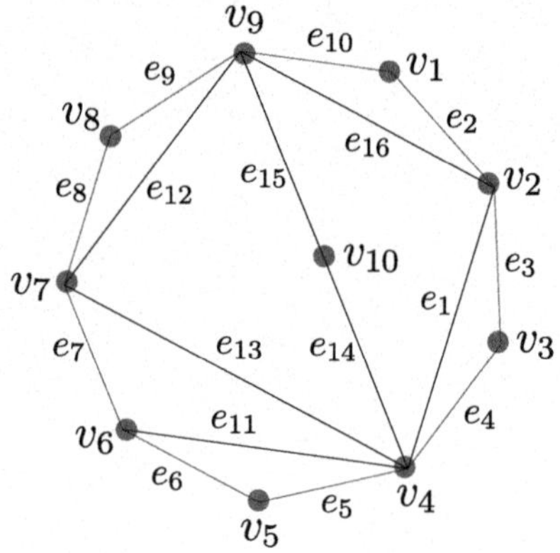


**Fig. 1.2** Graph $G$ with 10 vertices and 17 edges

known as a **loop**. A graph with no loops is a **multigraph.** A graph with at least one loop is a **pseudograph**. A **simple graph** is a graph with no parallel edges and loops.

A graph $G$ is **regular** of degree $r$ if $deg\ v = r$ for each vertex $v$ of $G$. Such graphs are called $r$**-regular** [13].

**Theorem 1.1.2** *For every graph G and every integer $r \geq \Delta(G)$, there exists an $r$-regular graph H containing G as an induced subgraph.*

The **complement** $\overline{G}$ of a simple graph $G$ is the simple graph with vertex set $V(G)$ defined by $uv \in E(\overline{G})$ if and only if $uv \notin E(G)$. A **clique** in a graph is a set of pairwise adjacent vertices. An **independent set** (or **stable set**) in a graph is a set of pairwise nonadjacent vertices.

**Example 1.1.3** Consider the graph in Fig. 1.2.

Here, $\{v_2, v_3, v_4\}$ is a clique and $\{v_1, v_3, v_5, v_7, v_{10}\}$ is an independent set.

A graph is **complete** if every two of its vertices are adjacent. A **complete** $(n, m)$ **graph** is, therefore, a regular graph of degree $n - 1$ having $m = n(n - 1)/2$. A complete graph of order $n$ is denoted by $K_n$. A **walk** is a list $v_0, e_1, v_1, \ldots, e_k, v_k$ of vertices and edges such that, for $1 < i < k$, the edge $e_i$ has endpoints $v_{i-1}$ and $v_i$. A **trail** is a walk with no repeated edge. A $u - v$ walk or $u - v$ trail has first vertex $u$ and last vertex $v$; these are its endpoints. A $u - v$ path is a path whose vertices of degree 1 (its endpoints) are $u$ and $v$; the others are internal vertices. The **length** of a walk, trail, path, or cycle is its number of edges. A walk or trail is **closed** if its endpoints are the same. Every $u - v$ walk contains a $u - v$ path [14]. In Fig. 1.2, $v_1, e_2, v_2, e_3, v_3, e_4, v_4, e_1, v_2, e_2, v_1, e_{10}, v_9$ is a walk, but not a path or trail since it has repeated vertices and edges. The walk $P : v_1, e_2, v_2, e_3, v_3, e_4, v_4, e_1, v_2$ is not a path but a trail. This is because $P$ has no repeated edges, but it has a repeated vertex $v_2$. The walk $Q : v_1, e_2, v_2, e_3, v_3, e_4, v_4, e_{14}, v_{10}$ is both a path and a trail.

A graph $G$ is $k$**-partite**, $k \geq 1$, if it is possible to partition $V(G)$ into $k$ subsets $V_1, V_2, \ldots, V_k$ such that every element of $E(G)$ joins a vertex of $V_i$ to a vertex in $V_j, i \neq j$. If $G$ is a 1-partite graph of order $n$, then $G = \overline{K}_n$. For $k = 2$, such graphs

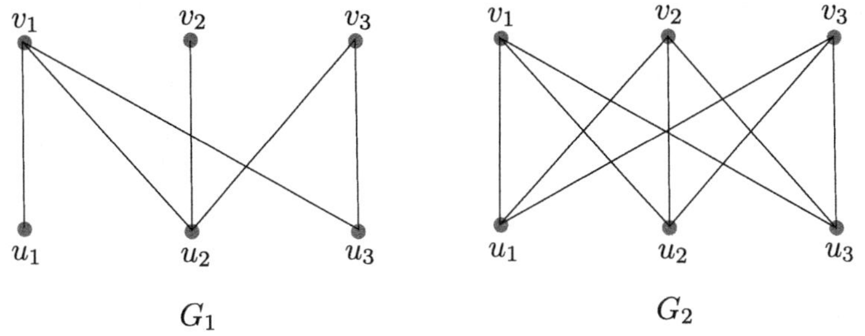


**Fig. 1.3** Bipartite graphs

are called **bipartite graphs** [13]. A **complete k-partite graph** $G$ is a $k$-partite graph with partite sets $V_1, V_2, \ldots, V_k$ having the added property that if $u \in V_i$ and $v \in V_j$, $i \neq j$, then $uv \in E(G)$. If $|V_i| = n_i$, then this graph is denoted by $K(n_1, n_2, \ldots, n_k)$ or $K_{n_1,n_2,\ldots,n_k}$. A complete $k$-partite graph is complete if and only if $n_i = 1$ for all $i$. When $k = 2$, it is known as **complete bipartite graph.** The graph $K_{1,s}$ is called a **star graph,** for the integer $k \geq 1$. In Fig. 1.3, both $G_1$ and $G_2$ are bipartite graphs. Moreover, $G_2$ is complete bipartite but $G_1$ is not.

**Theorem 1.1.4** ([13]) *A nontrivial graph is bipartite if and only if it has no odd cycles.*

A graph $G$ with vertex set $V(G) = \{v_1, v_2, \ldots, v_n\}$ and edge set $E(G) = \{e_1, e_2, \ldots, e_m\}$ can also be described by means of matrices. One such matrix is the $n \times n$ **adjacency matrix,** $A(G) = [a_{ij}]$, where

$$a_{ij} = \begin{cases} 1 \text{ if } v_i v_j \in E(G) \\ 0 \quad \text{otherwise.} \end{cases}$$

**Theorem 1.1.5** ([13]) *If $A$ is the adjacency matrix of a graph $G$ with vertex set $V(G) = \{v_1, v_2, \ldots, v_n\}$, then the $(i, j)^{th}$ entry of $A^k$, $k \geq 1$, is the number of different $v_i - v_j$ walks of length $k$ in $G$.*

A nontrivial closed trail of a graph $G$ is referred as a **circuit** of $G$, and a circuit $v_1, v_2, \ldots, v_n, v_1$ $(n \geq 3)$ whose $n$ vertices $v_i$ are distinct is called a **cycle**. An **acyclic graph** has no cycles. A cycle is **even** if its length is even; otherwise it is *odd*. A vertex $u$ is said to be **connected** to a vertex $v$ in a graph $G$ if there exists a $u - v$ path in $G$. A graph $G$ is **connected** if every two of its vertices are connected. A graph that is not connected is **disconnected.** A **component** of $G$ is a subgraph that is maximal with respect to the property of being connected. The number of components of $G$ is denoted by $k(G)$. $k(G) = 1$ if and only if $G$ is connected.

**Definition 1.1.6** ([13]) The distance $d(u, v)$ between vertices $u$ and $v$ in a connected graph $G$ is defined as the minimum of the lengths of $u - v$ paths of $G$.

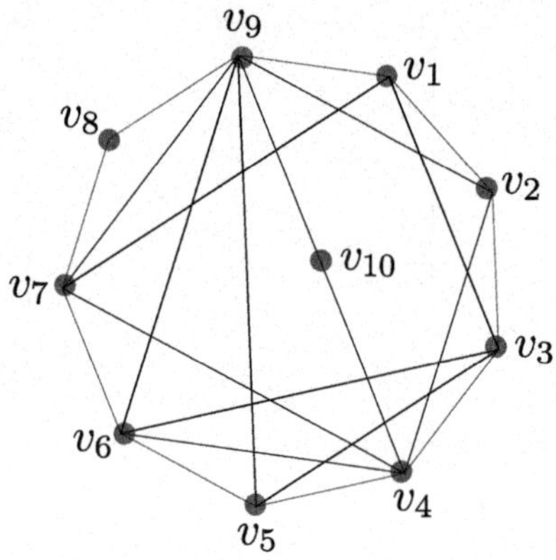


**Fig. 1.4** A simple graph $G$

Let $G$ be a connected graph and let $v$ be a vertex of $G$. The **eccentricity** $e(v)$ of $v$ is the distance to a vertex farthest from $v$. Thus $e(v) = \max\{d(v, u), u \in V\}$ and the vertex which has minimum eccentricity is known as a **central vertex**. The **radius** $r(G)$ of a graph $G$, is the minimum eccentricity of the vertices and the **diameter** $d(G)$ is the maximum eccentricity of the vertices. A graph $G$ is **self-centered** if every vertex is in the center of $G$, where the **center of** $G$ is the set of all central vertices [15].

**Example 1.1.7** In Fig. 1.4, $G$ has 10 vertices. $e(v_1) = 2$, $e(v_2) = 2$, $e(v_3) = 3$, $e(v_4) = 2$, $e(v_5) = 2$, $e(v_6) = 2$, $e(v_7) = 2$, $e(v_8) = 3$, $e(v_9) = 2$, $e(v_{10}) = 2$. Also, $r(G) = 2$ and $d(G) = 3$.

It can be shown easily that for every connected graph $G$, $r(G) \leq d(G) \leq 2r(G)$.

**Theorem 1.1.8** ([13]) *Every graph is the center of some connected graph.*

**Periphery** of a graph $G$ is the subgraph of $G$ induced by the vertices in $G$ whose eccentricity equals the diameter of $G$. We have the following result.

**Theorem 1.1.9** ([13]) *A graph $G$ is the periphery of some connected graph if and only if every vertex of $G$ has eccentricity* 1 *or no vertex of $G$ has eccentricity* 1.

**Definition 1.1.10** ([15]) The **status** $s(v)$ of a vertex $v$ in $G$ is the sum of the distances from $v$ to each other vertex in $G$. The **median** $M(G)$ of a graph $G$ is the set of vertices with minimum status.

**Example 1.1.11** In the graph $G$ given in Fig. 1.5, $s(a_1) = 8$, $s(a_2) = 8$, $s(a_3) = 8$, $s(a_4) = 8$, $s(a_5) = 5$ and $s(a_6) = 9$. Hence $M(G) = \{a_5\}$.

A vertex $v$ of a connected graph $G$ is called a **cut-vertex** of $G$ if $k(G) < k(G - v)$. In the graph given in Fig. 1.5, $k(G) = 1$ and $k(G - a_5) = 3$. So, $a_5$ is a cut-vertex.

**Theorem 1.1.12** *A vertex $v$ of a connected graph $G$ is a cut-vertex of $G$ if and only if there exist vertices $u$ and $w$ $(u,\ w \neq v)$ such that $v$ is on every $u - w$ path of $G$.*

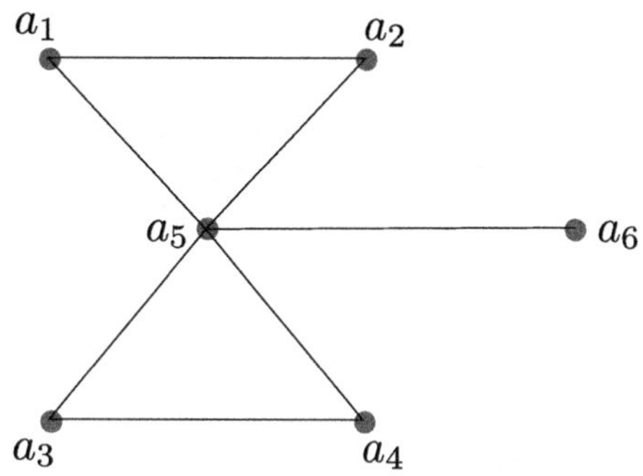


**Fig. 1.5** An illustration of status of vertices in a graph

Also it is not hard to see that every nontrivial connected graph contains at least 2 vertices that are not cut-vertices. A **bridge** of a graph $G$ is an edge $e$ such that $k(G-e) > k(G)$. If $e$ is a bridge of $G$ then $k(G-e) = k(G) + 1$. In the graph given in Fig. 1.5, the edge $a_5a_6$ is a bridge.

**Theorem 1.1.13** *An edge e of a connected graph G is a bridge of G if and only if there exist vertices u and w such that e is on every $u - w$ paths of G.*

**Theorem 1.1.14** *An edge e of a graph G is a bridge of G if and only if e lies on no cycle of G.*

A bridge incident with an end-vertex is called a **pendant edge.** The edge $a_5a_6$ is incident with the end-vertex $a_6$ in the graph given in Fig. 1.5. So it is a pendant edge. A nontrivial connected graph with no cut-vertices is called a **non-separable graph**. A **block** of a graph $G$ is a maximal non-separable subgraph of $G$. We can see that $K_n$ is a non-separable graph for any positive integer $n$.

**Theorem 1.1.15** *A graph G of order at least 3 is non-separable if and only if every two vertices of G lie on a common cycle of G.*

An **internal vertex** of a $u - v$ path $P$ is a vertex of $P$ different from $u$ and $v$. A collection $\{P_1, P_2, P_3, \ldots, P_k\}$ of paths is called **internally disjoint** if each internal vertex of $P_i$ $(i = 1, 2, \ldots, k)$ lies on no path $P_j$ $(j \neq i)$. In particular, two $u - v$ paths are internally disjoint if they have no vertices in common other than $u$ and $v$. **Edge-disjoint** $u - v$ paths have no edges in common.

**Theorem 1.1.16** *A graph G of order at least 3 is nonseparable if and only if there exist two internally disjoint $u - v$ paths for every two distinct vertices u and v of G.*

**Theorem 1.1.17** *The center of every connected graph G lies in a single block of G.*

A **tree** is an acyclic connected graph, while a **forest** is an acyclic graph. Thus every component of a forest is a tree. The graph given in Fig. 1.6 is an example for a forest in which each component is a tree.

We can observe that every nontrivial tree has at least 2 end-vertices.

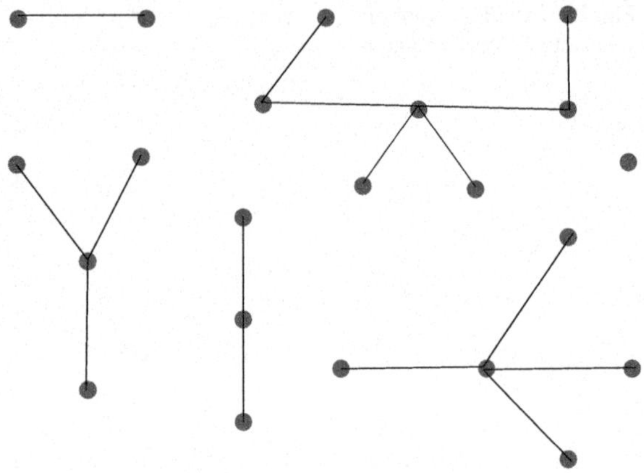

**Fig. 1.6** An example for a forest

**Theorem 1.1.18** *An $(n, m)$ graph $G$ is a tree if and only if $G$ is connected and $n = m + 1$.*

*Similarly we can show that an $(n, m)$ graph $G$ is a tree if and only if $G$ is acyclic and $n = m + 1$.*

**Theorem 1.1.19** ([16]) *A graph $G$ is a tree if and only if every two distinct vertices of $G$ are connected by a unique path of $G$.*

**Theorem 1.1.20** *Let $T$ be a nontrivial tree with $\Delta(T) = k$, and let $n_i$ be the number of vertices of degree $i$ in $T$ for $i = 1, 2, 3, \ldots, k$. Then,*

$$n_1 = n_3 + 2n_4 + 3n_5 + \cdots + (k - 2)n_k + 2. \tag{1.2}$$

**Theorem 1.1.21** *Let $n$ and $k$ be positive integers with $n \geq 2k$. Then every graph of order $n$ with $\delta(G) \geq k$ contains every forest of size $k$ without isolated vertices as a subgraph.*

We can prove that there are $n^{n-2}$ distinct labeled trees of order $n$. Let $G$ be a graph with $V(G) = \{v_1, v_2, \ldots, v_n\}$. The **degree matrix** $D(G) = [d_{ij}]$ is the $n \times n$ matrix with $d_{ii} = deg\ v_i$ and $d_{ij} = 0$ for $i \neq j$.

**Theorem 1.1.22** *If $G$ is a nontrivial labeled graph with adjacency matrix $A$ and degree matrix $D$, then the number of distinct spanning trees of $G$ is the value of any cofactor of the matrix $D - A$.*

A **vertex-cut** in a graph $G$ is a set $U$ of vertices in $G$ such that $G - U$ is disconnected. The **vertex-connectivity** or simply the **connectivity** $\kappa(G)$ of a graph $G$ is the minimum cardinality of a vertex-cut of $G$ if $G$ is not complete, and $\kappa(G) = n - 1$ if $G = K_n$ for some positive integer $n$. Hence $\kappa(G)$ is the minimum number of vertices whose removal results in a disconnected or trivial graph. A graph $G$ is said to be $k$-connected, $k \geq 1$, if $\kappa(G) \geq k$. Note that every cycle is 2-connected.

**Theorem 1.1.23** (Menger's theorem) *Let $u$ and $v$ be nonadjacent vertices in a graph $G$. Then the minimum number of vertices that separates $u$ and $v$ is equal to the maximum number of internally disjoint $u - v$ paths in $G$.*

**Theorem 1.1.24** *A nontrivial graph $G$ is $k$-connected if and only if for each pair $u, v$ of distinct vertices there are at least $k$ internally disjoint $u - v$ paths in $G$.*

**Theorem 1.1.25** *Let $G$ be a $k$-connected graph, $k \geq 2$. Then every $k$ vertices of $G$ lie on a common cycle of $G$.*

The **vertex independence number** or simply **independence number** $\beta(G)$ of a graph $G$ is the maximum cardinality among the independent sets of vertices of $G$. $\beta(K_{r,s}) = \max(r, s)$, $\beta(C_n) = \lfloor \frac{n}{2} \rfloor$ and $\beta(K_n) = 1$.

**Theorem 1.1.26** *Let $G$ be a graph of order $n \geq 3$ such that for all distinct nonadjacent vertices $u$ and $v$,*

$$deg\, u + deg\, v \geq n.$$

*Then $\kappa(G) \geq \beta(G)$.*

The line graph $L(G)$ of a graph $G$ is that graph whose vertices can be put in one-to-one correspondence with the edges of $G$ in such a way that two vertices of $L(G)$ are adjacent if and only if the corresponding edges of $G$ are adjacent. A graph $H$ is called a **line graph** if there exists a graph $G$ such that $H = L(G)$. A graph and its line graph are shown in Fig. 1.7.

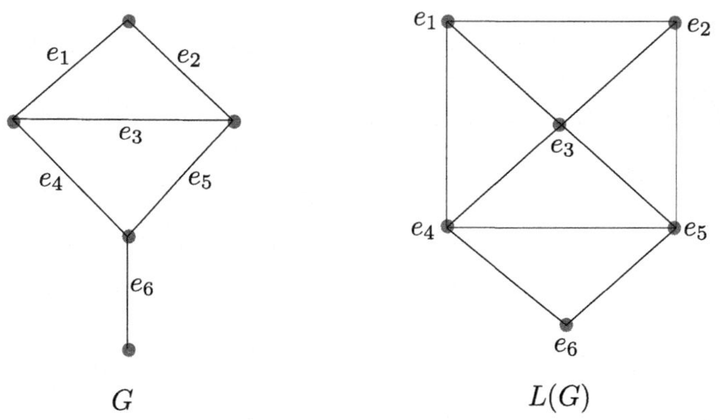


**Fig. 1.7** A graph and its line graph

## 1.2 Weighted Graph Theory

Earliest developments on weighted graph theory began with paths and cycles. In 1959, Paul Erdos studied about maximal paths and circuits in weighted graphs [17]. In 1990, Bondy and Fan discussed cycles in weighted graphs [18].

Distance and related concepts always have an indispensable role in the literature of graph theory [15]. Moreover, it is such a concept which spreads through all over graph theory. The important application of facility location on networks is based on various types of graphical centrality, all of which are defined using distance. Usually, there are two types of distances in weighted graph theory. One is the **distance between graphs** [19] and the other one is the **distance in a graph** [20, 21]. The second one is our focus of study.

There are different types of distances in a graph, such as detour distance [22], mean distance [23], average distance [24], etc. The concepts like **distance-regular graphs** [25], **distance-hereditary graphs** [26], **distance equienergetic graphs** [27], **distance-balanced graphs** [28], are also there in the literature.

In 1980, Mulder came up with a **transit function** called **interval function** [29]. Since **interval function** is something which can be used as an effective tool to describe the properties of graphs which depend on the distance between vertices, Nebesky along with Mulder worked more on this topics and gave some characterizations including axiomatic characterization of interval functions [30–32]. It is worth pointing out that intervals in graphs are extremely different from interval graphs [33]. An **interval graph** [34] is the intersection graph of a family of intervals on the real line. It has one vertex for each interval in the family, and an edge between every pair of vertices corresponding to intervals that intersect.

In 1978, Goldman and Witzgall had come up with the notion of gated sets in connection with geodesic in transportation science [35]. In several branches of mathematics, under various motivations, gated sets were introduced independently as **Chebychev sets** [36], **gated sets** [37] etc. Also, because of their role in the cartesian product of graphs, Tardif proposed the name **prefiber** for gated sets [38].

Graphs can be represented through matrices in a system's memory. Indeed, with a given graph, adequately labeled, there are associated several matrices, including the adjacency matrix, incidence matrix, distance matrix, cycle matrix and cocycle matrix. Techniques from linear algebra and group theory assist in studying the structure and enumeration of graphs. Related works can be seen in [39]. In certain situations, a reduction in flow is relevant than the total interruption. Motivated by this concept, Mathew and Sunitha introduced several connectivity concepts in weighted graph theory [40–45]. They categorized the edges of a weighted graph as $\alpha$-strong, $\beta$-strong and $\delta$-edges in connection with the strength of connectedness between its end vertices [41]. Also, they generalized the concepts of cutvertex, bridge, tree and block in usual graphs to weighted graphs as partial cutvertex, partial bridge, partial tree and partial block respectively. In addition, they obtained several results related to that concepts.

The study of distance related concepts in weighted graph theory is not yet found much in literature. This work was intended as an attempt to present the available work. Many concepts introduced in this work also generalizes the classical connectivity concepts.

**Definition 1.2.1** A graph $G$ is called a **weighted graph** if each edge $e$ is assigned a non-negative (preferably) weight $w(e)$ called the weight of $e$.

An unweighted graph can be regarded as a weighted graph in which each edge $e$ is assigned weight $w(e) = 1$. A **complete weighted graph** is a weighted graph whose underlying structure is complete.

**Definition 1.2.2** A **maximum spanning tree** (MST) of a weighted graph $G$ is a spanning graph of $G$, which is a tree and sum of weights of its edges, the largest among all such trees.

In a similar way minimum spanning tree can be defined. A graph is called **totally weighted,** if both its vertex set and edge set are weighted.

## 1.3 Heavy Paths and Optimal Cycles

The definitions and results in this section are from [18, 46–48].

**Definition 1.3.1** Let $G = (V, E)$ be a weighted graph with weight function $w$. Let $H$ be a subgraph of $G$ with vertex set $V(H)$ and edge set $E(H)$. The **weight of** $H$ is defined by $w(H) = \sum_{e \in E(H)} w(e)$.

**Definition 1.3.2** A cycle is called **optimal** if it is a cycle with maximum weight among all cycles of $G$.

**Definition 1.3.3** For every vertex $v \in V$, $N_H(v)$ denotes the neighbor set of $v$ in $H$, the **weighted degree** of $v$ in $H$ is defined as $d_H^w(v) = \sum_{u \in N_H(v)} w(vu)$. If there is no confusion, we denote $d_G^w(u)$ by $d^w(u)$.

An $(x, z)$-path is a path connecting the two vertices $x$ and $z$. For a given vertex $y$ of $G$, an $(x, z)$-path is called an $(x, y, z)$-path if it passes through the vertex $y$. A cycle is called a $y$-cycle if it passes through the vertex $y$. If $x$ and $z$ are two vertices on a path $P$, $P[x, z]$ denotes the segment of $P$ from $x$ to $z$. Let $C$ be a cycle in $G$ with a fixed orientation. For any two vertices $x$ and $z$ on $C$, by $C[x, z]$ we denote the segment of $C$ from $x$ to $z$ determined by this orientation. If $H$ is a subgraph of $G$, by $G - H$ we denote the induced subgraph $G[V(G) \setminus V(H)]$.

An unweighted graph can be regarded as a weighted graph in which each edge $e$ is assigned weight $w(e) = 1$. Thus, in an unweighted graph, $d^w(v) = d(v)$ for every vertex $v$, and an optimal cycle is simply a longest cycle.

In this section we provide generalizations of certain results, related to long paths and cycles.

**Theorem 1.3.4** *Let G be a graph and d an integer. If $d(v) \geq d$ for every vertex v in G, then G contains (i) a path of length at least d, and (ii) if $d \geq 2$, a cycle of length at least $d + 1$.*

Weighted generalization of Theorem 1.3.4 is as follows.

**Theorem 1.3.5** *Let G be a connected weighted graph on at least two vertices, z a specified vertex of G, and d a real number. Suppose that $w(zv) \geq d$ for every $v \in V(G) \setminus \{z\}$ and $w(e) > 0$ for every $e \in E(G)$.*

$(i)$ Then $G$ contains a $z$-path of weight at least $d$. Moreover, if $G$ contains no $z$-path of weight more than $d$, then each block $B$ of $G$ $(a)$ is complete, $(b)$ includes $z$, and $(c)$ is weighted so that $w(vz) = \alpha$ for all $v \in V(B) \setminus \{z\}$ and $w(uv) = \beta$ for all $u, v \in V(B) \setminus \{z\}$, where $\alpha + \beta(| V(B) | -2) = d$.

$(ii)$ If $d(v) \geq 2$ for every $v \in V(G) \setminus \{z\}$, then $G$ contains a cycle of weight more than $d$.

***Proof*** $(i)$ The proof is by induction on $| V(G) |$. If $| V(G) |= 2$, the result is trivially true. Suppose now that $| V(G) |\geq 3$. Let $G^{'} = G - z$. If $G^{'}$ is disconnected with components $h_1, h_2, \ldots, h_m$, then applying the induction hypothesis to the subgraph of $G$ induced by $V(H_i) \cup \{z\}$ for each $i$, $1 \leq i \leq m$, completes the proof. So we may suppose that $G^{'}$ is connected. Now let $\alpha = \max\{w(zv) : v \in N(z)\}$, and choose $z^{'} \in N(z)$ such that $w(zz^{'}) = \alpha$. Set $d^{'} = d - \alpha$. Then $w(v, G^{'}) \geq d^{'}$ for all $v \in V(G^{'})$. By the induction hypothesis, $G^{'}$ contains a $z^{'}$-path $P^{'}$ of weight at least $d^{'}$, and so $G$ contains the $z$-path $P = zz^{'}P^{'}$ of weight at least $\alpha + d^{'} = d$.

If $G$ contains no $z$-path of weight more than $d$, then the maximum weight of a $z^{'}$-path in $G^{'}$ is exactly $d^{'}$ Moreover by the induction hypothesis $G^{'}$ has the described structure. Let $B^{'}$ be a block of $G^{'}$. Thus $B^{'}$ is complete, includes $z^{'}$, and is weighted so that $w(vz^{'}) = \alpha^{'}$ for all $v \in V(B^{'}) \setminus \{z^{'}\}$, and $w(uv) = \beta^{'}$ for every $u, v \in V(B^{'}) \setminus \{z^{'}\}$, where

$$\alpha^{'} + \beta^{'}(| V(B^{'}) | -2) = d^{'} \text{......(1)}$$

If $x \in V(B^{'}) \setminus \{z^{'}\}$, then using (1), $w(x, G^{'}) = d^{'}$. Thus $w(xz) = w(x, G) - w(x, G^{'}) \geq d - d^{'} = \alpha$. Since $\alpha > 0$, $xz \in E(G)$. Moreover, by the choice of $z^{'}$, $w(xz) = \alpha$. Thus $w(vz) = \alpha$ for all $v \in V(B^{'})$. It follows that any vertex of $B^{'}$ could have been selected as the vertex $z^{'}$. This implies that $G^{'} = B^{'}$. The proof of $(i)$ is now complete by taking $\beta = \alpha^{'}$.

$(ii)$ Let $P = v_0 v_1 \cdots v_m$ be a path in $G$ of maximum weight starting at $v_0 = z$, and as long as possible subject to these conditions. Then, $v_m$ is adjacent only to vertices of $P$. Let

$$S = \{i : v_i v_m \in E(G)\} \text{ and } j = \min S.$$

Then for each $i \in S$, $v_o v_1 \cdots v_i v_m v_{m-1} \cdots v_{i+1}$ is another path starting at $v_0 = z$. By the choice of $P$, $w(v_i v_{i+1}) \geq w(v_i v_m)$. Summing this over all $i \in S$, we obtain

$$\sum_{i\in S} w(v_i v_{i+1}) \geq \sum_{i\in S} w(v_i v_m) = w(v_m) \geq d.$$

Let $C = v_j v_{j+1} \cdots v_{m-1} v_m v_j$. Then

$$w(C) \geq \sum_{i\in S} w(v_i v_{i+1}) + w(v_m v_j) \geq d + w(v_m v_j) > d.$$

Thus $G$ contains the cycle $C$ of weight more than $d$. ■

The condition $w(e) > 0$ in Theorem 1.3.5 can be even omitted for proving the existence of a cycle with minimum weight $d$ and a $z$-path of weight $d$. Theorem 1.3.6 discusses the corresponding extremal graphs.

A subgraph $H$ is said to be **contracted** if $B$ is deleted and replaced by a new vertex $b$ joined to every vertex $v \in V(G - B)$ for which $G$ contains an edge $uv$ for some $u \in V(B)$. Also a **pendant triangle** is an **endblock** which is a triangle.

**Theorem 1.3.6** *Let $G$ be a connected weighted graph on at least two vertices, $z$ a specified vertex of $G$, and $d$ a real number. Suppose that $w(zv) \geq d > 0$ for every $v \in V(G) \setminus \{z\}$.*

($i$) Then $G$ contains a $z$-path of weight at least $d$. Moreover, if $G$ contains no $z$-path of weight more than $d$, then the graph $G^{'}$ obtained from $G$ by recursively contracting blocks of weight zero until no such block remains has the following structure: each block $B$ of $G^{'}$ ($a$) includes $z$, and ($b$) is weighted so that $w(vz) = \alpha$ for all $v \in V(B) \setminus \{z\}$ with $vz \in E(G)$ and $w(uv) = \beta$ for all $u, v \in V(B) \setminus \{z\}$ with $uv \in E(G)$, where $\alpha + \beta(\mid V(B) \mid -2) = d$; moreover ($c$) if $\alpha > 0$, then $z$ is adjacent to every vertex of $B - z$, and if $\beta > 0$ then $B - z$ is complete.

($ii$) If $d(v) \geq 2$ for every $v \in V(G) \setminus \{z\}$, then $G$ contains a cycle of weight at least $d$. Moreover, if $G$ contains no cycle of weight more than $d$, then $G$ has a pendent triangle with edge weights zero, zero and $d$.

***Proof*** The first assertion easily follows from the above theorem. It is enough to prove the result for graphs that contain no zero-weight blocks. Let $G$ be such a graph. We use induction on $\mid E(G) \mid$. If $\mid E(G) \mid = \mid V(G) \mid -1$, then $G$ is a tree. Since $G$ contains no zero weight block, each edge of $G$ is of positive weight. The result follows from Theorem 1.3.5(i). Suppose now that $\mid E(G) \mid \geq \mid V(G) \mid$.

If $G$ contains an endblock $B$ which does not include $z$, let $b$ be the cut vertex of $G$ contained in $V(B)$. From the last assertion, $B$ contains a $b$-path $P_1$ of weight at least $d$. Moreover, since $G$ contains no zero-weight block and $b \neq z$, there is a $(z, b)$-path $P_2$ of positive weight. Then $P_1 \cup P_2$ is a $z$-path of weight more than $d$, a contradiction. This shows that every endblock of $G$ includes $z$ or equivalently every block of $G$ include $z$. Furthermore, we may suppose that $G$ contains a single block, for otherwise applying the induction hypothesis to each block would complete the proof.

If every edge of $G$ is of positive weight, then the result follows from Theorem 1.3.5(i), Suppose that $xy \in E(G)$ with $w(xy) = 0$. Let $H = G - xy$. We distinguish two cases.

**Case 1** $H$ contains no zero weighted block. By the induction hypothesis, each block of $H$ has the required structure.

**Subcase 1a** $H$ is a block. If $z \neq x, y$, then $H - z$ is not complete, and so by $(c)$ $\beta = 0$. Thus $w(xy) = 0 = \beta$. If $z = x$, then $\beta(\mid V(B) \mid -2) \geq w(y) \geq d$. This implies from $(b)$, that $w(xy) = 0 = \alpha$. The same conclusion holds if $z = y$. In either case, $G = H + xy$ has the required structure.

**Subcase 1b** If $H$ has more than one block, we must have that $w(e) = 0$ for every $e \in E(H - z)$, for otherwise an optimal $z$-path in $G$ would be of weight more than $d$. This shows that $G$ has the required structure (with $\alpha = d$ and $\beta = 0$).

**Case 2** $H$ contains a zero-weighted block. Since $G$ consists of a single block, $H$ is the union of blocks $B_1, B_2, \ldots, B_m, m \geq 2$, where $x \in B_1, y \in B_m$, $\mid V(B_i) \cap V(B_{i+1}) \mid= 1, 1 \leq i \leq m$, and $\mid V(B_i) \cap V(B_j) \mid= 0$ for $j \neq i \pm 1$. Without loss of generality, suppose that $z \notin V(B_1 - b)$ where $b$ is the cut vertex of $H$ contained in $B_1$. If $B_1$ contains a $b$-path $P_1$ of weight more than $d$, then extending $P_1$ to $z$ yields a $z$-path in $G$ of weight more than $d$. Thus $B_1$ contains no $b$-path of weight more than $d$. By the induction hypothesis, $B_1$ has the properties $(a)$, $(b)$ and $(c)$, which implies that $B_1$ has a $(b, x)$-path $P$ of weight $d$. If there is an edge $e$ of positive weight in some $B_i, 2 \leq i \leq m$, then $e$ and $P$ lie on a cycle in $G$ of weight more than $d$, and thus there is a $z$-path in $G$ of weight more than $d$. Therefore, $w(e) = 0$ for all $e \in E(B_1) \cup E(B_2) \cup \cdots \cup E(B_m)$. This implies that $m = 2$, $y = z$ and $G = B_1 \cup \{xy, by\}$, where $w(xy) = w(by) = 0$. The above argument, with the edge $xy$ replaced by the edge $by$, shows that $B_1$ is a uniformly weighted complete graph. Thus $G$ has the required structure with $\alpha = 0$ and $\beta = \frac{d}{|V(B_1)|-1}$.

$(ii)$ The first assertion follows from Theorem 1.3.5 To prove second assertion, we proceed by induction on $\mid E(G) \mid$. Since $d(v) \geq 2$ for all $v \in V(G) \setminus \{z\}$, we have $\mid E(G) \mid \geq \mid V(G) \mid$. If $\mid E(G) \mid = \mid V(G) \mid$, then each component of $G$ is a cycle and has a weight at least $d$, with equality only if it contains $z$. If $G$ contains no cycle of weight more than $d$, it follows that $G$ must consists of a single triangle with edge weights zero, zero and $d$ where $w(z) = 0$. Thus the result holds for $\mid E(G) \mid = \mid V(G) \mid$. Suppose now $\mid E(G) \mid > \mid V(G) \mid$.

If $G$ is not 2-connected, we apply the induction hypothesis to a component of $G$ or to an endblock $B$ of $G$ such that $z$ is not an internal vertex of $B$. This will complete the proof. We may therefore suppose that $G$ is 2-connected. Since $w(v) \geq d > 0$ for all $v \in V(G) \setminus \{z\}$ and $\mid E(G) \mid > \mid V(G) \mid$, there are at least two edges of positive weight. If there is an edge $e$ with $w(e) \geq d$, let $e'$ be any other edge with $w(e) > 0$. By the 2-connectedness of $G$, there is a cycle containing both $e$ and $e'$, which is thus of weight more than $d$. Therefore we may suppose that $w(e) < d$ for all $e \in E(G)$. If $w(e) > 0$ for all $e \in E(G)$, then by Theorem reftheo1.3.5(ii), $G$ contains a cycle of weight more than $d$. Thus consider $xy \in E(G)$ with $w(xy) = 0$. If $x \neq z$, then $d(x) \geq 3$, by assumption and the hypothesis that $w(x) \geq d$. The same conclusion holds for $y$. Hence $d(v) \geq 2$ for all $v \in V(G - xy) \setminus \{z\}$. By the induction hypothesis $G - xy$ contains a cycle of weight at least $d$. However, $G - xy$ does not contain a pendant triangle with weights zero, zero and $d$, by assumption. Thus $G - xy$ contains a cycle of weight more than $d$, and so does $G$. This completes the proof. ■

Consider the following theorem in [17].

**Theorem 1.3.7** *Let $G$ be a 2-connected graph and $d$ an integer. Let $x$ and $y$ be two distinct vertices of $G$. If $d(v) \geq d$ for all $v \in V(G) \setminus \{x, y\}$, then $G$ contains an $(x - y)$ path of length at least $d$.*

A cycle in a graph $G$ can be considered with two different orientations. Consider a cycle $C$ with a fixed orientation. For any two vertices $a, b \in V(C)$, $C[a, b]$ denote the segment of $C$ from $a$ to $b$ determined by the orientation.

The weighted generalization of Theorem 1.3.7 is as follows.

**Theorem 1.3.8** *Let $G$ be a 2-connected weighted graph and $d$ a real number. Let $x$ and $y$ be distinct vertices of $G$. If $w(v) \geq d$ for all $v \in V(G) \setminus \{x, y\}$, then $G$ contains an $(x - y)$-path of weight at least $d$.*

***Proof*** Let $\mid V(G) \mid= n$. We use induction on $n$. If $n = 3$, let $u$ be the third vertex other than $x$ and $y$. Then the path $xuy$ is of weight $w(u) \geq d$. Suppose now that $n \geq 4$ and the theorem is true for all graphs on $k$ vertices, $3 \leq k \leq n - 1$. Let $H = G - y$ be the graph obtained by deleting $y$ from G. We distinguish two cases:

**Case 1** $H$ is 2-connected. Choose $y' \in N(y) \setminus \{x\}$ (note that $\mid N(y) \mid\geq 2$ by the 2-connectedness of $G$ ) such that

$$w(y'y) = max\{w(vy) : v \in N(y) \setminus \{x\}\}.$$

Then for all $v \in V(H) \setminus \{x\}$,

$$w(v, H) = w(v, G) - w(vy) \geq d - w(y'y).$$

By the induction hypothesis, there is an $(x, y')$-path $Q$ in $H$ of weight at least $d - w(y'y)$. Then $P = Qy'y$ is an $(x, y)$-path in $G$ of weight at least $d$.

**Case 2** $H$ is not 2-connected. Choose an endblock $B$ of $H$ such that $x$ is not an internal vertex of $B$. Let $b$ be the unique cut vertex of $H$ contained in $B$ and let $B'$ be the subgraph of $G$ induced by $V(B) \cup \{y\}$. If $yb \in E(G)$, then $B'$ is 2-connected and for all $v \in V(B') \setminus \{y, b\}$, $w(v, B') = w(v, G) \geq d$. By the induction hypothesis, there is a $(y, b)$-path $P'$ in $B'$ of weight at least $d$. Extending $P'$ in $H$ from $b$ to $x$, we obtain an $(x, y)$-path in $G$ of weight at least $d$.

If $yb \notin E(G)$, we add $yb$ to $B'$ and set $w(yb) = 0$. Applying the above argument to the resulting graph, we obtain a $(y, b)$-path $P'$ of weight at least $d$. If $d > 0$, then $P' \neq yb$, since $w(yb) = 0$; if $d = 0$, then we can choose $P'$ such that $P' \neq yb$, since all we need is that $w(P') \geq d$. This shows that we still have $P' \subseteq B'$. Extending $P'$ as before completes the proof of the theorem. ■

**Corollary 1.3.9** *Let $G$ be a nonseparable weighted graph on at least two vertices and $d$ a real number. Let $x$ be a vertex of $G$. If $w(v) \geq d$ for all $v \in V(G) \setminus \{x\}$, then $G$ contains a path of weight at least $d$ from $x$ to any other vertex of $G$.*

**Lemma 1.3.10** *Let $C$ be an optimal cycle in a weighted graph $G$. If there exists a path $P$ in $G - C$, connecting vertices $x$ and $y$, such that $| N_C(x) |\geq 1$, $| N_C(y) |\geq 1$ and $| N_C(x) \cup N_C(y) |\geq 2$, then,*

$$w(C) \geq 2min\{w(x, C), w(y, C)\} + 2w(P).$$

Lemma 1.3.10 is a direct consequence of the following theorem.

**Theorem 1.3.11** *Let $C$ be an optimal cycle in a weighted graph $G$. Suppose that there exists a path $P$ in $G - C$, connecting vertices $x$ and $y$, such that $| N_C(x) |\geq 1$, $| N_C(y) |\geq 1$ and $| N_C(x) \cup N_C(y) |\geq 2$. Define $X = N_C(x) \setminus N_C(y)$, $Y = N_C(y) \setminus N_C(x)$ and $Z = N_C(x) \cap N_C(y)$. Then*

$$w(C) \geq 2w(x, X) + 2w(y, Y) + w(x, Z) + w(y, Z) + max\{2, | Z |\}w(P)$$

*unless $| Z |= 1$ and either $X = \phi$ or $Y = \phi$, in which cases, $w(C) \geq 2w(y, Y) + 2max\{w(x, Z) + w(P), w(y, Z)\}$ if $X = \phi$; $w(C) \geq 2w(x, X) + 2max\{w(y, Z) + w(P), w(x, Z)\}$ if $Y = \phi$.*

***Proof*** Let $A = X \cup Y \cup Z$ and suppose that $A = \{a_1, a_2, \ldots, a_k\}$, where $a_i$ are in order around $C$. For each pair of vertices $(a_i, a_{i+l})$, we shall construct from $C$ two new cycles by replacing the segment $C[a_i, a_{i+l}]$ with two $(a_i, a_{i+l})$-paths. These two paths are defined according to four cases:

$(i)$ $a_i, a_{i+1} \in Z$. The two paths are $a_i x P y a_{i+1}$ and $a_i y P x a_{i+1}$.

$(ii)$ $a_i \in Z$ ad $a_{i+1} \in X$ or $Y$. The two paths are, $a_i y P x a_{i+1}$ and $a_i x a_{i+1}$ or $a_i x P y a_{i+1}$ and $a_i y a_{i+1}$. If $a_{i+1} \in Z$ and $a_i \in X$ or $Y$, the paths are defined in the same way.

$(iii)$ $a_i \in X$ and $a_{i+1} \in Y$ or $a_i \in Y$ and $a_{i+1} \in X$. The two paths are two copies of $a_i x P y a_{i+1}$ or $a_i y P x a_{i+1}$.

$(iv)$ $a_i, a_{i+1} \in X$ or $a_i, a_{i+1} \in Y$. The two paths are two copies of $a_i x a_{i+1}$ or paths are two copies of $a_i x a_{i+1}$ or $a_i y a_{i+1}$.

In each case, we have defined two paths to replace the segment $C[a_i, a_{i+l}]$ and hence formed two cycles (in $(iii)$ and $(iv)$ these two cycles are identical). Since there are $k$ pairs of vertices $(a_i, a_{i+l})(i = 1, 2, \ldots, k)$, we obtain $2k$ cycles. In these cycles, every edge of $C$ is traversed exactly $2k - 2$ times; every edge from $x$ or $y$ to $Z$ is traversed twice, every edge from $x$ to $X$ is traversed four times and, similarly, every edge from $y$ to $Y$ is traversed four times. Now suppose that the path $P$ is traversed $l$ times (we determine $l$ later). Then the average weight of these $2k$ cycles is,

$$\frac{1}{2k}(2(k-1)w(C) + 2w(x, Z) + 2w(y, Z) + 4w(x, X) + 4w(y, Y) + lw(P)).$$

Since $C$ is an optimal cycle, its weight is no less than this average weight. Thus,

$$w(C) \geq w(x, Z) + w(y, Z) + 2w(x, X) + 2w(y, Y) + \frac{l}{2}w(P).$$

We now determine $l$. If $\mid Z \mid \geq 2$, then it is not difficult to see that $l \geq 2 \mid Z \mid$; if $\mid Z \mid = 1$, $X \neq \phi$ and $Y \neq \phi$, then $l \geq 4$ (at least twice from Case $(ii)$ and twice from Case $(iii)$); if $\mid Z \mid = 0$, then noting that $\mid N_C(x) \mid \geq 1$ and $\mid N_C(y) \mid \geq 1$, we see that $X \neq \phi$ and $Y \neq Y$, and so that $l \geq 4$ (from case $(iii)$). Therefore, $l \geq max\{4, 2 \mid Z \mid\}$, unless $\mid Z \mid = 1$ and either $X = \phi$ or $Y = \phi$. In the case $\mid Z \mid = 1$ and $X = \phi$, we suppose, without loss of generality that $Z = \{a_1\}$ and $Y = \{a_2, a_3, \ldots, a_k\}$. Since $C$ is optimal, $w(C[a_1, a_2]) \geq w(a_1x) + w(P) + w(a_2y)$ and $w(C[a_1, a_2]) \geq w(a_1y) + w(a_2y)$. Thus,

$$w(C[a_1, a_2]) \geq max\{w(a_1x) + w(P), w(a_1y)\} + w(a_2y)....(1)$$

Similarly,

$$w(C[a_k, a_1]) \geq max\{w(a_1x) + w(P), w(a_1y)\} + w(a_ky)....(2)$$

Also for $i \neq 1, k$

$$w(C[a_i, a_{i+1}]) \geq w(a_iy) + w(a_{i+1}y)....(3)$$

Combining (1), (2) and (3) we have

$$w(C) \geq 2w(y, Y) + 2max\{w(x, Z) + w(P), w(y, Z)\}$$

By the same argument for $\mid Z \mid = 1$ and $Y = \phi$, we have,

$$w(C) \geq 2w(x, X) + 2max\{w(y, Z) + w(P), w(x, Z)\}$$

This completes the proof. ■

**Theorem 1.3.12** *Let $G$ be a 2-connected graph and $d$ an integer. If $d(w) \geq d$ for every vertex $v$ in $G$, then $G$ contains either a cycle of length at least $2d$ or a Hamilton cycle.*

Theorem 1.3.12 can be generalized as follows.

**Theorem 1.3.13** *Let $G$ be a connected weighted graph and $d$ a real number. If $w(v) \geq d$ for every vertex $v$ in $G$, then either $G$ contains a cycle of weight at least $2d$ or every optimal cycle in $G$ is a Hamilton cycle.*

***Proof*** Suppose that there exists an optimal cycle $C$ in $G$ which is not a Hamilton cycle, and let $H$ be a component of $G - C$. We consider two cases:

**Case 1** H is nonseparable.

Choose distinct vertices $x$ and $y$ in $H$ (unless $| V(H) |= 1$, in which case necessarily $x = y$) such that

$(i)$ $d_C(x) \geq 1, d_C(y) \geq 1$, and

$(ii)$ $w(x, C) \geq w(y, C) \geq w(v, C)$ for all $v \in V(H) \setminus \{x, y\}$.

If $| N_C(x) \cup N_C(y) |\geq 2$, then by Lemma 1.3.10,

$$w(C) \geq 2w(y, C) + 2w(P)....(1)$$

where $P$ is an $(x, y)$ path in $H$. If $| V(H) |= 1$, then $w(P) = 0 \geq d - w(y) = d - w(y, C)$. Otherwise by the choice of $x$ and $y$, $w(v, H) = w(v) - w(v, C) \geq d - w(y, C)$ for all $v \in V(H) \setminus \{x\}$, and so by applying Corollary 1.3.9 to $H$, we have an $(x, z)$-path $Q$ in $H$ of weight $w(Q) \geq d - w(x, C) = d - w(xa)$. Thus $w(C) = w(C[a, b]) + w(C[b, a]) \geq 2(w(ax) + w(Q) + w(bz)) \geq 2d$. This completes Case 1.

**Case 2** H is separable.

Let $B_1$ and $B_2$ be two distinct endblocks of $H$, and let $b_i$ be the unique cut vertex of $H$ contained in $B_i (i = 1, 2)$. For $i = 1, 2$, we choose $x_i \in V(B_i) \setminus \{b_i\}$ such that $(i)$ $d_C(x_i) \geq 1$ and $(ii)$ $w(x_i, C) \geq w(v, C)$ for all $v \in V(B_i) \setminus \{b_i\}$. It follows that $w(v, B_i) = w(v) - w(v, C) \geq d - w(x_i, C)$ for all $v \in V(B_i) \setminus \{b_i\}$. For $i = 1, 2$, apply Corollary 1.3.9 to $B_i$ and obtain an $(x_i, b_i)$-path $P_i$ in $B_i$ of weight

$$w(P_i) \geq d - w(x_i, C).....(1)$$

If $| N_C(x_1) \cup N_C(x_2) |\geq 2$, then let $P$ be an $(x_1, x_2)$-path in $H$ of maximum weight. By (1),

$$w(P) \geq w(P_1) + w(P_2) \geq d - min\{w(x_1, C), w(x_2, C)\}.$$

By Lemma 1.3.10,

$$w(C) \geq 2min\{w(x_1, C), w(x_2, C)\} + 2w(P) \geq 2d.$$

If $N_C(x_1) = N_C(x_2) = \{u_1\}$, the there exists a vertex $u_2 \in V(C) \setminus \{u_1\}$ adjacent to some vertex $b \in V(H)$. As $(V(B_1) \setminus \{b_1\}) \cap (V(B_2) \setminus \{b_2\}) = \phi$, $b$ cannot belong to both $(V(B_1) \setminus \{b_1\})$ and $(V(B_2) \setminus \{b_2\})$. We suppose that $v \notin (V(B_2) \setminus \{b_2\})$. Extending the path $P_2$ in $H$ from $b_2$ to $b$, we obtain an $(x_2, b)$-path $P'$ in $H$ of weight

$$w(P') \geq w(P_2).....(2)$$

Since $C$ is optimal, $w(C) = w(C[u_1, u_2]) + w(C[u_2, u_1]) \geq 2(w(u_1x_2) + w(P') + w(u_2b)) \geq 2(w(u_1x_2) + w(P'))$.

Using (1) and (2) with $i = 1, 2$, we have

$$w(C) \geq 2(w(u_1x_2) + w(P_2)) \geq 2d + 2(w(u_1x_2) - w(x_2, C)) = 2d.$$

Hence the theorem is proved. ■

In 1989 Bondy and Fan [18] proposed two conjectures regarding weights of cycles in 2-connected graphs. Later they were proved by the same authors. We mention them for the completeness.

**Theorem 1.3.14** *Let G be a weighted graph on n vertices. Then G contains a path of weight at least* $\frac{2\omega(G)}{n}$

**Theorem 1.3.15** *Let G be a 2-edge connected weighted graph on n vertices. then G contains a cycle of weight at least* $\frac{2\omega(G)}{(n-1)}$.

In [18] the authors proved that Theorem 1.3.14 is a consequence of Theorem 1.3.15. The proof goes like this:

Let $G$ be a weighted graph on $n$ vertices, where $n \geq 2$ and suppose that $G$ is connected (otherwise consider the nontrivial components of $G$). Let $G'$ be the weighted graph obtained from $G$ by adding a new vertex $y$ and joining $y$ to every vertex of $G$ by an edge of weight $M$, where $M > w(G)$. Then $G'$ is 2-edge connected. Let $C$ be an optimal cycle of $G'$. Since $M > w(G)$ and $C$ is optimal, $C$ must contain the vertex $y$. If Theorem 1.3.15 is true, then

$$w(C) \geq \frac{2\omega(G')}{\mid V(G') \mid -1} = \frac{2(\omega(G) + nM)}{n} = \frac{2\omega(G)}{n} + 2M.$$

Thus $C - y$ is a path in $G$ of weight $w(C) - 2M \geq \frac{2\omega(G)}{n}$.

Before the formal proof of Theorem 1.3.15, they proved some special cases for complete weighted graphs. We give it here for completeness. we consider some basic definitions below.

A graph is called a **vertex weighted graph** if each vertex of the graph is assigned a nonnegative number, called the weight of the vertex. If $G$ is a graph on $n$ vertices $v_1, v_2, \ldots, v_n$, then $G(w_1, w_2, \ldots, w_n)$ denotes the vertex-weighted graph obtained by assigning to vertex $v_i$ the weight $w_i$, $1 \leq i \leq n$. If $w_1 = w_2 = \cdots = w_n = t$, then the notation is abbreviated to $G(t)$. Suppose that $G(w_l, w_2, \ldots, w_n)$ is a vertex-weighted graph. Then the induced (edge-) weighted graph $WG(w_1, w_2, \ldots, w_n)$ is the weighted graph obtained by setting $w(v_iv_j) = w_i + w_j$ for every edge $v_iv_j$. So $WK_n(t)$ is the uniformly-weighted complete graph on $n$ vertices in which each edge is of weight $2t$.

Recall that a set $S$ of vertices of a graph $G$ is a vertex cut of $G$ if the removal of $S$ leaves a graph with more components than $G$; $S$ is an $s$-vertex cut if $\mid S \mid= s$. A graph is separable if it has a 1-vertex cut; otherwise it is nonseparable. It follows from the definitions that every 2-connected graph is nonseparable, and that every nonseparable graph on at least three vertices is 2-connected.

**Theorem 1.3.16** *Let G be a weighted complete graph on n vertices, where $n \geq 3$. Then the maximum weight of a cycle of G is at least $\frac{2w(G)}{n-1}$ with equality if and only if $G = WK_n(w_1, w_2, \ldots, w_n)$ for some $w_1, w_2, \ldots w_n$ such that*

$$\sum_{i=1}^{n} w_i = \frac{2w(G)}{n-1}.$$

***Proof*** If $G = WK_n(w_1, w_2, \ldots, w_n)$ and $\sum_{i=1}^{n} w_i = \frac{w(G)}{n-1}$, then for any cycle $C$ of $G$,

$$w(C) = 2\sum\{w_i : v_i \in V(G)\} \leq 2\sum_{i=1}^{n} w_i = \frac{2w(G)}{n-1},$$

and equality holds when $C$ is a Hamiltonian cycle of $G$. This shows that the maximum weight of a cycle of $G$ is exactly $\frac{2w(G)}{n-1}$.

Conversely, let $G$ be any weighted complete graph on $n$ vertices. Let $\mathcal{C}$ be the set of Hamilton cycles of $G$. Since $G$ is complete, $|\mathcal{C}| = \frac{(n-1)!}{2}$ and each edge of $G$ is contained in exactly $(n-2)!$ Hamiltonian cycles. Thus the average weight of a Hamiltonian cycle of $G$ is

$$\frac{1}{|\mathcal{C}|}\sum_{C \in \mathcal{C}} w(C) = \frac{2}{(n-2)!}(n-2)!w(G) = \frac{2w(H)}{n-1}.$$

It follows that the maximum weight of a cycle of $G$ is at least $\frac{2w(G)}{n-1}$ with equality only if

$$w(C) = \frac{2w(G)}{n-1} \text{ for all } C \in \mathcal{C}.....(1)$$

Suppose that the value of $w(C)$ is $\frac{2w(G)}{n-1}$. We first note that the edge weights of $G$ satisfy the triangle inequality :

$$w(xy) + w(xz) \geq w(yz) \text{ for any } x, y, z \in V(G).....(2)$$

Let $P$ be a Hamiltonian path in $G - x$ connecting $y$ and $z$. Then $C = xyPzx$ and $C' = yPzy$ are cycles of $G$. Since $C$ is a hamiltonian cycle of $G$, we have by (1),

$$w(xy) + w(xz) + w(P) = w(C) = \frac{2w(G)}{n-1} \geq w(C') = w(yz) + w(P).$$

Thus (2) holds.

Net we observe that for any four vertices $u, x, y$ and $z$,

$$w(xu) + w(yz) = w(xy) + w(uz)...(3)$$

Let $P$ be a Hamiltonian path in $G - u, y$ connecting $x$ and $z$ and consider two Hamiltonian cycles $C = xuyzPx$ and $C' = xyuzPx$. By (1),

$$w(xu) + w(uy) + w(yz) + w(P) = w(C) = w(C') = w(xy) + w(yu) + w(uz) + w(P),$$

and so (3) holds. We now assign, to each vertex $x \in V(G)$, the weight $w_x = \frac{1}{2}(w(xy) + w(xz) - w(yz))$, where $y$ and $z$ are any two vertices other than $x$. By (2), $w_x \geq 0$. To verify that these vertex weights are well-defined (that is, that $w_x$ does not depend on the choice of $y$ and $z$), it suffices to show that, for any $u \in V(G) \setminus \{x, y, z\}$, the assignment determined by $u$ and $z$ is the same as that determined by $y$ and $z$, that is,

$$\frac{1}{2}(w(xu) + w(xz) - w(uz)) = \frac{1}{2}(w(xy) + w(xz) - w(yz)).$$

But this follows directly from (3) It remains to check that for any two distinct vertices $x$ and $y$,

$$w_x + w_y = w(xy)....(4)$$

Let $z$ be a third vertex. Then, by the definition,

$$w_x = \frac{1}{2}(w(xy) + w(xz) - w(yz)).$$

and

$$w_y = \frac{1}{2}(w(yx) + w(yz) - w(xz)).$$

and so (4) holds. Therefore, $G = WK_n(w_1, w_2, \ldots, w_n)$. Clearly, we have $\sum_{i=1}^{n} w_i = \frac{w(G)}{n-1}$, since, if $C$ is a Hamiltonian cycle of $G$,

$$2\sum_{i=1}^{n} w_i = w(C) = \frac{2w(G)}{n-1}.$$

This completes the proof. ■

**Corollary 1.3.17** *Let $G$ be a weighted complete graph on $n$ vertices. Then the maximum weight of a path of $G$ is at least $\frac{w(G)}{n}$, with equality holds if and only if $G = WH_n(t)$, where $t = \frac{w(G)}{n(n-1)}$.*

***Proof*** If $G = WK_n(t)$ and $t = \frac{w(G)}{n(n-1)}$, then for any path $P$,

$$w(P) = 2t \mid E(P) \mid \leq 2t(n-1) = \frac{2w(G)}{n},$$

and equality holds when $P$ is a Hamiltonian path of $G$. This shows that the maximum weight of a path of $G$ is exactly $\frac{2w(G)}{n}$.

Conversely, let $G$ be a weighted graph on $n$ vertices. Let $G'$ be the weighted graph obtained from $G$ by adding a new vertex $y$ and joining $y$ to every vertex of $G$ by an edge of weight $M$, where $M > w(G)$. Then $G'$ is 2-edge connected. Let $C$ be an optimal cycle of $G'$. Since $M > w(G)$ and $C$ is optimal, $C$ must contain the vertex $y$. By theorem,

$$w(C) \geq \frac{2w(G')}{|V(G')|-1} = \frac{2(w(G)+nM)}{n} = \frac{2w(G)}{n} + 2M.$$

Thus $C - y$ is a path in $G$ of weight $w(C) - 2M \geq \frac{2w(G)}{n}$. Thus it follows from Theorem 1.3.13 that $G$ contains a path of weight at least $\frac{2w(G)}{n}$, with equality only if $G' = WK_{n+1}(w_0, w_1, \ldots, w_n)$, where $w_0$ is the weight of $y$. Thus if $G$ contains no path of weight more than $\frac{2w(G)}{n}$, then $G = WK_n(t)$ where $t = M - w_0$. Moreover, since a Hamiltonian path of $G$ is of weight $\frac{2w(G)}{n}$, we have $t = \frac{w(G)}{n(n-1)}$, as required.
■

Now we proceed to the proof of Theorem 1.3.15. But we provide few definitions first.

**Definition 1.3.18** A spanning tree $T$ of a graph $G$ is called a **tritree** if every fundamental cycle of $T$ in $G$ is a triangle.

**Definition 1.3.19** Let $G$ be a weighted graph and $\mathbb{H}$ be a set of subgraphs (not necessarily distinct). If there is an assignment of a positive real number $\alpha_H$ to each $H \in \mathbb{H}$ such that, for every $e \in E(G)$, $w(e) = \sum\{\alpha_H : e \in H \in \mathbb{H}\}$, then we say that $G$ is a **weighted union** of the members of $H$, and write $G = \sum_{H \in \mathbb{H}} \alpha_H H$.

A 2-edge-connected weighted graph $G$ on $n$ vertices is **cycle-extremal** if its optimal cycles are of weight precisely $2\frac{w(G)}{(n-1)}$.

**Definition 1.3.20** Let $G$ be a weighted graph and $e \in E(G)$. Define $G_e$ to be the weighted graph obtained from $G$ by contracting the edge $e$ and, for each pair of multiple edges in the resulting graph, deleting the edge of smaller weight (or either, if they have equal weight).

**Lemma 1.3.21** *Let $G$ be a 2-connected weighted graph and $x \in V(G)$. If $|V(G)| > 3$, then there is $y \in N(x)$ such that $G_{xy}$ is 2-edge connected.*

***Proof*** Let $y \in N(x)$. If $G_{xy}$ is not 2-edge connected, then since $G$ is 2-connected, there must be $z \in N(x) \setminus \{y\}$ such that $d(z) = 2$. Then $G_{xz}$ is 2-edge connected. ■

**Lemma 1.3.22** *Let $G$ be a 2-connected weighted graph and $P$ an optimal path in $G$, with ends $x$ and $y$. Then there is a cycle $C$ in $G$ such that $w(C) > w(P)$ or $w(C) \geq w(x) + w(y)$.*

***Proof*** Let $P = v_0 v_1 \cdots v_l$, where $v_0 = x$ and $v_l = y$, and define $S = \{v_i : xv_i \in E\}$ and $T = \{v_i : v_{i-1}y \in E\}$. Note that by optimality of $P$,

$$w(v_{i-1}v_i) \geq w(xv_i), v_i \in S \text{ and} w(v_{i-1}v_i) \geq w(v_{i-1}y), v_i \in T....(1)$$

**Case. 1** $S \cap T \neq \phi$.

If $y \in S \cap T$ and $w(xy) > 0$, the cycle $C = Pyx$ has weight $w(C) = w(P) + w(xy) > w(P)$. Otherwise, for $v_i \in S \cap T$, define $C_i = v_0 v_1 \cdots v_{i-1} v_l v_{l-1} \cdots v_1 v_0$. Then the cycles $C_i$ together cover the edges $v_{i-1}v_i, v_i \notin S \cap T, \mid S \cap T \mid$ times the edges $v_{i-1}v_i, v_i \in S \cap T. \mid S \cap T \mid -1$ times, and the edges $xv_i$ and $v_{i-1}y, v_i \in S \cap T$, once. Therefore, $\sum\{w(C_i : v_i \in S \cap T\} = (\mid S \cap T \mid -1)w(P) + \sum\{w(v_{i-1}v_i) : v_i \notin S \cap T\} + \sum\{w(xv_i) + w(v_{i-1}y) : v_i \in S \cap T\}$. Using (1) it follows that $\sum\{w(C_i) : v_i \in S \cap T\} - (\mid S \cap T \mid -1)w(P) \geq \sum\{w(v_{i-1}v_i) : v_i \in S \setminus T\} + \sum \{w(v_{i-1}v_i) : v_i \in T \setminus S\} + \sum\{w(xv_i) : v_i \in S \cap T\} + \sum\{w(v_{i-1}y) : v_i \in S \cap T\}$ $\geq \sum\{w(xv_i) : v_i \in S\} + \sum\{w(v_{i-1}y) : v_i \in T\} = w(x) + w(y)$. Thus if $C$ is the cycle $C_i$ of maximum weight, $C$ has the required property.

**Case. 2** $S \cap T = \phi$.

A **vine** on a path $P$ is a set $\mathbb{Q} = \{Q_j : 1 \leq j \leq m\}$ of internally disjoint paths in $G$ so that (1) $P \cap Q_j = \{a_j, b_j\}, 1 \leq j \leq m$; and (2) $x = a_1 < a_2 < b_1 \leq a_3 < b_2 \leq a - 4 < \cdots \leq a_m < b_{m-1} < b_m = y$, where $a_j$ and $b_j$ are the ends of $Q_j$ and $<$ denotes the precedence on $P$. Since $G$ is 2-connected, it follows easily that there is a vine $\mathbb{Q}$ on $P$. We select $\mathbb{Q}$ so that $(i)$ $m$ is as small as possible; $(ii)$ subject to $(i)$, $\mid \cup_{j=1}^{m-1} V_j \mid$ as small as possible, where $V_j$ denotes the set of internal vertices of the segment $P[a_{j+1}, b_j]$ of $P$, $1 \leq j \leq (m-1)$.

We claim that the cycle $C = P \cup (\cup_{j=1}^{m}) - (\cup_{j=1}^{m-1})$ has the required property.

Denote the vertex following $a_m$ on $P$ by $a_m^+$. By the choice of $\mathbb{Q}$, the assumption that $S \cap T = \phi$ and (1), $w(C) = \sum\{w(v_{i-1}v_i) : v_i \notin V_j \cup \{b_j\}, 1 \leq j \leq (m-1)\} + \sum\{w(Q_j) : 1 \leq j \leq m\} \geq \sum\{w(v_{i-1}v_i) : v_i \in S \setminus \{b_1\} + \sum\{w(v_{i-1}v_i) : v_i \in T \setminus \{a_m^+\}\} + w(xb_1) + w(a_m y) \geq \sum\{w(xv_i) : v_i \in S \setminus \{b_1\}\} + \sum\{w(v_{i-1}y) : v_i \in T \setminus \{a_m^+\}\} + w(xb_1) + w(a_m y) \geq \sum\{w(xv_i) : v_i \in S\} + \sum\{w(v_{i-1}y) : v_i \in T\} = w(x) + w(y)$. Hence the proof. ■

**Theorem 1.3.23** *Let G be a 2-edge connected weighted graph on n vertices. Then G contains a cycle of weight at least* $\frac{2w(G)}{n-1}$.

***Proof*** We apply induction on $n$, and then on $\binom{n}{2-|E(G)|}$. If $n = 3$ or $\binom{|E(G)|=n}{2}$, the result follows from Theorem 1.3.14. Suppose now that $\binom{n>3, |E(G)|>n}{2}$ and that the result holds for all 2-edge connected graphs $G'$ with $\mid V(G') < n$ or with $\mid V(G') = n$ and $\mid E(G') \mid > \mid E(G) \mid$.

If $G$ is separable, let $G = G_1 \cup G_2$, where $\mid V(G_1) \cap V(G_2) \mid = 1$. Set $n_i = \mid V(G_i) \mid, i = 1, 2$. By induction hypothesis, $G_i$ contains a cycle of weight at least $\frac{2w(G_i)}{(n_i - 1)}, i = 1, 2$. Therefore if $C$ is an optimal cycle of $G$,

$$w(C) \geq w(C_i) \geq \frac{2w(G_i)}{(n_i - 1)}, i = 1, 2.$$

hence $2w(G) = 2w(G_1) + 2w(G_2) \leq (n_1 - 1)w(C) + (n_2 - 1)w(C) = (n - 1) w(C)$ which gives $w(C) \geq \frac{2w(G)}{(n-1)}$. If there is $x \in V(G)$ such that $w(x) \leq w(G)(n - 1)$, then by Lemma 1.3.21, there is $y \in N(x)$ such that $G_{xy}$ is 2-edge connected. Note that

$$w(G_{xy}) \geq w(G) - w(x) \geq \frac{n-2}{n-1} w(G).$$

By induction hypothesis, $G_{xy}$ contains a cycle $C'$ of weight

$$w(C') \geq \frac{2w(G_{xy})}{n-2} \geq \frac{2w(G)}{n-1}.$$

However, either $C'$ is a cycle in $G$ or it can be extended to a cycle in $G$. Therefore, we may assume that $G$ is 2-connected and that

$$w(v) > \frac{w(G)}{n-1} \text{ for every } v \in V(G)...(i)$$

Since $G$ is not complete, let $xy \notin E(G)$. Add $xy$ to $G$ with zero weight and let $G'$ denote the resulting weighted graph. By the induction hypothesis, $G'$ contains a cycle $C'$ of weight at least $\frac{2w(G')}{(n-1)} = \frac{2w(G)}{(n-1)}$. If $xy \notin C'$, then $C'$ is also a cycle of $G$, which completes the proof. Suppose that $xy \in C$. Since $w(xy) = 0$, the path $C - xy$ in $G$ is of weight at least $\frac{2w(G)}{(n-1)}$. Let $P$ be an optimal path in $G$, so $w(P) \geq \frac{2w(G)}{(n-1)}$. It follows from Lemma 1.3.22, using $(i)$, that $G$ contains a cycle of weight more than $\frac{2w(G)}{(n-1)}$. This completes the proof of the theorem. ■

A generalization of Theorem 1.3.12 can be found in [49] as given below.

**Theorem 1.3.24** ([49]) *Let $G$ be a 2-connected graph such that $d(u) + d(v) \geq s$ for every pair of nonadjacent vertices $u$ and $v$. Then $G$ contains either a cycle of length at least $s$ or a Hamilton cycle.*

**Lemma 1.3.25** *Let $G$ be a 2-connected weighted graph and $P = v_1 v_2 \cdots v_p$ a path in $G (p \geq 3)$. Define $S = \{v_i | v_1 v_i \in E(G)\}$ and $T = v_i | v_{i-1} v_p \in E(G)\}$. Assume each of the following conditions is satisfied.*

1. $N_G(v_1) \backslash V(P) = N_G(v_p) \backslash V(P) = \emptyset$;
2. $S \cap T = \phi$ *(in particular,* $v_1 v_p \notin E(G)$*);*
3. $w(v_{i-1} v_i) \geq \begin{cases} w(v_1 v_i), & \text{for all } v_i \in S \\ w(v_{i-1} v_p), & \text{for all } v_i \in T. \end{cases}$

*Then there is a cycle $C$ in $G$ of weight $w(C) \geq w(v_1) + w(v_p)$.*

Proof of lemma 1.3.25 follows from Lemma 1.3.22.

Note that in an unweighted graph, $w(v) = d(v)$ for every vertex $v$ and an optimal cycle is a longest cycle. A direct generalization of Theorem 1.3.8 is not possible, not even for complete graphs [47]. But a partial generalization is possible as in the following result.

**Theorem 1.3.26** *Let $G$ be a 2-connected weighted graph such that $w(u) + w(v) \geq s$ for every pair of nonadjacent vertices $u$ and $v$. Then $G$ contains either a cycle of weight at least $s$ or a Hamilton cycle.*

***Proof*** Let $G$ be a 2-connected weighted graph such that $w(u) + w(v) \geq s$ for every pair $u$, $v$ of nonadjacent vertices. Suppose $G$ does not contain a Hamilton cycle. We need to prove that $G$ contains a cycle of weight at least $s$.

Choose a path $P = v_1v_2 \cdots v_p$ in $G$ such that

1. (i) $P$ is as long as possible (i.e., $p$ is maximum);
2. (ii) $w(P)$ is as large as possible, subject to (i).

**Claim** There exists no cycle in $G$ containing all vertices of $P$.

Suppose there exists a cycle $C$ containing all vertices of $P$. If $V(C)\backslash V(P) \neq \phi$, then removing one edge from $C$ gives a path longer than $P$, contradicting the choice in $(i)$. So we have $V(C) = V(P)$. Since $C$ is not a Hamilton cycle and $G$ is connected, we can find a vertex $u \in V(G)\backslash V(C)$ and a path $Q$ from $u$ to $v_j$ for some $v_j \in V(C)$, such that $Q$ is internally disjoint from $C$. The subgraph $C \cup Q$ of $G$ contains a path longer than $P$, again contradicting the choice in $(i)$.

Define $S$ and $T$ as in Lemma 1.3.25. We complete the proof by showing that Lemma 1.3.25 applies, i.e., $(a)$, $(b)$, $(c)$ in Lemma 1.3.25 are all satisfied.

By the choice in $(i)$ we immediately see that $(a)$ holds.

To prove $(b)$, suppose $v_i \in S \cap T$. Then we can form the cycle $C = v_1v_iv_{i+1} \cdots v_pv_{i-1}v_{i-2} \cdots v_1$, which contradicts the claim. Hence $(b)$ holds.

If $v_i \in S$, then the path $P_0 = v_{i-1}v_{i-2} \cdots v_1v_iv_{i+1} \cdots v_p$ has the same length as $P$. So because of $(ii)$, we must have $w(P) \geq w(P')$, hence $w(v_{i-1}v_i) \geq w(v_1v_i)$. Similarly, if $vi \in T$, $w(v_{i-1}v_i) \geq w(v_{i-1}v_p)$. Thus $(c)$ holds. ■

As mentioned above Theorem 1.3.26 is not a direct generalization of Theorem 1.3.8. But there is a natural class of edge-weightings for which it is a generalization, namely the positive-induced edge-weightings defined next.

**Definition 1.3.27** Let $w$ be an edge-weighting of a graph $G = (V, E)$ and assume there exists a function $w' : V \to \mathbb{R}$ (a vertex-weighting of $G$) such that, for every edge $uv$ of $G$,

$$w(uv) = \frac{w'(u) + w'(v)}{2}.$$

Then we say that the edge-weighting $w$ is induced (by the vertex-weighting $w'$). If $w'$ can be chosen in such a way that $w'(v) > 0$ for all $v \in V$, then we call $w$ **positive-induced.** For every cycle $C$ of $G$ we have

$$w(C) = \sum_{e \in E(C)} w(e) = \sum_{v \in V(C)} w'(v).$$

Hence if $w$ is positive-induced, the statements every optimal cycle of $G$ is a Hamilton cycle and $G$ contains a Hamilton cycle are indeed equivalent.

In a similar way we can define an edge-weighting $w$ to be **nonnegative-induced** if $w$ is induced by a vertex-weighting $w'$ such that $w'(v) \geq 0$ for all $v \in V$. It follows easily that if $w$ is nonnegative-induced, then the statements $G$ contains a Hamilton cycle and there exists an optimal cycle that is a Hamilton cycle are equivalent, but neither of these statements implies that every optimal cycle of $G$ is a Hamilton cycle.

Let $G = (V, E)$ be a graph with an edge-weighting $w$. From the definition, $w$ is induced if and only if we can find a solution of the following set of $|E|$ linear equations in the $|V|$ unknowns $w'(v)$, $v \in V$ :

$$w'(x) + w'(y) = 2w(xy), \text{ for all } xy \in E \tag{1.3}$$

A walk in a graph is an alternating sequence $v_0e_1v_1e_2v_2 \cdots e_kv_k$ of vertices and edges such that $e_i = v_{i-1}v_i$, for $i = 1, \ldots, k$. The walk is called closed if $v_0 = v_k$. If $W = v_0e_1v_1e_2v_2 \cdots e_kv_k$ is a walk in $G$, then we define

$$\phi_w(W) = \sum_{i=1}^{k}(-1)^{i-1}w(e_i).$$

The following theorem gives a characterization of induced edge-weightings.

**Theorem 1.3.28** *Let $G = (V, E)$ be a graph with an edge-weighting $w$. Then $w$ is induced if and only if for every closed walk $W$ of even length we have $\phi_w(W) = 0$.*

***Proof*** Without loss of generality we may assume that $G$ is connected. First suppose that $w$ is induced by $w'$. Then by $(i)$ we have for every closed walk $W = v_0e_1v_1 \cdots e_{2k}v_{2k}(= v_0)$ of even length:

$$\begin{aligned}
\phi_w(W) &= w(e_1) - w(e_2) + w(e_3) - \cdots - w(e_{2k}) \\
&= \frac{1}{2}((w'(v_0) + w'(v_1)) - \frac{1}{2}((w'(v_1) + w'(v_2)) + \cdots - \frac{1}{2}((w'(v_{2k-1}) + w'(v_{2k})) \\
&= \frac{1}{2}((w'(v_0) + w'(v_{2k}))
\end{aligned}$$

which proves one side of the theorem.

Now suppose that $\phi_w(W) = 0$ for every closed walk $W$ of even length. If $G$ is bipartite, choose a vertex $u \in V$ and a real number a and set $w'(u) = a$. If $G$ is not bipartite, choose a cycle $C = u_0e_1u_1 \cdots e_{2k-1}u_0$ of odd length in $G$, and set $u = u_0$ and $w'(u) = \phi_w(C)$. Now for any vertex $v$, consider a walk from $v$ to $u$, $W = vf_1v_1 \cdots f_lu$, and set

$$w'(v) = \begin{cases} 2\phi_w(W) + w'(u), & \text{if } l \text{ is even} \\ 2\phi_w(W) - w'(u), & \text{if } l \text{ is odd}. \end{cases} \tag{1.4}$$

We claim that this definition of $w'(v)$ is independent of the choice of the walk $W$. To see this, let $W = vf_1v_1 \cdots f_lu$ and $W' = vf_1'v_1' \cdots f_m'$ two walks from

$v$ to $u$. If $G$ is bipartite, then $l$ and $m$ must have the same parity. So $X = vf_1v_1 \cdots f_l u f_m' v_{m-1}' \cdots f_1' v$ is a closed walk of even length. Since $\phi_w(X) = 0$, we immediately get that $\phi_w(W) = \phi_w(W')$. If $G$ is not bipartite, one of the closed walks $vf_1v_1 \cdots f_l u f_m' v_{m-1}' \cdots f_1' v$ and $vf_1v_1 \cdots f_l u e_1 u_1 \cdots e_{2k-1} u f_m' v_{m_1}' \cdots f_1' v$ is of even length, the cycle $C = ue_1u_1 \cdots e_{2k-1}u$ being of odd length. Denote this walk of even length by $Y$. By assumption we have $\phi_w(Y) = 0$. Using the definition of $w'(u)$ and equation 1.4 we reach the desired conclusion.

Finally, we prove that the edge-weighting $w$ is indeed induced by the vertex-weighting $w'$ defined by equation 1.4. To see this, let $e = xy$ be an edge in G and $W = yf_1v_1 \cdots f_l u$ a walk from $y$ to $u$. Then $W' = xeyf_1v_1 \cdots f_l u$ is a walk from $x$ to $u$, and $\phi_w(W') = w(e) - \phi_w(W)$. If $l$ is even, then by (2),

$$w'(x) + w'(y) = 2\phi_w(W') - w'(u) + 2\phi_w(W) + w'(u) = 2w(e).$$

The same is true if $l$ is odd. This completes the proof of the theorem. ■

Next we characterize positive-induced edge-weightings. We omit the proof, which goes along similar lines to the proof of Theorem 1.3.28.

**Theorem 1.3.29** *Let $G = (V, E)$ be a connected graph with an induced edge-weighting $w$. If $G$ is nonbipartite, then $w$ is positive-induced if and only if for every closed walk $W = v_0e_1v_1 \cdots e_{2k-1}v_{2k-1} (= v_0)$ of odd length in $G$ we have $\phi_w(W) > 0$. If $G$ is bipartite, then $w$ is positive-induced if and only if for every walk $W = v_0e_1v_1 \cdots e_{2k-1}v_{2k-1}$ of odd length in $G$ we have $\phi_w(W) > 0$.*

It follows from the proof of Theorem 1.3.28 that all induced edge-weightings on a connected graph $G = (V, E)$ (including ones with negative values) can be obtained by assigning arbitrary weights to the edges of a spanning tree (if $G$ is bipartite) or to the edges of a connected spanning unicyclic subgraph whose cycle is odd (if $G$ is not bipartite). Thus the dimension of the vector space of induced edge-weightings is $|V| - 1$ if $G$ is bipartite, and $|V|$ if $G$ is not bipartite. A natural basis of this vector space can be derived from the vertex-deleted subgraphs of $G$. Each such subgraph $G - v$ can be identified with the edge-weighting $w_v$ defined by

$$w_v(e) = \begin{cases} 1, & \text{if } e \in E(G - v), \\ 0, & \text{otherwise.} \end{cases}$$

and this edge-weighting $w_v$ is induced by the vertex-weighting $w_v'$ given by

$$w_v'(u) = 1 \qquad (u \neq v), \qquad w_v'(v) = -1.$$

Because the vertex-weightings $w_v'$, $v \in V(G)$, are linearly independent, they generate the space of all vertex-weightings. It follows that the edge- weightings

$w_v$, $v \in V(G)$, generate the space of all induced edge-weightings; thus, when $G$ is nonbipartite, they form a basis. In the bipartite case, any $|V| - 1$ of them constitute a basis.

A graph $G$ is said to be **triangle-free** if it contains no cycle of length 3. Also a path $P$ is a **longest heaviest path** of $G$ if $(i)$ $w(P)$ is maximum, and, $(ii)$ $P$ is a longest path of $G$ subject to $(i)$. If we consider a weighted complete graph with unit edge weight, the second conclusion of Theorem 1.3.13 can be dropped. But there are lot of graphs in which both conditions of Theorem 1.3.13 hold and we cannot obtain the existence of a heavy cycle by this theorem. The following results are from [50]. It is shown that if $G$ is a triangle free weighted graphs, we can find a heavy cycle even if $G$ is hamiltonian.

**Lemma 1.3.30** ([50]) *Let $G$ be a weighted graph and let $P$ be a longest heaviest path of $G$ with end vertices $x$ and $y$. Assume that $d(x) + d(y) - w(xy) \leq| E(P) |$. Then if $xy \notin E(G)$, then $P$ has weight at least $w(x) + w(y)$, and if $xy \in E(G)$, then the cycle $xPyx$ has weight at least $w(x) + w(y)$.*

***Proof*** Let $P = a_1a_2...a_p$ be a longest heaviest path of $G$ where $a_1 = x$ and $a_p = y$. Then we have $N(a_1) \subseteq V(P)$ and $N(a_p) \subseteq V(P)$. Let

- $N_1 = \{a_i | a_i \in N_G(a_1), a_{i-1} \notin N_{G-a_1}(a_p)\}$,
- $N_2 = \{a_i | a_i \in N_G(a_1), a_{i-1} \in N_{G-a_1}(a_p)\}$,
- $N_3 = \{a_i | a_i \in N_{G-a_1}(a_p), a_{i+1} \notin N_G(a_1)\}$ and
- $N_4 = \{a_i | a_i \in N_{G-a_1}(a_p), a_{i+1} \in N_G(a_1)\}$.

Moreover, let

$E_1 = \{a_1v | v \in N_1\}, E_2 = \{a_1v | v \in N_2\}, E_3 = \{va_p | v \in N_3\}$ and $E_4 = \{va_p | v \in N_4\}$. Now we define a mapping $\varphi_1$ of $\bigcup_{i=1}^{3} E_i$ to $E(P)$ such that

- for $e = a_1a_i \in E_1 \bigcup E_2$, $\varphi_1(e) = a_{i-1}a_i$ and
- for $e = a_ia_p \in E_3$, $\varphi_1(e) = a_ia_{i+1}$,

and let $F_i = \{\varphi_1(e) | e \in E_i\}$ for $i = 1, 2, 3$. Now it is easy to see that $F_1 \bigcap F_2 = \emptyset$. And by definition of $E_3$, $a_{i+1} \notin N(a_1)$ if $a_ia_p \in E_3$, hence $F_1$, $F_2$ and $F_3$ are disjoint.

It follows from the fact $d(x) + d(y) - \varepsilon(xy) \leq |E(P)|$ that

$$\begin{aligned}\sum_{i=1}^{3} |F_i| &= |E_1| + |E_2| + |E_3| \\ &= |N_1| + |N_2| + |N_3| \\ &= |N(a_1)| + |N(a_p)\backslash\{a_1\}| - |N_4| \\ &\leq |E(P)| - |N_4| \\ &= |E(P)| - |E_4|.\end{aligned}$$

Thus $|E(P)\backslash \bigcup_{i=1}^{3} F_i| \geq |E_4|$. Let $\varphi_2$ be an injection of $E_4$ to $E(P)\backslash \bigcup_{i=1}^{3} F_i$ and let $F_4 = \{\varphi_2(e|e \in E_4)\}$. Note that $F_1$, $F_2$, $F_3$ and $F_4$ are disjoint.

Assume that $a_1a_i \in E_1$ and $Q_1 = a_{i-1}a_{i-2}...a_1a_ia_{i+1}...a_p$. Then, since $w(Q_1) \leq w(P)$, $w(a_1a_i) \leq w(\varphi_1(a_1a_i))$. By similar argument as above, we have $w(e) \leq w(\varphi_1(e))$ for all $e \in E_1 \bigcup E_3$. Suppose $a_j \in N_2$. Then we have $a_{j-1}a_p \in E_4$. Let $C$ be a cycle $a_ja_1a_2...a_{j-1}a_pa_{p-1}...a_j$ and $e = \varphi_2(a_{j-1}a_p)$. Since $e \in E(C)$, $Q_2 = C - e$ is a path in $G$. Then it follows from the fact $w(Q_2) \leq w(P)$ that $w(a_1a_j) + w(a_{j-1}a_p) \leq w(\varphi_1(a_1a_j)) + w(\varphi_2(a_{j-1}a_p))$ for all $a_j \in N_2$. Therefore, if $a_1a_p \notin E(G)$,

$$\begin{aligned}
d^w(a_1) + d^w(a_p) &= \sum_{v \in N_1} w(a_1v) + \sum_{v \in N_2} w(a_1v) + \sum_{v \in N_3} w(va_p) + \sum_{v \in N_4} w(va_p) \\
&= \sum_{e \in E_1} w(e) + \sum_{e \in E_3} w(e) + \sum_{a_j \in N_2} (w(a_1a_j) + w(a_{j-1}a_p)) \\
&\leq \sum_{e \in F_1} w(e) + \sum_{e \in F_3} w(e) + \sum_{e \in F_2} w(e) + \sum_{e \in F_4} w(e) \\
&\leq w(P),
\end{aligned}$$

which implies the assertion. And in case of $a_1a_p \in E(G)$,

$$\begin{aligned}
d^w(a_1) + d^w(a_p) &= \sum_{v \in N_1} w(a_1v) + \sum_{v \in N_2} w(a_1v) + \sum_{v \in N_3} w(va_p) + \sum_{v \in N_4} w(va_p) + w(a_1a_p) \\
&= \sum_{e \in E_1} w(e) + \sum_{e \in E_3} w(e) + \sum_{a_j \in N_2} (w(a_1a_j) + w(a_{j-1}a_p)) + w(a_1a_p) \\
&\leq \sum_{e \in F_1} w(e) + \sum_{e \in F_3} w(e) + \sum_{e \in F_2} w(e) + \sum_{e \in F_4} w(e) + w(a_1a_p) \\
&\leq w(P) + w(a_1a_p).
\end{aligned}$$

Hence the cycle $a_1Pa_pa_1$ has weight at least $d^w(a_1) + d^w(a_p)$, which implies the assertion. ■

**Theorem 1.3.31** *Let $G$ be a 2-connected triangle-free weighted graph and let $d$ be a nonnegative real number. If $d^w(v) \geq d$ for every vertex $v$ in $G$, then $G$ has a cycle of weight at least $2d$.*

***Proof*** Let $P$ be a longest heaviest path in $G$, and let $x, y$ be end vertices of $P$. Since $G$ is triangle-free and $N(x), N(y) \subseteq V(P)$, $|N(x)| \leq |V(P)|/2$ and $|N(y)| \leq |V(P)|/2$. Moreover, if $xy \notin E(G)$, $|N(x)| \leq (|V(P)| - 1)/2$ and $N(y) \leq (|V(P)| - 1)/2$. Hence, whether $x$ and $y$ are adjacent or not, we have $d(x) + d(y) - \varepsilon(xy) \leq |E(P)|$. In case of $xy \in E(G)$, Lemma 1.3.30 implies the existence of a cycle of weight at least $d^w(x) + d^w(y) \geq 2d$, which is a required cycle. Thus we may assume $xy \notin E(G)$, then Lemma 1.3.30 implies that $w(P) \geq d^w(x) + d^w(y) \geq$

$2d$. Now it follows from Lemma 1.3.22 that there exists a cycle $C$ in $G$ such that $w(C) > w(P) \geq 2d$ or $w(C) \geq d^w(x) + d^w(y) \geq 2d$, which is a required cycle. ■

Theorem 1.3.7 can be generalized as follows.

**Theorem 1.3.32** ([51]) *Let $G$ be a 2-connected graph and $d$ an integer. Let $x$ and $z$ be two distinct vertices of $G$. Suppose that $d(v) \geq d$ for all $v \in V(G)\backslash\{x, z\}$.*

*(1) Then for any given vertex $y$ of $G$, $G$ contains an $(x, y, z)$-path of length at least $d$.*
*(2) If for some vertex $y \in V(G)\backslash\{x, z\}$, $G$ contains no $(x, y, z)$-path of length more than $d$; then the connected component $H_y$ of $G - x - z$ that contains $y$ is isomorphic to $K_{d-1}$ and $V(H_y) \subseteq N(x) \cap N(z)$. If $y \in \{x, z\}$, then the assertion holds for any connected component of $G - x - z$.*

Similar to the generalization of Theorem 1.3.7, we can generalize Theorem 1.3.32 to weighted graphs as follows. Note that this generalizes Theorem 1.3.8 also.

**Theorem 1.3.33** *Let $G$ be a 2-connected weighted graph and $d$ a real number. Let $x$ and $z$ be two distinct vertices of $G$. Suppose that $d^w(v) \geq d$ for all $v \in V(G)\backslash\{x, z\}$.*

*(1) Then for any given vertex $y$ of $G$, $G$ contains an $(x, y, z)$-path of weight at least $d$.*
*(2) If $w(e) > 0$ for all $e \in E(G)$ and for some vertex $y \in V(G)\backslash\{x, z\}$, $G$ contains no $(x, y, z)$-path of weight more than $d$, then (a) the connected component $H_y$ of $G - x - z$ that contains $y$ is complete (b) $V(H_y) \subseteq N(x) \cap N(z)$ (c) $w(xv) = \alpha_x$, $w(zv) = \alpha_z$ for all $v \in V(H_y)$ and $w(uv) = \beta_y$ for all $u, v \in V(H_y)$ so that $\alpha_x + \beta_y(|V(H_y)| - 1) + \alpha_z = d$. If $y \in \{x, z\}$, then the assertion holds for any connected component of $G - x - z$.*

***Proof*** If $y \in \{x, z\}$, then the result in (1) follows from Theorem 1; The assertions in (2) can be proved by choosing any connected component of $G - x - z$ as $H_y$ in the following proof. So we may assume that $y \notin \{x, z\}$.

Let $|V(G)| = n$. We use induction on $n$. If $n = 3$, let $y$ be the third vertex other than $x$ and $z$, then the path $xyz$ is an $(x, y, z)$-path of weight $d^w(y) \geq d$. Suppose now $n \geq 4$ and the theorem is true for all graphs on $k$ vertices with $3 \leq k \leq n - 1$. Let $G' = G - z$ be the graph obtained by deleting $z$ from $G$. We consider two cases:

**Case 1** $G'$ is 2-connected.

(1) Since $G$ is 2-connected, we can choose $z' \in N(z)\backslash\{x\}$ such that

$$w(zz') = max\{w(zv) : v \in N(z)\backslash\{x\}\}.$$

Then for all $v \in V(G')\backslash\{x\}$,

$$d^w_{G'}(v) = d^w(v) - w(zv) \geq d - w(zz').$$

By the induction hypothesis, for any given vertex $y \in V(G')\backslash\{x\}$, $G'$ contains an $(x, y, z')$-path $Q$ of weight at least $d - w(zz')$. Then the path $P = Qz'z$ is an $(x, y, z)$-path of weight at least $d$.

(2) If for some vertex $y \in V(G)\backslash\{x, z\}$, $G$ contains no $(x, y, z)$-path of weight more than $d$, then the maximum weight of an $(x, y, z')$-path in $G'$ is exactly $d' = d - w(zz')$. Moreover, by the induction hypothesis, $G'$ has the described structure. Let $H'_y$ be the connected component of $G' - x - z'$ that contains $y$. (If $y = z'$, take any connected component of $G' - x - z'$ as $H'_y$). Thus, $H'_y$ is complete, $V(H'_y) \subseteq N_{G'}(x) \cap N_{G'}(z')$ and $G'$ is weighted so that

$$w(xv) = \alpha'_x,\ w(z'v) = \alpha'_z, \text{ for all } v \in V(H'_y)$$

and

$$w(uv) = \beta'_y \text{ for all } u, v \in V(H'_y),$$

where

$$\alpha'_x + \beta'_y(|V(H'_y)| - 1) + \alpha'_{z'} = d'.$$

If $v \in V(H'_y)$, then $d^w_{G'}(v) = d'$. Thus

$$w(zv) = d^w(v) - d^w_{G'}(v) \geq d - d' = w(zz').$$

Since $w(zz') > 0$, we have that $zv \in E(G)$. Moreover, by the choice of $z'$, it is clear that $w(zv) = w(zz')$ for all $v \in V(H'_y)$. It follows that any vertex in $V(H'_y) \cup \{z\}$ could have been selected as the vertex $z'$. This implies that $\alpha'_{z'} = \beta'_y$.

Suppose that there exists another connected component $H^*$ of $G' - x - z'$. By the induction hypothesis, there must be an $(x, z')$-path of weight at least $d - w(zz')$ in $G[V(H^*) \cup \{x, z'\}]$. On the other hand, there is a $(z, y, z')$-path of weight $w(zz') + \beta'_y|V(H'_y)|$ in $G[V(H'_y) \cup \{z, z'\}]$. Combining these two paths, we get an $(x, y, z)$-path of weight at least $d + \beta'_y|V(H'_y)| > d$, which contradicts the assumption. Hence $G - x - z = G[V(H'_y) \cup \{z'\}]$ and

$$\begin{aligned} w(xz') &= d^w(z') - w(zz') - \beta'_y|V(H'_y)| \\ &\geq d - w(zz') - \beta'_y|V(H'_y)| \\ &= d' - \beta'_y|V(H'_y)| \\ &= \alpha'_x + \beta'_y(|V(H'_y)| - 1) + \alpha'_{z'} - \beta'_y|V(H'_y)| \\ &= \alpha'_x. \end{aligned}$$

Furthermore, by the assumption that $G$ contains no $(x, z)$-path of weight more than $d$ we know that $w(xz') \leq \alpha'_x$. So $xz' \in E(G)$ and $w(xz') = \alpha'_x$. Now let $H_y$ denote the connected component of $G - x - z$ that contains $y$ and set $\alpha_z = w(zz')$, $\alpha_x = \alpha'_x$ and $\beta_y = \beta'_y$. Then $H_y$ is complete, $V(H_y) \subseteq N(x) \cap N(z)$ and $G$ is weighted so that

$$w(xv) = \alpha_x,\ w(zv) = \alpha_z \text{ for all } v \in V(H_y)$$

and

$$w(uv) = \beta_y \text{ for all } u, v \in V(H_y),$$

where

$$\alpha_x + \beta_y(|V(H_y)| - 1) + \alpha_z = d.$$

**Case 2** $G'$ is not 2-connected.

(1) Since $G$ is 2-connected, $G'$ must be connected. We shall frequently make use of the following claim.

**Claim** Suppose $B$ is an end-block of $G'$ and $b$ is the unique cut-vertex of $G'$ contained in $B$. Let $B'$ be the subgraph of $G$ induced by $V(B) \cup \{z\}$. Then for any given vertex $y$ of $B'$, $B'$ contains a $(b, y, z)$-path $P'$ of weight at least $d$.

**Proof of Claim** If $zb \in E(G)$, then $B'$ is 2-connected and for all $v \in V(B') \backslash \{b, z\}$, we have

$$d^w_{B'}(v) = d^w(v) \geq d.$$

By the induction hypothesis, for any given vertex $y$ of $B'$, $B'$ contains a $(b, y, z)$-path $P'$ of weight at least $d$.

If $zb \notin E(G)$, add $zb$ to $B'$ and set $w(zb) = 0$. Applying the induction hypothesis to the resulting graph, we know that for any given vertex $y$ of $B'$, the resulting graph contains a $(b, y, z)$-path of weight at least $d$. If $d > 0$, then $P' \neq zb$, since $w(zb) = 0$. If $d = 0$, then we can choose $P'$ in $B'$ such that $P' \neq zb$, since all we need is that $w(P') \geq d$. This shows that we always have a $(b, y, z)$-path $P'$ in $B'$ of weight at least $d$.

**Case 2.1** $y$ is contained in a block of $G'$ with two or more cut-vertices.

Choose an end-block $B$ in $G'$ with cut-vertex $b$ such that there is an $(x, y, b)$-path $Q$ in $G' - (B - b)$. Let $B'$ be the subgraph of $G$ induced by $V(B) \cup \{z\}$. By the above claim, we have that there is a $(b, z)$-path $P'$ in $B'$ of weight at least $d$. Combining these two paths $Q$ and $P'$, we get an $(x, y, z)$-path of weight at least $d$.

**Case 2.2** $y$ is contained in an end-block $B$ of $H$ with a cut-vertex $b$ and $x \notin V(B)$.

Let $B'$ be the subgraph of $G$ induced by $V(B) \cup \{z\}$. It is easy to see that there exists an $(x, b)$-path $Q$ in $H - (B - b)$. By the above claim we have that there is a $(b, y, z)$-path $P'$ in $B'$ of weight at least $d$. Combining these two paths $Q$ and $P'$, we get an $(x, y, z)$-path of weight at least $d$.

**Case 2.3** $y$ and $x$ are contained in an end-block $B_1$ of $G'$.

If $x$ is the unique cut-vertex of $B_1$, let $B_1'$ be the subgraph of $G$ induced by $V(B_1) \cup \{z\}$. Then from the above claim we know that there is an $(x, y, z)$-path $P_1'$ in $B_1'$ of weight at least $d$. Otherwise, since $G'$ has at least two distinct end-blocks, we can choose an end-block $B_2$ in $G'$ other than $B_1$. Let $b_2$ be the unique cut-vertex of $G'$ contained in $B_2$ and $B_2'$ be the subgraph of $G$ induced by $V(B_2) \cup \{z\}$. Then there is a $(b_2, z)$-path $P_2'$ in $B_2'$ of weight at least $d$ by the above Claim, and there is also an $(x, y, b_2)$-path $Q$ in $G' - (B_2 - b_2)$. Combining these two paths $Q$ and $P_2'$, we get an $(x, y, z)$-path of weight at least $d$.

(2) From the above proof, we need only consider the case in which $y$ is contained in an end-block $B_1$ of $G'$ with $x$ as its unique cut-vertex. In this case, the result follows from the induction hypothesis by considering the graph $G[V(B_1) \cup \{z\}]$. This completes the proof. ■

Now we generalize another result involving long cycles as we did in Theorem 1.3.13.

**Theorem 1.3.34** ([52]) *Let $G$ be a 2-connected graph and $d$ an integer. If $d(v) \geq d$ for every vertex $v$ in $G$, then for any given vertex $y$ of $G$, $G$ contains either a $y$-cycle of length at least $2d$ or a Hamilton cycle.*

To generalize this result, we need the following result.

**Theorem 1.3.35** *Let $C$ be an optimal cycle in a weighted graph $G$. Suppose that there is an $(x, y, z)$-path in $G - C$ such that $|N_C(x)| \geq 1$, $|N_C(z)| \geq 1$ and $|N_C(x) \cup N_C(z)| \geq 2$. Define*

$$X = N_C(x) \backslash N_C(z),\ Z = N_C(z) \backslash N_C(x) \text{ and } Y = N_C(x) \cap N_C(z).$$

*If $|Y| = 1$ and either $X = \emptyset$ or $Z = \emptyset$, then there exists a $y$-cycle $C'$ in $G$ such that*

$$w(C') \geq \tfrac{w(C)}{2} + min\{d^w_C(x), d^w_C(z)\} + w(P).$$

*Otherwise, there exist $l (l \geq 4)$ $y$-cycles $C_1, C_2, \cdots, C_l$ in $G$ such that*

$$\sum_{i=1}^{l} w(C_i) \geq (l-2)w(C) + 2d^w_Y(x) + 2d^w_Y(z) + 4d^w_X(x) + 4d^w_Z(z) + lw(P).$$

***Proof*** If $|Y| = 1$ and either $X = \emptyset$ or $Z = \emptyset$, we have two cases. In the case $|Y| = 1$ and $X = \emptyset$, we can assume that $Y = \{a_1\}$ and $Z = \{a_2, \cdots, a_k\}$. Without loss of generality, we suppose that the segment $C[a_2, a_1]$ is of weight at least $w(C)/2$. So the cycle $C' = xPza_2C[a_2, a_1]a_1x$ is a $y$-cycle of weight

$$\begin{aligned} w(C') &\geq \frac{w(C)}{2} + w(xa_1) + w(za_2) + w(P) \\ &\geq \frac{w(C)}{2} + min\{d^w_C(x), d^w_C(z)\} + w(P). \end{aligned}$$

The case $|Y| = 1$ and $Z = \emptyset$ can be discussed by the same argument.

Otherwise, let $A = X \cup Y \cup Z$ and suppose that $A = \{a_1, a_2, \cdots, a_k\}$, where $a_i$ are in order around $C$. For each pair of vertices $(a_i, a_{i+1})$, we shall construct two new cycles from $C$ by replacing the segment $C[a_i, a_{i+1}]$ with two $(a_i, a_{i+1})$-paths. These two paths are defined according to four cases:

(1) $a_i, a_{i+1} \in Y$. The two paths are

$$a_i x P z a_{i+1} \text{ and } a_i z P x a_{i+1}.$$

(2) $a_i \in Y$ and $a_{i+1} \in X$ or $Z$. The two paths are

$$a_i z P x a_{i+1} \text{ and } a_i x a_{i+1}, \text{ or } a_i x P z a_{i+1} \text{ and } a_i z a_{i+1}.$$

If $a_{i+1} \in Y$ and $a_i \in X$ or $Z$, the paths are defined in the same way.
(3) $a_i \in X$ and $a_{i+1} \in Z$ or $a_i \in Z$ and $a_{i+1} \in X$. The two paths are two copies of

$$a_i x P z a_{i+1} \text{ or } a_i z P x a_{i+1}.$$

(4) $a_i, a_{i+1} \in X$ or $a_i, a_{i+1} \in Z$. The two paths are two copies of

$$a_i x a_{i+1} \text{ or } a_i z a_{i+1}.$$

In each case, we have defined two paths to replace the segment $C[a_i, a_{i+1}]$ and hence formed two cycles. Since there are $k$ pairs of vertices $(a_i, a_{i+1})(i = 1, \ldots, k)$, we obtain $2k$ cycles. In these cycles, every edge of $C$ is traversed $2k - 2$ times; every edge from $x$ or $z$ to $Y$ is traversed twice, every edge from $x$ to $X$ is traversed four times and, similarly, every edge from $z$ to $Z$ is traversed four times. Now suppose that the path $P$ is traversed $l$ times (we determine $l$ later). Then the weight sum of these $2k$ cycles is

$$2(k-1)w(C) + 2d_Y^w(x) + 2d_Y^w(z) + 4d_X^w(x) + 4d_Z^w(z) + lw(P).$$

Without loss of generality, we can denote the $l$ cycles which pass through the path $P$ (also pass through the vertex $y$) by $C_1, C_2, \ldots, C_l$. Since $C$ is an optimal cycle of $G$, we get $l$ $y$-cycles $C_1, C_2, \ldots, C_l$ such that

$$\sum_{i=1}^{l} w(C_i) \geq (l-2)w(C) + 2d_Y^w(x) + 2d_Y^w(z) + 4d_X^w(x) + 4d_Z^w(z) + lw(P).$$

Now we determine $l$. If $|Y| \geq 2$, then it is not difficult to see that $l \geq 2|Y|$; if $|Y| = 1$, $X \neq \emptyset$, and $Z \neq \emptyset$, then $l \geq 4$; if $|Y| = 0$, then noting that $|N_C(x)| \geq 1$ and $|N_C(z)| \geq 1$, we have that $X \neq \emptyset$ and $Z \neq \emptyset$, and $l \geq 4$. Therefore for all the cases we have that $l \geq 4$. ■

**Theorem 1.3.36** *Let $G$ be a 2-connected weighted graph and $d$ a real number. If $d^w(v) \geq d$ for every vertex $v$ in $G$, then for any given vertex $y$ of $G$, either $G$ contains a $y$-cycle of weight at least $2d$ or every optimal cycle in $G$ is a Hamilton cycle.*

***Proof*** Suppose that there exists an optimal cycle $C$ in $G$ which is not a Hamilton cycle. From Theorem 1.3.31, we have that $w(C) \geq 2d$. If $y$ is contained in the cycle $C$, then we are done. Otherwise, let $H$ be the component of $G - C$ which contains $y$. We consider two cases:

**Case 1** $H$ is nonseparable.

**Case 1.1** $V(H) = \{y\}$.

Suppose that $N_C(y) = \{a_1, a_2, \ldots, a_k\}(k \geq 2)$, where $a_i$ are in order around $C$. For each pair of vertices $(a_i, a_{i+1})$, we shall construct a $y$-cycle $C_i$ from $C$ by replacing the segments $C[a_i, a_{i+1}]$ with the path $a_i y a_{i+1}$. Since there are $k$ pairs of vertices $(a_i, a_{i+1})(i = 1, 2, \ldots, k)$, we obtain $k$ cycles, and,

$$\begin{aligned}\sum_{i=1}^{k} w(Ci) &= (k-1)w(C) + 2d_C^w(y)\\ &\geq 2(k-1)d + 2d\\ &= 2kd.\end{aligned}$$

Then, among these $k$ cycles there must be a $y$-cycle $C'$ with weight at least $2d$.

**Case 1.2** $|V(H)| \geq 2$.

Choose distinct vertices $x$ and $z$ in $H$ such that
(1) $|N_C(x)| \geq 1$, $|N_C(z)| \geq 1$, and
(2) $d_C^w(x) \geq d_C^w(z) \geq d_C^w(v)$ for all $v \in V(H)\backslash\{x, z\}$.

**Case 1.2.1** $|N_C(x) \cup N_C(z)| \geq 2$.

By the choice of $x$ and $z$, we have

$$d_H^w(v) = d^w(v) - d^w C(v) \geq \max\{0, d - d_C^w(z)\} \text{ for all } v \in V(H)\backslash\{x\}.$$

If $|V(H)| = 2$, it is easy to find an $(x, y, z)$-path $P$ in $H$ of weight at least $\max\{0, d - d_C^w(z)\}$. Otherwise, we can choose an $(x, y, z)$-path $P$ in $H$ such that

$$w(P) \geq \max\{0, d - d_C^w(z)\}.$$

Now denote $N_C(x)\backslash N_C(z)$, $N_C(x) \cap N_C(z)$ and $N_C(z)\backslash N_C(x)$ by $X$, $Y$ and $Z$, respectively. If $|Y| = 1$ and $X = \emptyset$ or $Z = \emptyset$. By Theorem 1.3.35 we know that there is a $y$-cycle $C'$ in $G$ such that

$$w(C') \geq \frac{w(C)}{2} + min\{d_C^w(x), d_C^w(z)\} + w(P) \geq 2d.$$

Otherwise, from Theorem 1.3.35 we know that $G$ contains $l(l \geq 4)$ $y - cycles$ $C_1, C_2 \cdots C_l$ such that

$$\begin{aligned}\sum_{i=1}^{l} w(C_i) &\geq (l-2)w(C) + 2d_Y^w(x) + 2d_Y^w(z) + 4d_X^w(x) + 4d_Z^w(z) + lw(P)\\ &= (l-2)w(C) + 2d_C^w(x) + 2d_C^w(z) + 2d_X^w(x) + 2d_Z^w(z) + lw(P)\\ &\geq (l-2)w(C) + 4d_C^w(z) + l\max\{0, d - d_C^w(z)\}\\ &\geq 2ld.\end{aligned}$$

Then, among these $l$ $y$-cycles in $G$ there must be one with weight at least $2d$.

**Case 1.2.2** $N_C(x) = N_C(z) = \{a\}$.

Since $G$ is 2-connected, there exists a vertex $b \in V(C)\backslash\{a\}$ which is adjacent to some vertex $u \in V(H)\{x, z\}$. By the choice of $x$ and $z$, we have

$$d_H^w(v) = d^w(v) - d_C^w(v) \geq d - d_C^w(x) \text{ for all } v \in V(H).$$

Applying Theorem 1.3.33 to $H$, we have an $(x, y, u)$-path $Q$ in $H$ of weight

$$w(Q) \geq d - d_C^w(x) = d - w(xa).$$

then the path $axQub$ is of weight at least $d$. It is easy to see that we can form a $y$-cycle of weight at least $2d$.

**Case 2** H is separable.

**Case 2.1** $y$ is contained in a block of $H$ with two or more cut-vertices.

Let $B_1$ and $B_2$ be two distinct end-blocks of $H$, and let $b_i$ be the unique cut-vertex of $H$ contained in $B_i(i = 1, 2)$. For $i = 1, 2$, we choose $x_i \in V(B_i)\backslash\{b_i\}$ such that
(1) $|N_C(x_i)| \geq 1$, and
(2) $d_C^w(xi) \geq d_C^w(v)$ for all $v \in V(B_i)\backslash\{b_i\}$.
It follows that
$d_{B_i}^w(v) = d^w(v) - d_C^w(v) \geq \max\{0, d - d_C^w(x_i)\}$ for all $v \in V(B_i)\backslash\{b_i\}$, $(i = 1, 2)$.
Applying Theorem 1.3.33 to $B_i$ we obtain an $(x_i, b_i)$-path $P_i$ in $B_i$ of weight

$$w(P_i) \geq \max\{0, d - d_C^w(x_i)\}.$$

If $|N_C(x_1) \cup N_C(x_2)| \geq 2$, then let $P$ be an $(x_1, y, x_2)$-path in $H$ of maximum weight; then

$$w(P) \geq w(P_1) + w(P_2) \geq \max\{0, d - \min\{d_C^w(x_1), d_C^w(x_2)\}\}.$$

Denote $N_C(x_1)\backslash N_C(x_2)$, $N_C(x_2)\backslash N_C(x_1)$ and $N_C(x_1) \cap N_C(x_2)$ by $X_1$, $X_2$ and $Y$, respectively. If $|Y| = 1$ and $X_1 = \emptyset$ or $X_2 = \emptyset$, then by Theorem 1.3.35 we know that there is a $y$-cycle $C'$ in $G$ such that

$$w(C') \geq \frac{w(C)}{2} + \min\{d_C^w(x_1), d_C^w(x_2)\} + w(P) \geq 2d.$$

Otherwise, from Theorem 1.3.35 we know that $G$ contains $l(l \geq 4)$ $y$-cycles $C_1, C_2, \ldots, C_l$ such that

$$\begin{aligned}
\sum_{i=1}^{l} w(C_i) &\geq (l-2)w(C) + 2d_Y^w(x_1) + 2d_Y^w(x_2) \\
&\quad + 4d_{X_1}^w(x_1) + 4d_{X_2}^w(x_2) + lw(P) \\
&\geq 2(l-2)d + 4\min\{d_C^w(x_1), d_C^w(x_2)\} \\
&\quad + l\max\{0, d - \min\{d_C^w(x_1), d_C^w(x_2)\}\} \\
&\geq 2ld.
\end{aligned}$$

So, among these $l$ $y$-cycles there must be one with weight at least $2d$.

If $N_C(x_1) = N_C(x_2) = \{a\}$, then there exists a vertex $b \in V(C)\backslash\{a\}$ which is adjacent to some vertex $u \in V(H)\backslash\{x_1, x_2\}$. As $(V(B_1)\backslash\{b_1\}) \cap (V(B-2)\backslash\{b_2\}) = \emptyset$, $u$ cannot belong to both $(V(B_1)\backslash\{b_1\})$ and $(V(B_2)\backslash\{b_2\})$. Without loss of generality, we suppose that $u \notin V(B_2)\{b_2\}$. It is easy to see that there is a $(u, y, b_2)$-path $Q$ In $H-(B_2-b_2)$. So the path $P = buQb_2P_2x_2a$ is of weight

$$w(P) \geq w(P_2) + w(x_2a) \geq d.$$

Therefore, it is easy to construct a $y$-cycle of weight at least $2d$.

**Case 2.2** $y$ is contained in an end-block $B_1$ of $H$.

Choose another end-block $B_2$ of $H$ and let $b_i$ be the unique cut-vertex of $H$ contained in $B_i (i = 1, 2)$. For $i = 1, 2$, choose $x_i \in V(B_i)\backslash\{b_i\}$ such that
(1) $|N_C(x_i)| \geq 1$, and
(2) $d_C^w(x_i) \geq d_C^w(v)$ for all $v \in V(B_i)\backslash\{b_i\}$.

Applying Theorem 1.3.33 to $B_1$ and $B_2$, we obtain an $(x_1, y, b_1)$-path $P_1$ in $B_1$ of weight at least $\max\{0, d - d_C^w(x_1)\}$, and an $(x_2, b_2)$-path $P_2$ in $B_2$ of weight at least $\max\{0, d - d_C^w(x_2)\}$. It is also easy to know that there is a $(b_1, b_2)$-path $Q$ in $H-(B_1-b_1)-(B_2-b_2)$. So the path $P = P_1QP_2$ is an $(x_1, y, x_2)$-path with weight

$$w(P) \geq w(P_1) + w(P_2) \geq \max\{0, d - \min\{d_C^w(x_1), d_C^w(x_2)\}\}.$$

If $|N_C(x_1) \cup N_C(x_2)| \geq 2$, using the similar argument in Case 2.1, we can get a $y$-cycle of weight at least $2d$.

If $N_C(x_1) = N_C(x_2) = \{a\}$, there exists a vertex $b \in V(C)\backslash\{a\}$ which is adjacent to some vertex $u \in V(H)\backslash\{x_1, x_2\}$.

If $u \in V(B_1)$ and $u = b_1$, the path $bb_1P_1x_1a$ is of weight at least $d$; If $u \neq b_1$, we can choose a $(u, y, b_2)$-path $Q$, then the path $P = buQb_2P_2x_2a$ is of weight at least $d$. So in both cases we can form a $y$-cycle of weight at least $2d$.

If $u \notin V(B_1)$, we can choose a $(b_1, u)$-path $Q$ in $H-(B_1-b_1)$, and therefore the path $P = ax_1P_1b_1Qub$ is of weight at least $d$. It is easy to form a $y$-cycle with weight at least $2d$. The proof is now complete. ■

## 1.4 Tritrees and Cyclic Extremal Graphs

Tritrees were introduced by Bondy in [18]. A **tritree** of a graph G is a spanning tree T of G such that every fundamental cycle of G with respect to T is a triangle. A **trigraph** is a graph which has a tritree. For example, a spanning star is a tritree. The square of a tree is a trigraph.

We have two results taken from [18] given below.

**Proposition 1.4.1** *Let $T$ be a tritree of a graph $G$. Then any cycle of $G$ contains at most two edges of $T$.*

**Proposition 1.4.2** *Let G be a graph with no 2-vertex cut. Then each tritree of G is a spanning star.*

**Proposition 1.4.3** *Let G be a weighted graph and $G'$ a weighted graph obtained by adding to G edges of weight zero. If* $G' = \sum_{T \in \mathcal{T}} \alpha_T T$ *for some set* $\mathcal{T}$ *of tritrees of* $G'$, *then* $\mathcal{T}$ *is also a set of tritrees of G and* $G = \sum_{T \in \mathcal{T}} \alpha_T T$.

***Proof*** Since $\alpha_T$ is positive and the new edges in $G'$ have zero weight, none of the new edges belongs to any tritree in $\mathcal{T}$. Hence each tritree in $\mathcal{T}$ is also a tritree of $G$, and $G = \sum_{T \in \mathcal{T}} \alpha_T T$. ■

**Proposition 1.4.4** *Let G be a weighted graph on n vertices. If* $G = \sum_{T \in \mathcal{T}} \alpha_T T$ *for some set* $\mathcal{T}$ *of tritrees of G, then* $\sum_{T \in \mathcal{T}} \alpha_T = \dfrac{w(G)}{(n-1)}$.

***Proof*** By definition,

$$w(G) = \sum_{T \in \mathcal{T}} \alpha_T \mid E(T) \mid = \sum_{T \in \mathcal{T}} \alpha_T (n-1) = (n-1) \sum_{T \in \mathcal{T}} \alpha_T,$$

which gives the required equality. ■

**Proposition 1.4.5** *Let G be a weighted graph on n vertices. if* $G = \sum_{T \in \mathcal{T}} \alpha_T T$ *for some set* $\mathcal{T}$ *of tritrees of G, then, for any* $e \in E(G)$, $w(e) \leq \frac{w(G)}{n-1}$, *with equality if and only if every tritree in* $\mathcal{T}$ *contains e.*

***Proof*** let $T' = \{T \in \mathcal{T} : T \text{ contains } e\}$. Then using Theorem 1.3.4, we have

$$w(e) = \sum_{T \in \mathcal{T}'} \alpha_T \leq \sum_{T \in \mathcal{T}} \alpha_T = \frac{w(G)}{n-1},$$

with equality if and only if $\mathcal{T}' = \mathcal{T}$ as required. ■

**Proposition 1.4.6** *Let G be a weighted graph on n vertices and C an optimal cycle in G. If* $G = \sum_{T \in \mathcal{T}} \alpha_T T$ *for some set* $\mathcal{T}$ *of tritrees of G, then* $w(C) \leq \frac{2w(G)}{(n-1)}$ *with equality if and only if* $\mid E(C) \cap E(T) \mid = 2$ *for every* $T \in \mathcal{T}$.

***Proof*** By Proposition 1.4.1, $C$ contains at most two edges of $T$, for any $T \in \mathcal{T}$. Thus using Proposition 1.4.4,

$$w(C) \leq \sum_{T \in \mathcal{T}} 2\alpha_T = 2 \sum_{T \in \mathcal{T}} \alpha_T = \frac{2w(G)}{n-1},$$

with equality if and only if $| E(C) \cap E(T) |= 2$ for every $T \in \mathcal{T}$, as required. ■

**Proposition 1.4.7** *If $G$ is a cyclic extremal graph, then $w(v) \geq \frac{w(G)}{(n-1)}$ for all $v \in V(G)$.*

***Proof*** Let $v \in V(G)$. If $w(v) < \frac{w(G)}{(n-1)}$, by Lemma 1.3.22, there is $y \in N(v)$ such that $G_{vy}$ is 2-edge connected, where

$$w(G_{vy} \geq w(G) - w(y) > \frac{(n-2)w(G)}{n-1}.$$

By Theorem 1.3.15, $G_{vy}$ has a cycle $C'$ of weight $w(C') \leq \frac{2w(G_{vy}}{n-2} > \frac{2w(G)}{n-1}$. But either $C'$ is a cycle in $G$ or it can be extended to a cycle in $G$. This contradicts the fact that $G$ is cyclic extremal. Therefore, $w(v) \geq \frac{w(G)}{(n-1)}$ for all $v \in V(G)$. ■

**Definition 1.4.8** Let $G_1$ and $G_2$ be two weighted graphs such that $| V(G_1) \cap V(G_2) |= 1$. Set $n_i =| V(G_i) |, i = 1, 2$. If $\frac{w(G_1)}{(n_1-1)} = \frac{w(G_2)}{(n_2-1)}$, then $G_1 \cup G_2$ is called the 1-sum of $G_1$ and $G_2$.

**Proposition 1.4.9** *Let $G$ be a separable cycle-extremal graph. Then $G$ is a 1-sum of two cyclic-extremal graphs.*

***Proof*** Since $G$ is separable, let $G = G_1 \cup G_2$, where $| G_1 \cap G_2 |= 1$. Set $n_i =| V(G_i) |, i = 1, 2$. We prove that $\frac{w(G_1)}{(n_1-1)} = \frac{w(G_2)}{(n_2-1)}$ and that $G_i$ is cyclic-extremal for $i = 1, 2$. Let $C$ be an optimal cycle in $G$ and $C_i$ an optimal cycle in $G_i, i = 1, 2$. Since $G$ is cyclic-extremal,

$$\begin{aligned} 2w(G) &= (n-1)w(C) = (n_1-1)w(C) + (n_2-1)w(C) \\ &\geq (n_1-1)w(C_1) + (n_2-1)w(C_2) \end{aligned}$$

By Theorem 1.3.15, $(n_i - 1)w(C_i) \geq 2w(G_i), i = 1, 2$, and so $2w(G) \geq 2w(G_1) + 2w(G_2) = 2w(G)$. Thus all the above inequalities are equalities, and $\frac{2w(G_1)}{n_1-1} = w(C_1) = w(C) = w(C_2) = \frac{2w(G_2)}{n_2-1}$, which completes the proof. ■

**Proposition 1.4.10** *Let $G$ be a 2-connected cycle-extremal graph on $n$ vertices. Then,*

(*i*) *each edge of $G$ lies in an optimal cycle, and*
(*ii*) *any two nonadjacent vertices are connected by a path of weight at least $\frac{2w(G)}{(n-1)}$.*

***Proof*** $(i)$ Let $e \in E(G)$. Replace $w(e)$ by $w(e) + \epsilon$. By Theorem 1.3.15, the resulting weighted graph $G$ has a cycle $C_\epsilon$ of weight,

$$w(C_\epsilon) \geq \frac{2w(C_\epsilon)}{n-1} = \frac{2w(G)}{n-1} + \frac{2\epsilon}{n-1}.$$

$G$ has no cycle of weight more than $\frac{2w(G)}{(n-1)}$. Thus $C_\epsilon$ must pass through $e$. Letting $\epsilon \to 0$, and noting that the number of cycles through $e$ is finite, we deduce that some optimal cycle of $G$ must pass through $e$.
$(ii)$ Let $uv \notin E(G)$. Join $u$ and $v$ by an edge $e$ of weight $\epsilon$, and denote the resulting graph by $G_\epsilon$. The above argument shows that $u$ and $v$ are connected in $G$ by a path of weight at least $\frac{2w(G)}{(n-1)}$. ■

**Proposition 1.4.11** *Let $G$ be a 2-connected cycle-extremal graph of positive weight with a 2-vertex cut $\{x, y\}$. Then $xy \in E(G)$ and $w(xy) = \frac{w(G)}{(n-1)}$.*

***Proof*** Let $G = H_1 \cup H_2$, where $V(H_1) \cap V(H_2) = \{x, y\}$ and $E(H_1) \cup E(H_2) = E(G) \setminus \{xy\}$. So,

$$w(H_1) + w(H_2) = w(G) - w(xy).....(i)$$

Note that $w(xy) = 0$ if $xy \notin E(G)$. Let $n_i =| V(H_i) |$ and $P_i$ an $(x, y)$-path of maximum weight in $H_i, i = 1, 2$. Since $H_i + xy$ is 2-connected and, by Proposition 1.4.7, $w(v) \geq \frac{w(G)}{(n-1)}$ for all $v \in V(H_i) \setminus \{x, y\}, i = 1, 2$. It follows from Theorem 1.3.8, that

$$w(P_i) \geq \frac{w(G)}{n-1}, i = 1, 2........(ii)$$

Put $G_i = H_i + xy$ with $w_G(xy) = w(P_{3-i}), i = 1, 2$. So $w(G_i) = w(H_i) + w(P_{3-i}), i = 1, 2$. Let $C$ be an optimal cycle in $G$ and $C_i$ an optimal cycle in $G_i, i = 1, 2$. By Theorem 1.3.15,

$$(n_i - 1)w(C_i) \geq 2w(G_i) = 2w(H_i) + 2w(P_{3-i}), i = 1, 2........(iii)$$

Since either $C_i$ is a cycle in $G$ or it can be converted to be a cycle in $G$ by replacing $xy$ with $P_{3-i}$, we have

$$w(C) \geq w(C_i), i = 1, 2.........(iv)$$

Thus,

$$nw(C) = (n_1 - 1)w(C) + (n_2 - 1)w(C) \geq (n_1 - 1)w(C_1) + (n_2 - 1)w(C_2).$$

By $(i)$ and $(iii)$,

$$nw(C) \geq 2(w(H_1) + w(H_2) + w(P_1) + w(P_2)) = 2(w(G) - w(xy) + w(P_1) + w(P_2)).$$

Hence,

$$w(xy) \geq w(G) + w(P_1) + w(P_2) - \frac{nw(C)}{2}.$$

Using $(ii)$, and noting that $w(C) = \frac{2w(C)}{(n-1)}$, we have,

$$w(xy) \geq \frac{w(G)}{n-1} .......(v)$$

Since $w(G) > 0$, this implies that $xy \in E(G)$. Now $P_1 \cup \{x, y\}$ is a cycle in $G$, which gives that

$$w(P-1) + w(xy) \leq w(C) = \frac{2w(G)}{n-1}.$$

This together with $(ii)$ and $(v)$ implies that $w(xy) = \frac{w(G)}{(n-1)}$, and the proof is complete. ■

**Definition 1.4.12** Let $G_1$ and $G_2$ be two nonseparable weighted graphs such that $V(G_1) \cap V(G_2) = \{x, y\}$ and $E(G_1) \cap E(G_2) = \{xy\}$. if

$$\frac{w(G_1)}{n_1 - 1} = w(xy) = \frac{w(G_2)}{n_2 - 1},$$

where $n_i =| V(G_i) |, i = 1, 2$, then $G_1 \cup G_2$ is called the **2-sum** of $G_1$ and $G_2$.

**Proposition 1.4.13** *Let $G$ be a 2-connected cycle-extremal graph of positive weight with a 2-vertex cut $\{x, y\}$. Then $G$ is a 2-sum of two cycle-extremal graphs.*

***Proof*** Define $G_i, n_i, C_i, i = 1, 2$ as in the proof of Proposition 1.4.9. Since equalities hold in $(iii)$ and $(iv)$ of Proposition 1.4.11,

$$\frac{2w(G_1)}{n_1 - 1} = w(C_1) = w(C) = w(C_2) = \frac{2w(G_2)}{n_2 - 1}.$$

This implies that $G_i$ is a cyclic extremal graph for $i = 1, 2$. Moreover, from Proposition 1.4.11, $xy \in E(G)$ and $w(xy) = \frac{w(G)}{(n-1)} = \frac{w(C)}{2}$. Hence

$$\frac{w(G_1)}{n_1 - 1} = w(xy) = \frac{w(G_2)}{n_2 - 1}.$$

Therefore $G$ is the 2-sum of the cyclic extremal graphs $G_1$ and $G_2$. ■

**Theorem 1.4.14** *A 2-edge connected weighted graph is cyclic extremal if and only if it is a weighted union of tritrees.*

The proof for this major theorem follows from Proposition 1.4.6, Theorem 1.3.23 and the following theorem.

**Theorem 1.4.15** *If $G$ is a cyclic-extremal graph, then $G = \sum_{T \in \mathcal{T}} \alpha_T T$ for some set $\mathcal{T}$ of tritrees of $G$.*

***Proof*** Proof is by by induction on $n$ and then on $\binom{n}{2} - \mid E(G) \mid$. For $n = 3$ or $\mid E(G) \mid = \binom{n}{2}$, the result follows from Theorem 1.3.16. Suppose now that $n > 3$ and $\mid E(G) \mid < \binom{n}{2}$. If $w(G) = 0$, the result is trivially true by taking $\mathcal{T} = \phi$. Thus we may assume that $w(G) > 0$.

If $G$ is separable, then by Proposition 1.4.9, $G$ is a 1-sum of cycle extremal graphs, $G_1$ and $G_2$. By the induction hypothesis,

$$G_1 = \sum_{i=1}^{m_1} \alpha'_j T'_j \text{ and } G_2 = \sum_{k=1}^{m_2} \alpha''_k T''_k.$$

By Proposition 1.4.4 and the definition of 1-sum,

$$G_1 = \sum_{i=1}^{m_1} \alpha'_j = \sum_{k=1}^{m_2} \alpha''_k.$$

Denote this common value by $\alpha$, and let

$$\alpha_{jk} = \frac{\alpha'_j \alpha''_k}{\alpha} \text{ and } T_{jk} = T'_j \cup T''_k,\ 1 \le j \le m_1 \text{ and } 1 \le k \le m_2.$$

It is clear that $T_{jk}$ is a tritree of $G$. We claim that

$$G = \sum_{j=1}^{m_1} \sum_{k=1}^{m_2} \alpha_{jk} T_{jk} ......(i)$$

Let $e \in E(G)$. We may suppose without loss of generality that $e \in E(G_1)$. Then, $w(e) = \sum\{\alpha'_j : e \in E(T'_j), 1 \le j \le m_1\} = \frac{1}{\alpha}(\sum_{k=1}^{m_1} \alpha''_k) \sum\{\alpha'_j : e \in E(T'_j), 1 \le j \le m_1\} = \sum\{\frac{\alpha'_j \alpha''_k}{\alpha} : e \in E(T'_j \cup T''_k), 1 \le j \le m_1, 1 \le k \le m_2\} = \sum\{\alpha_{jk} : e \in E(T_{jk}), 1 \le m_1, 1 \le k \le m_2\}$. Therefore $(i)$ holds as claimed.

If $G$ is 2-connected, and has a 2-vertex cut $\{x, y\}$ then by Proposition 1.4.13, $G$ is a 2-sum of cycle-extremal graphs $G_1$ and $G_2$. By induction hypothesis,

$$G_1 = \sum_{i=1}^{m_1} \alpha'_j T'_j \text{ and } G_2 = \sum_{k=1}^{m_2} \alpha''_k T''_k.$$

By Theorem 1.3.4 and the definition of 2-sum,

$$\sum_{j=1}^{m_1} \alpha'_j = w(xy) = \sum_{k=1}^{m_1} \alpha''_k \text{....}(ii)$$

$xy$ being the common edge of $G_1$ and $G_2$, by Propositions 1.4.5 and 1.4.11, each of the tritrees $T'_j$ and $T''_k$, $1 \le j \le m_1$ and $1 \le k \le m_2$, includes the edge $xy$. Furthermore, since $\{x, y\}$ is a 2-vertex cut of $G$, $T'_j \cup T''_k$ is a tritree of $G$, $1 \le j \le m_1$ and $1 \le k \le m_2$. As before, denote the common value in $(ii)$ by $\alpha$, and let

$$\alpha_{jk} = \frac{\alpha'_j \alpha''_k}{\alpha} \text{ and } T_{jk} = T'_j \cup T''_k,\ 1 \le j \le m_1,\ 1 \le k \le m_2.$$

Then,

$$G = \sum_{j=1}^{m_1} \sum_{k=1}^{m_2} \alpha_{jk} T_{jk},$$

as required by the theorem. Therefore, we may assume, noting that $n > 3$ that $G$ is 3-connected...$(iii)$.

Since $G$ is not complete, and by Proposition 1.4.10, the optimal paths of $G$ are of weight at least $\frac{2w(G)}{(n-1)}$. If $w(v) > \frac{w(G)}{(n-1)}$ for all $v \in V(G)$, then by Lemma 1.3.22, $G$ contains a cycle of weight more than $\frac{2w(G)}{(n-1)}$. This is impossible. Therefore, we may assume that there is a vertex $u \in V(G)$ such that $w(u) \le \frac{w(G)}{(n-1)}$. This implies by Proposition 1.4.7, that

$$w(u) = \frac{w(G)}{n-1} \text{.....}(iv)$$

Let $G' = G - u$. Then,

$$w(G') = w(G) - w(u) = \frac{n-2}{n-1} w(G).$$

Now let $C'$ be an optimal cycle in $G'$. Since $C' \subseteq G$,

$$w(C') \le \frac{2w(G)}{n-1}.$$

On the other hand by $(iii)$, $G'$ is 2-connected. It follows from Theorem 1.3.23 that

$$w(C') \ge \frac{2w(G')}{n-2} \ge \frac{2w(G)}{n-1}.$$

Consequently, $G'$ is 2-connected cyclic extremal graph......$(v)$.

Let $x, y \in N(u)$. If $xy \notin E(G)$, then by $(v)$ and proposition 1.4.10(ii), $x$ and $y$ are connected by a path $P'$ of weight

$$w(P') \geq \frac{2w(G')}{n-2} = \frac{2w(G)}{n-1}.$$

However, $xP'yux$ is a cycle in $G$, and $G'$ has no cycle of weight more than $\frac{2w(G)}{n-1}$. Therefore, $w(ux) = w(uy) = 0$. This shows that for all $x, y \in N(u)$,

$$xy \notin E(G) \to w(ux) = w(uy) = 0.......(vi)$$

Let $x, y \in N(u)$. If $xy \in E(G)$, then by $(v)$, and Proposition 1.4.10(i), there is an optimal cycle $C'$ in $G'$ passing through $xy$. Since $G'$ is cyclic extremal,

$$w(C') = \frac{2w(G')}{n-2} = \frac{2w(G)}{n-1}.$$

Let $C$ be the cycle obtained from $C'$ by replacing the edge $xy$ with the path $xuy$. Then $C \subseteq G$, and hence $w(ux) + w(uy) \leq w(xy)$, for otherwise $C$ is of weight more than $\frac{2w(G)}{(n-1)}$. Therefore, for all $x, y \in N(u)$, and $xy \in E(G)$,

$$w(ux) + w(uy) \leq w(xy).......(vii)$$

Set $X = \{v \in N(u) : w(uv) > 0\}$. By $(vi)$, $X$ induces a complete graph $K$ in $G$. If $| X |= 2$, let $C$ be the triangle $xyux$; if $| X |> 2$, let $C$ be a Hamiltonian cycle of $K$. In either case, by $(vii)$, $w(C) \geq 2w(u)$. It follows from $(iv)$ that $w(C) \geq \frac{2w(G)}{(n-1)}$. But, $G$ contains no cycle of weight more than $\frac{2w(G)}{(n-1)}$. So all inequalities in $(vii)$ are equalities. Furthermore, if we set $w(xy) = 0$ for $x, y \in N(u)$ and $xy \notin E(G)$, then we also have, by $(vi)$, that $w(ux) + w(uy) = 0 = w(xy)$. Therefore,

$$w(ux) + w(uy) = w(xy) \text{ for all } x, y \in N(u).......(viii)$$

Our next goal is to prove that

$$G' = \sum_{T' \in \mathcal{T}'} \alpha_{T'} T'......(ix)$$

where each $T' \in \mathcal{T}'$ is a spanning star of $G'$.

If $G'$ has no 2-vertex cut, then $(ix)$ follows from the induction hypothesis and Proposition 1.4.2. Suppose now, that $G'$ has a 2-vertex cut. Let $\{x, y\}$ be a 2-vertex cut of $G'$. By $(v)$ and Proposition 1.4.11, $xy \in E(G)$. Let $G' = G_1 \cup G_2$, where $V(G_1) \cap V(G_2) = \{x, y\}$ and $E(G_1) \cap E(G_2) = \{xy\}$. Choose the 2-vertex cut $\{x, y\}$ so that $| V(G_1) |$ is as small as possible. Then, any 2-vertex cut of $G'$ is contained in $V(G_2)$........$(x)$.

To see this, let $\{z_1, z_2\}$ be a 2-vertex cut of $G'$. If $\{z_1, z_2\} \nsubseteq V(G_2)$, suppose, without loss of generality, that $z_1 \in V(G_1) \setminus \{x, y\}$. By the choice of $\{x, y\}$, it must be that $z_2 \in V(G_2) \setminus \{x, y\}$. But $G_1 - z_1$ is a connected graph containing $xy$ and $G_2 -$

$z_2$ is a connected graph containing $xy$. Therefore $G - \{z_1, z_2\}$ is connected.This contradiction proves $(x)$.

Since $G$ is 3-connected, there is $z \in V(G_1) \setminus \{x, y\}$ and $zu \in E(G)$. Let $H$ be the graph obtained from $G'$ by adding edges of weight zero to join $z$ to all vertices in $N(u)$ which are not adjacent to $z$ in $G$. It is clear that any cycle in $H$ is either a cycle in $G$ or can be converted to a cycle in $G$ by replacing at most two edges of weight zero with two edges incident with $u$. This implies that no cycle in $H$ is of weight more than

$$\frac{2w(G)}{n-1} = \frac{2w(G')}{n-2} = \frac{2w(H)}{n-2}.$$

It follows from Theorem 1.3.23 that $H$ is cycle-extremal. By the induction hypothesis,

$$H = \sum_{T \in T_H} \alpha_T T,$$

where $T_H$ is a set of tritrees of $H$. Moreover, by $(iii)$, and the structure of $H$, $H$ has no 2-vertex cuts because any 2-vertex cut would contain $z$ and so contradicts $(x)$. Therefore, by Proposition 1.4.2, each tritree in $T_H$ is a spanning star, and $(ix)$ follows from Proposition 1.4.3.

Recall $X = \{v \in N(u) : w(uv) > 0\}$. If $| X | \leq 2$, then by $(viii)$ and $(iv)$, there is an edge $e \in E(G')$ incident with a vertex in $X$ with $w(e) = w(u) = \frac{w(G)}{(n-1)}$. By Proposition 1.4.6, each tritree $T'$ in $(ix)$ contains the edge $e$, and hence $T'$ is a star centered at a vertex in $X$. If $| X | > 2$, let $C$ be a Hamiltonian cycle of the graph induced by $X$. Then as mentioned before, $w(C) = 2w(u) = 2w(G)/(n-1)$. consequently, by Proposition 1.4.6, $| E(C) \cap E(T') | = 2$ for every tritree $T'$ in $(ix)$. This also yields that $T'$ is centered at a vertex in $X$. Therefore setting $V(G') = \{v_1, v_2, \ldots, v_{n-1}\}$ and $X = \{v_1, v_2, \ldots, v_m\}$, we may rewrite $(ix)$ as

$$G' = \sum_{i=1}^{m} \alpha_i T_i',$$

where $T_i'$ is the star tritree centered at vertex $v_i$, $1 \leq i \leq m$. Put $\alpha_i = 0, m+1 \leq i \leq n-1$. Then by Definition 1.3.19, for any $i$ and $j$ $1 \leq i < j \leq n-1$,

$$w(v_i v_j)\alpha_i + \alpha_j, \ldots.(xi)$$

where $w(v_i v_j) = 0$ if $v_i v_j \notin E(G)$. By $(iii)$, $| N(u) | \geq 3$. Let $v_i, v_j, v_k \in N(u)$. We have ,

$$w(uv_i) = \frac{1}{2}[(w(uv_i) + w(uv_j)) + (w(uv_i) + w(uv_k)) - (w(uv_j) + w(uv_k))].$$

By $(viii)$,

$$w(uv_i) = \frac{1}{2}[w(v_iv_j) + w(v_iv_k) - w(v_jv_k)].$$

It follows from $(xi)$ that

$$w(uv_i) = \frac{1}{2}[(\alpha_i + \alpha_j) + (\alpha_i + \alpha_k) - (\alpha_j + \alpha_k)] = \alpha_i.$$

Therefore, defining $T_i = T_i' + uv_i$, $1 \le i \le m$, we have,

$$G = \sum_{i=1}^{m} \alpha_i T_i,$$

where $T_i$ is a star tritree of $G$ centered at $v_i$, $1 \le i \le m$. This completes the proof. ■

**Definition 1.4.16** A weighted graph $G$ on $n$ vertices is path-extremal if its optimal paths are of weight precisely $\frac{2w(G)}{n}$.

**Theorem 1.4.17** *A weighted graph is path extremal if and only if it is a complete graph in which all edges have the same weight.*

***Proof*** Add a new vertex $v_0$ and join it to each vertex of $G$ by an edge of weight $M$, where $M > w(G)$. The resulting graph $G'$ is 2-connected and has weight $w(G') = w(G) + nM$. Let $C'$ be an optimal cycle in $G'$. Then, by Theorem 1.3.23,

$$w(C') \ge \frac{2w(G')}{n} = \frac{2w(G)}{n} + 2M.$$

Since $M > w(G)$, vertex $v_0$ lies on $C'$. Let $P = C' - v_0$. Then

$$w(P) = w(C') - 2M \ge \frac{2w(G)}{n}.$$

but equality holds here because $G$ is path extremal. Consequently,

$$w(C') = \frac{2w(G')}{n}$$

and $G'$ is cyclic extremal. By Theorem 1.3.24,

$$G' = \sum_{i=0}^{m} \alpha_i T_i' .....(i)$$

If $G'$ has a 2-vertex cut $\{x, y\}$, then either $x = v_0$ or $y = v_0$. Thus by Proposition 1.4.11,

$$M = w(xy) = \frac{w(G')}{n} = \frac{w(G)}{n} + M,$$

and $w(G) = 0$, a contradiction. If $G$ has no 2-vertex cut, then by Proposition 1.4.2, $T_i'$ is a spanning star of $G'$, $0 \leq i \leq m$. Since $M > w(G)$, there must be some $T_i'$ centered at $v_0$. We may suppose that $T_0'$ is the star tritree centered at $v_0$. If $\alpha_0 = M$, then $m = 0$ and $w(G) = 0$, again a contradiction. Otherwise, $m = n$ and $\alpha_i = M - \alpha_0 > 0$, $1 \leq i \leq n$. Let $\alpha = \alpha_i$ and $T_i = T_i' - v_0 v_i$, $1 \leq i \leq n$. It follows from $(i)$ that $G = \sum_{i=1}^{n} \alpha T_i$, and so $G$ is a uniformly weighted complete graph. ■

## 1.5 Weighted Digraphs

We shall consider only loopless digraphs; in other words, if $\langle x, y \rangle$ is a (directed) edge (or *arc*) of $G$ then $x \neq y$. We shall write $xy$ for $\langle x, y \rangle$. An **oriented graph** $G$ is a digraph with no cycles of length two: thus if $xy \in E(G)$ then $yx \notin E(G)$. An **edge-weighting** of a graph or digraph $G$ is a function $w : E(G) \to \mathbb{R}$. We shall only consider edge-weightings with non-negative weights. For $x \in V(G)$, the **inweight** of $x$ is

$$w_{\text{in}}(x) = \sum_{y \in \Gamma^-(x)} w(yx)$$

and the **outweight** of $x$ is

$$w_{\text{out}}(x) = \sum_{y \in \Gamma^+(x)} w(xy).$$

We shall assume that graphs and digraphs have at least one vertex.

**Theorem 1.5.1** *Let $G$ be a digraph with edge-weighting $w$ such that every vertex $v$ in $G$ satisfies $w_{\text{out}}(v) \geq 1$. Then $G$ contains a path $P$ such that*

$$w(P) \geq 1. \tag{1.5}$$

***Proof*** The key idea of our proof is that, in order to make induction easy, we prove a stronger assertion. Indeed, let $G$ be a digraph with edge-weighting $w$ and let $v_0 \in V(G)$. We prove by induction on $n = |G|$ that if every $v \in V(G) \backslash \{v_0\}$ satisfies $w_{\text{out}}(v) \geq 1$ then $G$ contains a path $P$ such that $w(P) \geq 1$. If $n = 2$, then the result is clear. Suppose $n > 2$ and $v_0 \in V(G)$.

If $v_0$ has no in edges, then consider the graph $G^* = G \backslash \{v_0\}$ with the same edge-weighting. Every vertex has outweight at least 1, so picking any $u \in V(G^*)$, the conditions of the inductive hypothesis are satisfied, so we can find a path $P$ in $G^*$ with $w(P) \geq 1$, which can also be considered as a path in $G$.

Otherwise, $d^-(v_0) > 0$. Let $uv_0$ be an edge with $w(uv_0)$ maximal, i.e., $uv_0$ is the heaviest in edge. Let $G^*$ be the digraph $G \backslash \{u, v_0\}$ with an extra vertex $x$ and, for

$v \in V(G)\backslash\{u, v_0\}$, an edge from $v$ to $x$ iff $vu \in E(G)$ or $vv_0 \in E(G)$. Thus $\Gamma^+_{G^*}(x) = \emptyset$ and $\Gamma^-_{G^*}(x) = (\Gamma^-_G(u) \cup \Gamma^-_G(v_0))\backslash\{u, v_0\}$. Let $w^*$ be the weighting obtained by setting $w^* = w$ on $G^*\backslash x$ and, for $vx \in E(G^*)$,

$$w^*(vx) = \begin{cases} w(vu) + w(uv_0) & vu \in E(G) \\ w(vv_0) & \text{otherwise.} \end{cases} \tag{1.6}$$

It is easily checked that $G^*$, $w^*$ and $x$ satisfy the conditions of the inductive hypothesis. Indeed, for $v \in V(G^*)\backslash\{x\}$, we have $w^*_{\text{out}}(v) = w_{\text{out}}(v) - w(vu) - w(vv_0) + w^*(vx)$. If $vu \in E(G)$, then

$$\begin{aligned} w^*_{\text{out}} &= w_{\text{out}}(v) - w(vu) - w(vv_0) + w(vu) + w(uv_0) \\ &= w_{\text{out}}(v) - w(vv_0) + w(uv_0) \\ &\geq w_{\text{out}}(v) \end{aligned}$$

by maximality of $w(uv_0)$. If $vu \notin E(G)$, then clearly $w^*_{\text{out}}(v) = w_{\text{out}}(v)$. Thus $w^*_{\text{out}} \geq 1$ for all $v \neq x$. Now $G^*$ has fewer vertices than $G$. Therefore, by our inductive hypothesis, there is a path $P^*$ contained in $G^*$ such that $w^*(P^*) \geq 1$. Now if $x \notin V(P^*)$ then $P^*$ can also be thought of as a path $P$ in $G$, where $w(P) = w^*(P^*)$, so we have the required path. Otherwise, $P^*$ must end in $x$, since $d^+_{G^*}(x) = 0$. Suppose the last edge in $P^*$ is $vx$. We use $P^*$ to define a path $P$ contained in $G$ as follows. $P^*$ is the same as $P$ except for the last vertex. If $vu \in E(G)$ then replace $vx$ with $vuv_0$; otherwise replace $vx$ with $vv_0$. In either case, it follows immediately from (1.6) that $w(P) \geq w^*(P^*) \geq 1$, so we have found the required path. ■

For strongly connected digraphs, we can say slightly more about our heavy paths.

**Corollary 1.5.2** *Let $G$ be a strongly connected digraph with edge-weighting $w$ such that every vertex $v$ in $G$ satisfies $w_{\text{out}}(v) \geq 1$. Then, for every vertex $v$ in $G$, there is a path $P$ such that $w(P) \geq 1$ and $P$ ends in $v$.*

***Proof*** As in the proof of Theorem 1.5.1, we prove a stronger assertion. Let $G$ be a digraph with edge-weighting $w$ and let $v \in V(G)$. We prove that if, for every vertex $v' \neq v$, $w_{\text{out}}(v') \geq 1$ and there is a path from $v'$ to $v$, then there is a path $P$ ending in $v$ such that $w(P) \geq 1$. It is easily checked that this condition is stable under the contraction used in the proof of Proposition 1.4.2; the result follows by a similar induction ■

**Theorem 1.5.3** *Let $G$ be a digraph with edge-weighting $w$, such that every vertex $v$ in $V(G)$ satisfies $w_{\text{out}}(v) \geq 1$. Then $G$ contains a cycle $C$ with $w(C) \geq (24n)^{-1/3}$.*

***Proof*** Let $c = 24^{-1/3}$ . We prove the assertion of the theorem by induction on $n = |G|$. As noted above, we may assume that $G$ is strongly connected (by considering a strongly connected component with no out edges). If there is an edge weighing more than $cn^{-1/3}$ then we can extend it to a cycle and we are done. Suppose then that no edge weighs more than $cn^{-1/3}$, and that $G$ contains no cycle of weight $cn^{-1/3}$.

Suppose first that some $v \in V(G)$ satisfies $d^+(v) \geq 6cn^{2/3}$. Starting with the triple $(G_0, w_0, v_0) = (G, w, v)$, consisting of our graph $G$, edge-weighting $w$ and special vertex $v$, we shall perform a sequence of contractions to obtain triples $(G_1, w_1, v_1)$, $(G_2, w_2, v_2), \ldots$ where each $G_i$ is a strongly connected digraph with edge weighting $w_i$ such that every vertex except $v_i$ has outweight at least 1.

Given $(G_i, w_i, v_i)$, if there is an edge weighing at least $cn^{-1/3}$ then we can extend it to a cycle of weight at least $cn^{-1/3}$. As we shall note below, this corresponds to a cycle in $G$ with weight at least $cn^{-1/3}$, which is a contradiction. Thus we may assume that no edge of $G_i$ has weight more than $cn^{-1/3}$. Let $vv_i$ be the heaviest edge into $v_i$ ($d^-_{G_i}(v_i) > 0$ since $G_i$ is strongly connected). We define $G^*$ by contracting the edge $vv_i$ : $G^*$ is obtained from $G$ by deleting $v$ and $v_i$ and adding a vertex $v_{i+1}$ with edges from $v_{i+1}$ to $y \in V(G^*)$ iff $v_i y \in E(G_i)$ and from $y$ to $v_{i+1}$ iff $yv \in E(G_i)$ or $yv_i \in E(G_i)$. We define the weighting $w^*$ by $w^* = w_i$ on $G^* \backslash \{v_{i+1}\}$, and $w^*(v_{i+1}y) = w_i(v_i y)$ for $v_{i+1}y \in E(G^*)$ and, for $yv_{i+1} \in E(G^*)$,

$$w^*(yv_{i+1}) = \begin{cases} w_i(yv) + w_i(vv_i) & \text{if } vu \in E(G_i) \\ w_i(vv_i) & \text{if } vu \notin E(G_i). \end{cases} \tag{1.7}$$

Clearly, no edge in $G^*$ weighs more than $2cn^{-1/3}$, since no edge in $G_i$ weighs more than $cn^{-1/3}$. Furthermore, a cycle in $G^*$ corresponds to a cycle of equal or greater weight in $G_i$, where we replace an edge $yv_{i+1}$ by $yvv_i$ or $yv_i$ as appropriate. Since all our operations will be contractions of this form and taking subgraphs, any cycle in $G_i$ corresponds to a cycle of equal or greater weight in $G$. Now let $H$ be a strongly connected component of $G_i \backslash \{v_{i+1}\}$ (this is well-defined, since $w^*(xy) \leq 2cn^{-1/3} < 1$ for $xy \in E(G^*)$). We define $G_{i+1}$ to be the subgraph of $G^*$ induced by $H \cup \{v_{i+1}\}$ and $w_{i+1}$ to be $w^*$ restricted to this graph. For $y \neq v_{i+1}$, the outweight of $y$ in $G_{i+1}$ is equal to the outweight of $y$ in $G^*$; it follows from (1.7) that this is at least as large as the outweight of $y$ in $G_i$, which is at least 1.

We claim that $G_{i+1}$ is also strongly connected. Indeed, it is enough to show that $d^-(v_{i+1}) > 0$ and $d^+(v_{i+1}) > 0$. If $d^-(v_{i+1}) = 0$, then consider the digraph $G' = G_{i+1} \backslash \{v_{i+1}\}$. Every vertex has outweight at least 1, so $G'$ contains a cycle of weight at least $c|G'|^{-1/3} > cn^{-1/3}$, which corresponds to a cycle in $G$ of weight at least $cn^{-1/3}$, which is a contradiction. Thus $d^-(v_{i+1}) > 0$. If $d^+(v_{i+1}) = 0$, then consider the same digraph $G'$. Each $y \in V(G')$ has outweight at least $1 - w_{i+1}(yv_{i+1}) \geq 1 - 2cn^{-1/3}$. However, $G'$ contains no vertices from $\Gamma^+_G(v)$ (if $vy \in E(G)$ and $y \in V(G_{i+1})$ then we would have $v_j y \in E(G_j)$ for $j = 0, \ldots, i+1$), and so $|G'| \leq n - d^+_G(v) - 1 < n - 6cn^{2/3}$. Thus, by our inductive hypothesis, $G'$ contains a cycle of weight at least

$$\frac{c(1 - 2cn^{-1/3})}{(n - 6cn^{2/3})^{1/3}} > cn^{-1/3},$$

which corresponds to a cycle of weight at least $cn^{-1/3}$ in $G$, a contradiction. Thus $d^+(v_{i+1}) > 0$, and so $G_{i+1}$ is strongly connected. However, clearly $|G| = |G_0| > |G_1| > \cdots$, so at some point we reach a contradiction.

Therefore, every vertex $v \in V(G)$ must satisfy $d^+(v) < 6cn^{2/3}$. Let $G'$ be the graph obtained from $G$ by deleting every edge weighing less than $n^{-2/3}/12c$. Then every vertex still has outweight at least $1 - 6cn^{2/3}(n^{-2/3}/12c) = 1/2$. Now no edge weighs more than $cn^{-1/3}$, so every vertex must satisfy

$$d^+_{G'}(v) \geq (1/2)/(cn^{-1/3}) = n^{1/3}/2c.$$

Therefore, $G'$ contains a cycle of length at least $n^{1/3}/2c$, which must weigh at least $(n^{1/3}/2c)(n^{-2/3}/12c) = n^{-1/3}/24c^2 = cn^{-1/3}$, which is a contradiction. ■

## References

1. König, D.: Theorie der endlichen und unendlichen Graphen. Akademische Verlagsgesellschaft, Leipzig (1936)
2. Berge, C.: The Theory of Graphs. Methuen (1962)
3. Ore, O.: Theory of Graphs, vol. 38. American Mathematical Society (1962)
4. Harary, F., Graph Theory. Addison-Wesley, Reading (1969)
5. Roberts, F., S.: Discrete Mathematical Models, with Applications to Social, Biological, and Environmental Problems. Prentice-Hall (1976)
6. Coulson, C. A.: On the calculation of the energy in unsaturated hydrocarbon molecules. In: Mathematical Proceedings of the Cambridge Philosophical Society, vol. 36, pp. 201–203. Cambridge University Press (1940)
7. Cayley, A.: On the theory of the analytical forms called trees. London, Edinburgh, Dublin Philos. Mag. J. Sci. **13**(85), 172–176 (1857)
8. Bandelt, H.J., Chepoi, V.: Metric graph theory and geometry: a survey. Contemp. Math. **453**, 49–86 (2008)
9. Biggs, N., Biggs, N. L., Biggs, E.N.: Algebraic Graph Theory, vol. 67. Cambridge University Press (1993)
10. Gibbons, A.: Algorithmic Graph Theory. Cambridge University Press (1985)
11. Chung, F.R.K., Graham, F.C.: Spectral Graph Theory, vol. 92. American Mathematical Society (1997)
12. Chartrand, G., Zhang, P.: Introduction to Graph Theory. Tata McGraw-Hill Edition (2006)
13. Chartrand, G., Lesniak, L.: Graphs and Digraphs. Chapman & Hall/CRC, USA (2000)
14. West, D.B.: Introduction to Graph Theory, vol. 2. Prentice Hall (2001)
15. Buckley, F., Harary, F.: Distance in Graphs. Addision-Wesley, Redwood City (1990)
16. Chartrand, G., Zhang, P.: The Steiner number of a graph. Disc. Math **242**, 41–54 (2002)
17. Erdos, P., Gallai, A.: On maximal paths and circuits in graphs. Acta Math. Sci. Hung **10**, 337–356 (1959)
18. Bondy, J.A., Fan, G.: Cycles in weighted graphs. Combinatorica **11**, 191–205 (1991)
19. Chartrand, G., Saba, F., Zou, H.B.: Edge rotations and distance between graphs. Casopis pro pestovani matematiky **110**(1), 87–91 (1985)
20. Entringer, R.C., Jackson, D.E., Snyder, D.A.: Distance in graphs. Czechoslov. Math. J. **26**(2), 283–296 (1976)
21. Goddard, W., Oellermann, O.R.: Distance in graphs. Struct. Anal. Complex Netw. 49–72 (2011)
22. Chartrand, G., Escuadro, H., Zhang, P.: Detour distance in graphs. J. Combin. Math. Combin. Comput. **53**, 75–94 (2005)

23. Doyle, J.K., Graver, J.E.: Mean distance in a graph. Discret. Math. **17**(2), 147–154 (1977)
24. Dankelmann, P., Entringer, R.: Average distance, minimum degree and spanning trees. J. Graph Theory **33**(1), 1–13 (2000)
25. Brouwer, A.E., Haemers, W.H.: Distance Regular Graphs, Spectra of Graphs, pp. 177–185. Springer, New York (2012)
26. Bandelt, H.J., Mulder, H.M.: Distance hereditary graphs. J. Combin. Theory, Ser. B **41**(2), 182–208 (1986)
27. Ramane, H.S., Gutman, I., Revankar, D.S.: Distance equi energetic graphs. MATCH Commn. Math. Comput. Chem. **60**, 473–484 (2008)
28. Jerebic, J., Klavzer, S., Rall, D.: Distance balanced graphs. Ann. Comb. **12**(1), 71–79 (2008)
29. Mulder, H.: The Interval Function of Graphs, MCT132. Mathematics Centrum, Amsterdam (1980)
30. Mulder, H., Nebesky, L.: Axiomatic characterization of the interval function of a graph. Eur. J. Combin. **30**(5), 1172–1185 (2009)
31. Nebesky, L.: A characterization of the interval function of a connected graph. Czechoslov. Math. J. **44**(1), 173–178 (1994)
32. Nebesky, L.: Characterizing the interval function of a connected graph. Math. Bohem. **123**(2), 137–144 (1998)
33. Fishburn, P.C.: Interval graphs and interval orders. Discret. Math. **55**(2), 135–149 (1985)
34. McKee, T.A., McMorris, F.R.: Topics in Intersection Graph Theory. SIAM Monographs on Disc. Mathematics and Applications, Philadelphia (1999)
35. Goldman, V.V., Horton, G.K., Klein, M.L.: Phonon energies and lifetimes in Ne and He in the first floor self consistent approximation. Phys. Rev. Lett. **24**(25), 14–24 (1970)
36. Hedlikova, J.: Ternary spaces, media and Chebychev sets. Czechoslov. Math. J. **33**(3), 373–389 (1983)
37. Dress, A.W., Scharlau, R.: Gated sets in metric spaces. Aequat. Math. **34**(1), 112–120 (1987)
38. Tardif, C.: Prefibers and cartesian product of metric spaces. Discret. Math. **109**(1–3), 283–288 (1992)
39. Brualdi, R.A., Ryser, H.J.: Combinatorial Matrix Theory, vol. 39. Cambridge University Press, Cambridge (1991)
40. Mathew, S.: On totally weighted interconnection networks. J. Interconnect. Netw. **14**, 1350004 (1–16) (2013)
41. Mathew, S., Sunitha, M.S.: Some connectivity concepts in weighted graphs. Adv. Appl. Discrete Math. **6**(1), 45–54 (2010)
42. Mathew, S., Sunitha, M.S.: Cycle connectivity in weighted graphs. Proyecciones J. Math. **30**(1), 1–17 (2001)
43. Mathew, S., Sunitha, M.S.: Partial trees in weighted graphs -1. Proyecciones J. Math. **30**(2), 163–174 (2011)
44. Mathew, S., Sunitha, M.S.: A characterization of partial blocks in weighted graphs. Inf. Process. Lett. **112**(17–18), 706–710 (2012)
45. Mathew, S., Sunitha, M.S.: A generalization of classical connectivity parameters for graphs. World Appl. Sci. J. **22**, 18–22 (2013)
46. Bondy, J.A., Fan, G.: Optimal paths and cycles in weighted graphs. Ann. Discret. Math. **41**, 53–69 (1989)
47. Bondy, J.A., Broersma, H.J., Van den Heuvel, J., Veldman, H.J.: Heavy cycles in weighted graphs. Discuss. Math. Graph Theory **22**, 7–15 (2002)
48. Fujisawa, J.: Heavy fans, cycles and paths in weighted graphs of large connectivity. Discret. Math. **307**(1), 38–53 (2007)
49. Posa, L.: On the circuits of finite graphs. Magyar Tud. Akad. Mat. Kutato Int. Kozl. **8**, 355–361 (1963)
50. Fujisawa, J.: Weighted degrees and heavy cycles in weighted graphs. Discret. Math. **309**, 6483–6495 (2009)
51. Enomoto, H.: Long paths and large cycles in finite graphs. J. Graph Theory **8**, 287–301 (1984)
52. Grotschel, M.: Graphs with cycles containing given paths. Ann. Discret. Math. **1**, 233–245 (1977)

# Chapter 2
# Connectivity

Connectivity is one of the major topics in graph theory. Concepts like cut vertex and bridge were introduced by Euler in his basic papers of graph theory. Most of the applications of graph theory, especially in the field of networking are based on connectivity. This chapter focusses on the weighted analogues of basic connectivity concepts in unweighted graphs. Note that in classical sense, a graph without weights is working in the principles of Aristotle's logic; for example, the removal of a vertex from a graph can disconnect the graph or not. Two vertices are adjacent or not, etc. But when the graph becomes weighted, different possibilities like, reduction of strength of paths, increase in total connectivity for subgraphs, etc. may arise. As a consequence we discuss different types of new structures in weighted graph theory. The first attempt in this direction was made in [1]. This chapter is based on [1–4].

## 2.1 Strength of Connectedness

Problems like maximum bandwidth problem, widest path problem, bottleneck problem, etc. are well known in Mathematics and Computer Science. Basically they are problems related with networks or weighted graphs. In internet routing, the minimum weight of edges in a path is referred as the bandwidth of the path and the maximum among all such paths between two routers as the maximum bandwidth between the routers. In this section, we discuss this concepts in detail. We take $w$ as the weight function on the edge set.

**Definition 2.1.1** Let $G = (V, E, w)$ be a weighted graph. The **strength of a path** $P = v_0e_1v_1 \cdots e_nv_n$, from $v_0$ to $v_n$ denoted by $s(P)$ is defined as $s(P) = w(e_1) \wedge w(e_2) \wedge w(e_3) \wedge \cdots \wedge w(e_n)$, where $\wedge$ denote the minimum.

 
S. Mathew et al., *Weighted and Fuzzy Graph Theory*, Studies in Fuzziness and Soft Computing 429, https://doi.org/10.1007/978-3-031-39756-1_2

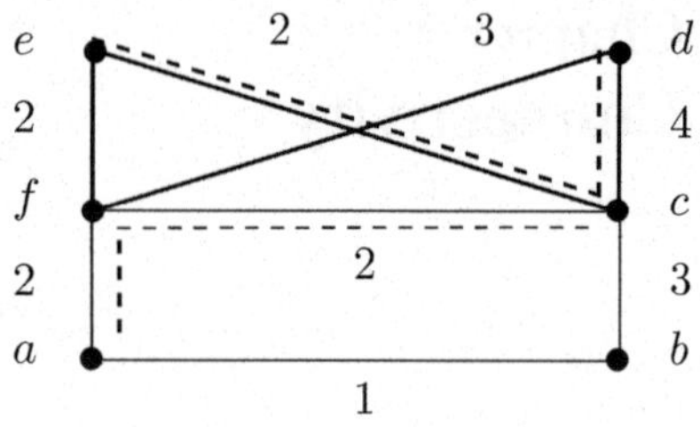


**Fig. 2.1** Strength of connectedness

Clearly strength of paths and cycles in unweighted graph is one. Also strength of a cycle can be similarly defined in weighted graphs.

One of the most important concept in weighted graphs, namely strength of connectedness is defined as follows.

**Definition 2.1.2** Let $G = (V, E, w)$ be a weighted graph. The **strength of connectedness** of a pair of vertices $u, v \in V$, denoted by $CONN_G(u, v)$ is defined as $CONN_G(u, v) = \vee\{s(P) : P \text{ is a } u - v \text{ path in } G\}$. Here $\vee$ denote the maximum. A $u - v$ path with strength equal to $CONN_G(u, v)$ is called a **strongest path.**

Note that $CONN_G(u, v) = 0$ if $u$ and $v$ are in different components. A strongest path is also called a **widest path** or **maximum bandwidth path**.

**Example 2.1.3** Consider the weighted graph in Fig. 2.1.

In Fig. 2.1, the strength of the $a - d$ path $abcd$ is one and that of the $a - d$ path $afcd$ is 2. Indeed, all paths from $a$ to $d$ passing through $b$ are of strength one and all paths from $a$ to $d$ passing through $f$ are of strength 2. Hence, $CONN_G(a, d) = 2$. Paths $afecd$ and $afcd$ are strongest $a - d$ paths. Next we have an obvious result.

**Proposition 2.1.4** *Let $G = (V, E, w)$ be a weighted graph and $H$, a weighted subgraph of $G$. Then for any pair of vertices $u, v \in V$, we have $CONN_H(u, v) \leq CONN_G(u, v)$.*

Now we generalize the concept of a cut vertex and a bridge in a graph. Naturally, when we delete a cut vertex, the graph becomes disconnected. That is, the strength of connectedness between some pair of vertices reduces from 1 to 0. In weighted graphs there exist vertices, whose removal reduces the strength of connectedness between some other pair of vertices. We call them partial cut vertices. Note that a cut vertex is clearly a partial cut vertex. The term partial is used in the sense that the maximum flow between some pair of vertices in $G - v$ is partially affected on the removal of $v$ if $v$ is such a vertex. A similar definition holds for partial bridges also.

**Definition 2.1.5** Let $G$ be a weighted graph. A vertex $w$ is called a **partial cut vertex** or a $p$**-cut vertex** of $G$ if there exists a pair of vertices $u, v$ in $G$ such that $u \neq v \neq w$ and $CONN_{G-w}(u, v) < CONN_G(u, v)$.

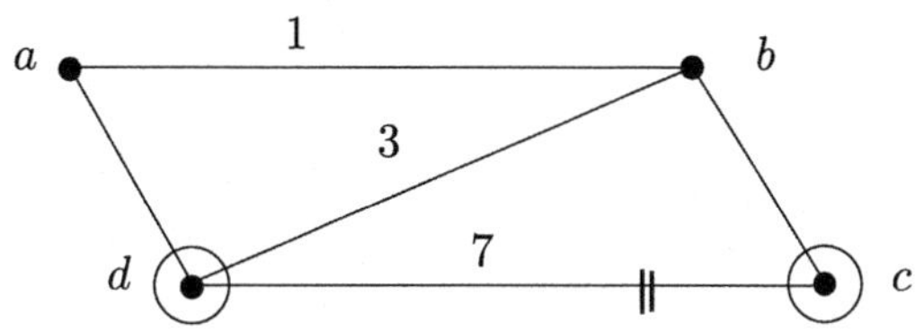


**Fig. 2.2** $p$-cut vertices and $p$-bridges

**Definition 2.1.6** Let $G$ be a weighted graph. An edge $e = uv$ is called a **partial bridge** or a $p$**-bridge** if $CONN_{G-e}(u, v) < CONN_G(u, v)$.

Similar to graphs without weights, we can give simple characterizations for $p$-cut vertices and $p$-bridges in weighted graphs as in the following propositions.

**Proposition 2.1.7** *Let $G = (V, E, w)$ be a weighted graph and $v$ be a vertex of $G$. Then $v$ is a $p$-cut vertex if and only if $v$ is an internal vertex of every strongest $x - y$ path for some pair of vertices $x, y \in V$ different from $v$.*

**Proposition 2.1.8** *Let $G = (V, E, w)$ be a weighted graph and let $e$ be an edge of $G$. Then, $e$ is a $p$-bridge if and only if $e$ is an edge in every strongest $x - y$ path for some pair of vertices $x, y \in V$.*

The proofs of both these propositions follow from the fact that the removal of every strongest path between some pair of vertices $a$ and $b$ reduces the strength of connectedness between $a$ and $b$. The proof is left to the reader. Note that every cut vertex of $G$ is a $p$-cut vertex and every bridge of $G$ is a $p$-bridge. The converse is not true as seen from the following example.

**Example 2.1.9** Consider the weighted graph in Fig. 2.2. Here $dc$ is a $p$-bridge since its deletion from $G$ reduces the strength of connectedness between $d$ and $c$ from 7 to 3. Similarly $ad$ also is a $p$-bridge. Also $c$ and $d$ are $p$-cut vertices of $G$ since $CONN_{G-c}(b, d) = 1 < 3 = CONN_G(b, d)$ and $CONN_{G-d}(a, c) = 1 < 4 = CONN_G(a, c)$. Note that $G$ is 2-connected and hence has no cut vertices or bridges.

Now we have some equivalent conditions for an edge to be a $p$-bridge, which is a weighted graph version of a fuzzy graph result by Rosenfeld [5].

**Theorem 2.1.10** *Let $G = (V, E, w)$ be a weighted graph and $e = xy \in E$. Then the following statements are equivalent.*

$(i)$ $e$ is a $p$-*bridge.*

$(ii)$ $CONN_{G-e}(x, y) < w(e)$.

$(iii)$ *$e$ is not a weakest edge of any cycle in $G$.*

***Proof*** $(ii) \Rightarrow (i)$ Assume the condition in $(ii)$. If $e$ is not a $p$-bridge, then $CONN_{G-e}(x, y) = CONN_G(x, y) \geq w(e)$, which violates $(ii)$. Hence $(i)$ holds.

$(i) \Rightarrow (iii)$ Assume $(i)$ and suppose that $e = xy$ is a weakest edge of a cycle say $C$. Then any path $P$ containing the edge $xy$ can be converted into a path not containing the edge $xy$, whose strength is at least equal to that of the path $P$ (using the path $C - xy$). Thus edge $xy$ cannot be a $p$-bridge.

$(iii) \Rightarrow (ii)$ If $CONN_{G-e}(x, y) \geq w(e)$, then there exists a path say $P$ from $x$ to $y$ not involving edge $e$ that has strength at least $w(e)$. Then $P \cup \{e\}$ is a cycle of which $e$ is a weakest edge. ■

A maximum spanning tree of a weighted graph is a spanning tree of maximum weight. The next result gives a method to construct a maximum spanning tree.

**Theorem 2.1.11** *Let $G = (V, E, w)$ be a connected weighted graph. If $T_0$ is a spanning tree obtained by successively removing weakest edges from cycles in $G$, then $T_0$ is a maximum spanning tree of $G$.*

***Proof*** Let $T$ be a maximum spanning tree of $G$. If $E(T_0) \subseteq E(T)$, then $T_0$ is a maximum spanning tree. Assume $E(T_0) \nsubseteq E(T)$. There exists at least one edge (say, $e_0 = u_0v_0$) in $T_0$ which is not there in $T$. $e_0$ together with the $u_0 \ldots v_0$ path in $T$ form a cycle $C$ in $G$. In the construction of $T_0$, a weakest edge $e$ is removed from $C$ and it is not $e_0$. So, $w(e_0) \geq w(e)$. $T + e_0$ contains exactly one cycle. Removing $e$ from it gives a spanning tree $T_1$ of $G$. If $w(e_0) > w(e)$, then weight of $T_1$ is greater than $T$, a maximum spanning tree. This is not possible. Therefore, $w(e_0) = w(e)$ and $T_1$ is also a maximum spanning tree of $G$. $T_1$ is a maximum spanning tree containing $e_0$ where $e_0$ is an edge in $T_0$, not in $T$. Suppose $E = \{e_0, e_1, \ldots, e_k\}$ be the edges in $T_0$ not present in $T$. Similarly, every edge in $T_0$ not in $T$ (say, $e_i$; $1 \leq i \leq k$) can be successively inserted into a maximum spanning tree $T_i$ giving another maximum spanning tree, $T_{i+1}$. Let $T_{k+1}$ is the maximum spanning tree obtained by inserting $e_k$ into $T_k$. $T_{k+1}$ contains all the edges in $T_0$. Therefore, $T_{k+1} = T_0$. Hence, $T_0$ is a maximum spanning tree. ■

**Theorem 2.1.12** *Let $G = (V, E, w)$ be a connected weighted graph and $T$ be a spanning tree of $G$. If every path in $T$ is strongest, then $T$ is a maximum spanning tree of $G$.*

***Proof*** If $G$ is a weighted tree, then $T = G$ and is the only spanning tree of $G$. Hence, $T$ is the only maximum spanning tree of $G$. Suppose $G$ contains cycles. Let $e = uv$ be an edge of $G$, not present in $T$. $e$ together with the strongest $u - v$ path in $T$ form a cycle $C(e)$ in $G$. Clearly, $e$ is a weakest edge of $C(e)$. Apply the algorithm in Theorem 2.1.11 on $G$, starting with $C(e)$. When weakest edge is removed from each of $C(e)$, we get $T$. From Theorem 2.1.11, $T$ is a maximum spanning tree. ■

Next we characterize $p$-bridges of a connected weighted graph using the concept of maximum spanning trees. Note that every path in a maximum spanning tree is a strongest path, by definition.

**Theorem 2.1.13** *Let $G = (V, E, w)$ be a connected weighted graph. Then an edge $e \in E$ is a p-bridge if and only if e is in every maximum spanning tree of G.*

***Proof*** Let $e = xy$ be a $p$-bridge of a weighted graph $G$. Then by Theorem 2.1.10 edge $xy$ is the unique strongest $x - y$ path in $G$ and hence is in every MST of $G$.

Conversely suppose that $e = xy$ is an edge of $G$ which is in every MST of $G$. If $e$ is not a $p$-bridge, by Theorem 2.1.10, $e$ must be a weakest edge of a cycle say $C$ in $G$ and hence $CONN_G(x, y) \geq w(e)$. Thus there exists at least one MST of $G$ not containing the edge $e$. ∎

If $G$ is a weighted graph with $n$ vertices, then $G$ can have at most $n - 1$ $p$-bridges. The equality occurs when $G$ is a weighted tree with $n$ vertices.

Next we characterize $p$-cut vertices of a weighted graph using the concept of maximum spanning trees.

**Theorem 2.1.14** *A vertex $v$ in a weighted graph $G = (V, E, w)$ is a p-cut vertex if and only if $v$ is an internal vertex of every MST.*

***Proof*** Let $v$ be a $p$-cut vertex of a weighted graph $G$. Then for some vertices $x, y \in V$, $v$ is in every strongest $x - y$ path. Now every MST contains a unique strongest $x - y$ path and hence $v$ is an internal vertex of each MST of $G$.

Conversely suppose that $v$ is an internal vertex in every MST. Let $T$ be any MST and let $xv$ and $vy$ are edges incident on $v$ in $T$. Clearly $xvy$ is a strongest $x - y$ path in $T$. If possible, suppose that $v$ is not a $p$-cut vertex. Then between every pair of vertices $u, u' \in V$, there exists a strongest path which avoids $w$. Consider one such $x - y$ path $P$. Clearly $P$ contains edges which are not in $T$. $CONN_G(x, y)$ is either weight of the edge $xv$ or weight of $vy$. If $CONN_G(x, y) = w(xv)$, then edges of $P$ will have weight at least $w(xv)$. Removal of edge $xv$ and addition of $P$ in $T$ gives another MST in which $v$ is a pendent vertex, a contradiction. ∎

From the above theorem it follows that in any weighted graph, there always exist at least two vertices which are non $p$-cut vertices.

## 2.2 Strong Edges and Strong Paths

In a graph without weights, all pairs of vertices have strength of connectedness one. But in weighted graphs, the strength of connectedness is different for different pairs of vertices. In this section we classify the edges in a weighted graph as strong and non strong. Further, we classify strong edges as $\alpha$-strong and $\beta$-strong as in the following definition. This section is based on [1].

**Definition 2.2.1** Let $G = (V, E, w)$ be a weighted graph. An edge $e = xy \in E$ is said to be $\alpha$**-strong** if $CONN_{G-e}(x, y) < w(e)$, $\beta$**-strong** if $CONN_{G-e}(x, y) = w(e)$ and a $\delta$**-edge** if $CONN_{G-e} > w(e)$. A $\delta$—edge $e$ is called a $\delta^*$**-edge** if $e$ is not a weakest edge of $G$.

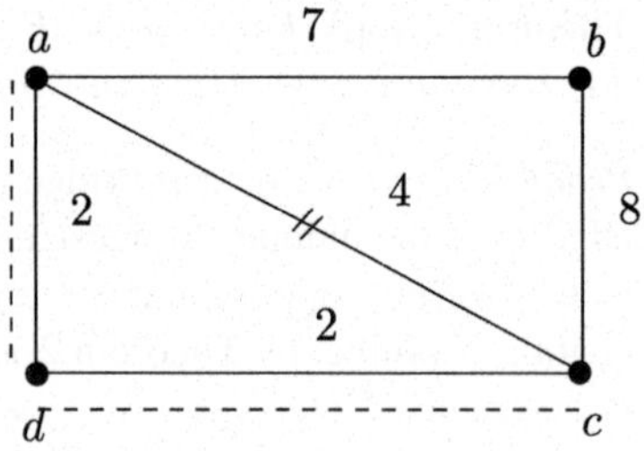


**Fig. 2.3** Graph in Example 2.2.4

Clearly an edge $e$ is strong if it is either $\alpha$-strong or $\beta$-strong. That is edge $xy$ is strong if its weight is at least equal to the strength of connectedness between $x$ and $y$ in $G$.

**Definition 2.2.2** A $u-v$ path $P$ in $G$ is called a **strong** $u-v$ **path** if all edges in $P$ are strong. In particular if all edges of $P$ are $\alpha$-strong, then $P$ is called an **$\alpha$-strong path** and if all edges of $P$ are $\beta$-strong, then $P$ is called a **$\beta$-strong path**.

**Definition 2.2.3** Let $G$ be a weighted graph and $C$, a cycle in $G$. $C$ is called a **strong cycle** if all edges in $C$ are strong.

**Example 2.2.4** Consider the weighted graph $G(V, E, w)$ given in Fig. 2.3.

Here $ab$ and $bc$ are $\alpha$-strong, $cd$ and $da$ are $\beta$-strong and edge $ac$ is a $\delta$-edge. Clearly edge $ac$ is $\delta^*$ since it is not a weakest edge in $G$. Also $P_1 = abc$ is an $\alpha$-strong path, $P_2 = cda$ is a $\beta$-strong path. In $G$, $C_1 = abcda$ is a strong cycle, but $C_2 = abca$ is not a strong cycle.

**Proposition 2.2.5** *An edge $xy$ of a weighted graph $G = (V, E, w)$ is strong if and only if $w(e) = CONN_G(x, y)$.*

***Proof*** Suppose $e = xy$ is strong. Then $CONN_G(x, y) \geq w(e)$. If $CONN_G(x, y) > w(e)$, then there exists a path say $P$ from $x$ to $y$ such that $s(P) > w(e)$. Then $e$ becomes the unique weakest edge of a cycle and hence $e$ cannot be strong, a contradiction.

Conversely suppose $w(e) = CONN_G(x, y)$. Then, $e$ itself is a strongest $x-y$ path. If $e$ is the unique strongest path, then $CONN_{G-e}(x, y) < w(e)$ and hence $e$ must be $\alpha$-strong. If there exists an alternate strongest $x-y$ path, then $CONN_{G-e}(x, y) = w(e)$ and hence $e$ must be $\beta$-strong. ■

Next we characterize $p$-bridges in weighted graphs using $\alpha$-strong edges.

**Theorem 2.2.6** *An edge $e$ in a weighted graph $G = (V, E, w)$ is a partial bridge if and only if $e$ is $\alpha$-strong.*

***Proof*** Suppose that $G$ is a weighted graph and $e = uv$ is a $p$-bridge of $G$. By Theorem 2.1.10, $CONN_{G-e}(u, v) < w(uv)$. By definition, it follows that $e$ is $\alpha$-strong.

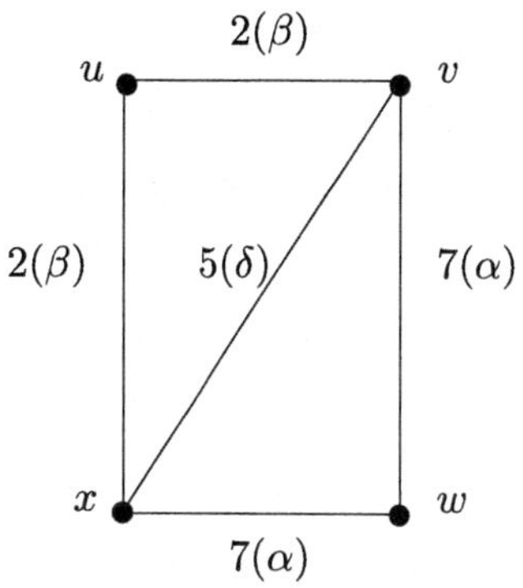


**Fig. 2.4** Types of edges—illustration

Conversely suppose that $e = uv$ is an $\alpha$-strong edge of $G$. Then $CONN_{G-e}(u, v) < w(e)$. That is edge $uv$ is the unique strongest path from $x$ to $y$ in $G$. Hence the removal of $e$ from $G$ reduces the strength of connectedness between $u$ and $v$. Thus it follows that $e$ is a $p$-bridge of $G$. ■

The concepts of **strong path** and strongest path are independent in a weighted graph. A strongest path can contain all types of edges as seen from the following example.

**Example 2.2.7** Let $G(V, E, w)$ be a the weighted graph given in Fig. 2.4.

In this graph, $CONN_G(u, w) = 2$. Hence $P = uvxw$ is a strongest $u - w$ path. $P$ contains all types of edges. Edge $uv$ is $\beta$—strong, $vx$ is a $\delta$—edge and edge $xw$ is $\alpha$—strong.

Note that in a graph without weights, if we assume that every edge is given a weight one, then each of its edges is strong and each path is strongest. In weighted graphs any strongest path not containing a $\delta$-edge is a strong path. We discuss the converse in the following proposition.

**Proposition 2.2.8** *Let $G$ be connected weighted graph and let $x, y$ be any two vertices in $G$. Then there exists a strong path from $x$ to $y$.*

We leave the proof to the reader. The proof lies in the fact that every $\delta$-edge $e$ can be replaced by a path $P$ of strength $s(P) > w(e)$.

**Proposition 2.2.9** *Let $G$ be a weighted graph. Let $P$ be a strong $x - y$ path in $G$. Then $P$ becomes a strongest $x - y$ path in the following cases.*

*$(i)$ $P$ contains only $\alpha$ -strong edges.*
*$(ii)$ If $P$ is the unique strong $x - y$ path.*
*$(iii)$ If all $x - y$ path in $G$ are of equal strength.*

***Proof*** $(i)$ Let $G$ be a weighted graph. and let $P$ be a strong $x - y$ path in $G$ containing only $\alpha$-strong edges. If possible suppose that $P$ is not a strongest $x - y$ path. Let $Q$

be a strongest $x - y$ path in $G$. Then $P \bigcup Q$ will contain at least one cycle $C$ in which every edge of $C - P$ will have strength greater than strength of $P$. Thus a weakest edge of $C$ is an edge of $P$. Let $uv$ be such an edge of $C$. Let $C'$ be the $u - v$ path in $C$, not containing the edge $uv$. Then, $w(uv) \leq$ Strength of $C' \leq CONN_{G-(u,v)}(u, v)$, which implies that $uv$ is not $\alpha$-strong, a contradiction. Thus $P$ is a strongest $x - y$ path.

$(ii)$ Let $G$ be a weighted graph and let $P$ be the unique strong $x - y$ path in $G$. If possible suppose that $P$ is not a strongest $x - y$ path. Let $Q$ be a strongest $x - y$ path in $G$. Then, strength of $Q >$ strength of $P$. That is, for every edge $uv$ in $Q$, $w(uv) > w(x'y')$ where $x'y'$ is a weakest edge of $P$.

**Claim** $Q$ is a strong $x - y$ path.
For; otherwise, if there exists an edge $uv$ in $Q$ which is a $\delta$-edge, then, $w(uv) < CONN_{G-(u,v)}(u, v) \leq CONN_G(u, v)$ and hence $w(uv) < CONN_G(u, v)$.

Thus, there exists a path from $u$ to $v$ in $G$ whose strength is greater than $w(uv)$. Let it be $P'$. let $w$ be the last vertex after $u$, common to $Q$ and $P'$ in the $u - w$ sub path of $P'$ and $w'$ be the first vertex before $v$, common to $Q$ and $P'$ in the $w' - v$ sub path of $P'$ (If $P'$ and $Q$ are disjoint $u - v$ paths then $w = u$ and $w' = v$). Then the path $P''$ consisting of the $x - w$ path of $Q$, $w - w'$ path of $P'$, and $w' - y$ path of $Q$ is an $x - y$ path in $G$ such that strength of $P'' >$ strength of $Q$, contradiction to the assumption that $Q$ is a strongest $x - y$ path in $G$. Thus $uv$ cannot be a $\delta$-edge and hence $Q$ is a strong $x - y$ path in $G$.

Thus we have another strong path from $x$ to $y$, other than $P$, which is a contradiction to the assumption that $P$ is the unique strong $x - y$ path in $G$. Hence $P$ should be a strongest $x - y$ path in $G$.

$(iii)$ If every path from $x$ to $y$ have the same strength, then each such path is a strongest $x - y$ path. In particular a strong $x - y$ path is a strongest $x - y$ path. ■

Next theorem gives a way to determine the partial cut vertices of a weighted graph.

**Theorem 2.2.10** *If $w$ is a common vertex of two or more $p$-bridges, then $w$ is a $p$-cut vertex.*

***Proof*** Let $w$ be a common vertex of two $p$-bridges $uw$ and $wv$. Then by Proposition 2.2.9, $uwv$ is a unique strongest path in $G$. Hence $CONN_{G-w}(u, v) < CONN_G(u, v)$. Thus $w$ is a $p$-cut vertex of $G$. ■

The converse of the above result is not true as seen from the following example.

**Example 2.2.11** Let $G = (V, E, w)$ be a weighted graph (Fig. 2.5) with $V = \{a, b, c, d\}$ and $E = \{e_1 = ab, e_2 = bc, e_3 = cd, e_4 = da, e_5 = ac\}$ with $w(e_1) = 5, w(e_2) = 5, w(e_3) = 2, w(e_4) = 7, w(e_5) = 2$. Clearly the vertex $b$ is a $p$-cut vertex, as $CONN_{G-\{b\}}(a, c) = 2 < 5 = CONN_G(a, c)$. But $da$ is the only $\alpha$-strong edge in $G$.

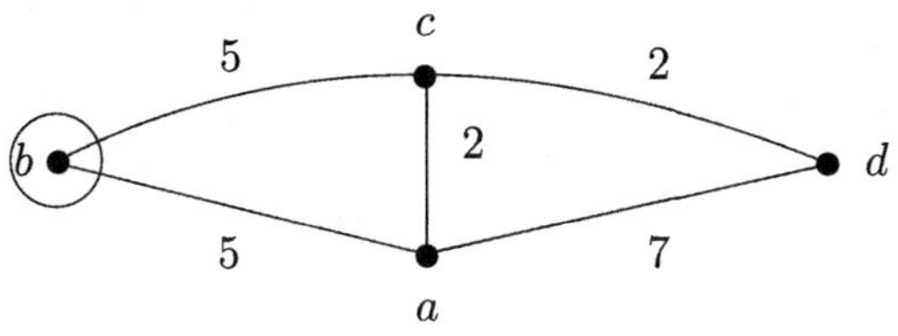


**Fig. 2.5** Graph in Example 2.2.11

## 2.3 Partial Blocks

A block is a graph without cut vertices. Similar to blocks, we define partial blocks in weighted graphs as follows. Most of the contents in this section are from [6].

**Definition 2.3.1** A connected weighted graph $G$ is called a **partial block** ($p$-block) if it has no partial cut vertices.

It is seen from Theorem 2.2.10 that a weighted block need not be a weighted $p$-block. But the converse is true since every cut vertex is a $p$-cut vertex.

**Example 2.3.2** Two partial blocks on 3 and 5 vertices are given in Fig. 2.6a and b respectively.

**Theorem 2.3.3** *A weighted graph $G = (V, E, w)$ is a partial block if and only if any two vertices $u, v \in V$ such that $uv$ is not $\alpha$-strong are joined by two internally disjoint strongest paths.*

***Proof*** Suppose that $G = (V, E, w)$ is a partial block. Let $u, v \in V$ be such that $uv$ is not an $\alpha$-strong edge. To show that there exist two internally disjoint strongest $u - v$ paths. Suppose not. That is, there exist exactly one strongest $u - v$ path in $G$. Since $uv$ is not $\alpha$-strong, strengths of all strongest $u - v$ paths must be at least two (note that if $uv$ is $\beta$-strong, then there exist two internally disjoint $u - v$ paths, which is not possible). Also for all strongest $u - v$ paths in $G$, there must be a vertex in common. Let $x$ be such a vertex in $G$. Then, $CONN_{G-x}(u, v) < CONN_G(u, v)$, which contradicts the fact that $G$ has no partial cut vertices.

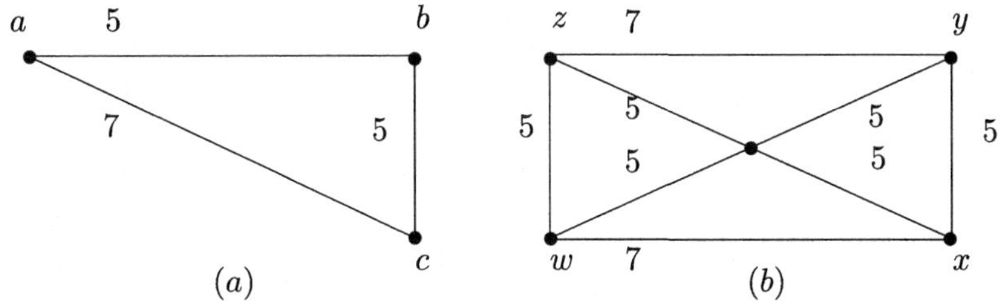


**Fig. 2.6** Partial blocks

Conversely suppose that any two vertices of $G$ are joined by two internally disjoint strongest paths. Let $t$ be a vertex in $G$. For any pair of vertices, $x, y \in V$ such that $x \neq y \neq t$, there always exists a strongest path not containing $t$. So $t$ cannot be a partial cut vertex and hence the theorem is proved. ∎

**Theorem 2.3.4** *If G is a partial block with at least three vertices then the following conditions hold and are equivalent.*

*(i) Every two vertices of G lie on a common strong cycle.*
*(ii) Each vertex and a strong edge of G lie on a common strong cycle.*
*(iii) Any two strong edges of G lie on a common strong cycle.*
*(iv) For two given vertices and a strong edge in G there exists a strong path joining the vertices containing the edge.*
*(v) For every three distinct vertices of G there exist strong paths joining any two of them containing the third.*
*(vi) For every three vertices of G there exist strong paths joining any two of them which does not contain the third.*

***Proof*** $(i)$ Suppose that $G$ is a partial block with at least three vertices. Let $u$ and $v$ be any two vertices in $G$ such that there exists a unique strong path between $u$ and $v$. Now two cases arise. (1) $uv$ is a strong edge. (2) $uv$ is either a $\delta$-edge or there exists a $u - v$ path of length more than two in $G$.

**Case 1** $uv$ is a strong edge.
By assumption, there exists only one strong path between $u$ and $v$ namely the edge $uv$ itself. By Proposition 2.2.8, edge $uv$ is the unique strongest path between $u$ and $v$. Thus $uv$ is a common edge of every maximum spanning tree of $G$ and hence by Theorem 2.1.14, it is a partial bridge of $G$. If $u$ is an end vertex ($d(u) = 1$) in all maximum spanning trees, then $v$ is an internal vertex in all maximum spanning trees and hence $v$ is a cut vertex for all the maximum spanning trees. Thus it follows that $CONN_{G-v}(w, u) < CONN_G(w, u)$ for some vertex $w$ of $G$ other than $u$ and $v$ and hence $v$ is a partial cut vertex of $G$ contradicting the assumption that $G$ is a partial block. If $v$ is an end vertex in all maximum spanning trees, then the case is similar. Now suppose that $u$ is an end vertex in some MST $T_1$ and $v$ is an end vertex in some MST $T_2$. Let $u'$ be a strong neighbor of $u$ in $T_2$. Since $u$ is an end vertex and $v$ is an internal vertex in $T_1$, there exists a strong path $P$ in $T_1$ from $u$ to $u'$ passing through $v$. The path $P$ together with the strong edge $uu'$ forms a strong cycle in $G$, a contradiction.

**Case 2** Either $uv$ is a $\delta$-edge or there exists a strong $u - v$ path of length more than two in $G$.
If $e = uv$ is a $\delta$-edge, then $w(e) < CONN_G(u, v)$. By assumption, there exists a unique strong path between $u$ and $v$ say $P$. By Proposition 2.2.8, $P$ is the unique strongest $u - v$ path and hence it belongs to all maximum spanning trees. Thus all internal vertices in $P$ are internal vertices in all the maximum spanning trees and hence by Theorem 2.1.14, all of them are partial cut vertices of $G$, a contradiction to the assumption that $G$ is a partial block.

$(i) \Rightarrow (ii)$ Suppose that any two vertices of $G$ lie on a common strong cycle. To prove that any given vertex and strong edge lie on a common strong cycle. Let $u$ be a vertex and $vw$ be an edge in $G$. Let $C$ be a strong cycle containing $u$ and $v$. If $w$ is a neighbor of $v$ in $C$, then there is nothing to prove. Now suppose that $w$ is not a neighbor of $v$ in $C$. Let $C_1$ be a strong cycle containing $u$ and $w$. Let $P_1$ and $P_2$ be the strong $u - v$ paths in $C$ and $P_1'$ and $P_2'$ the strong $u - w$ paths in $C_1$. Let $x_1$ be the vertex at which $P_1'$ leaves $P_1$. Then, clearly, $u...(P_1)...x_1...(P_1')...wv....(P_2)u$ is a strong cycle containing $u$ and $vw$. If $x = u$ then, $u...(P_1')...wv...(P_2)...u$ is the required cycle. If $x_1 = v$, let $x_2$ be the vertex at which $P_2'$ leaves $P_2$. Then $u...(P_1)...vw...(P_2')...x_2...(P_2)u$ is the required strong cycle. If $x_2 = u$ then $u...(P_2')...wv...(P_1)...u$ is the required strong cycle. Since $P_1$ and $P_2$ are internally disjoint, both $x_1$ and $x_2$ cannot be the same as $v$.

The remaining implications can be proved similarly. Also the conditions given in Theorem 2.3.4 are only necessary, not sufficient for a weighted graph to be a $p-$block. ■

**Theorem 2.3.5** (Characterization of $p-$blocks in weighted graphs)
Let $G$ *be a weighted graph with at least three vertices and having no partial bridges. Then the following statements are equivalent.*

$(i)$ $G$ is a $p-$ block.

$(ii)$ For any two vertices $x, y$ of $G$ there exists a cycle containing the vertices $x$ and $y$ which is formed by two strongest strong $x - y$ paths.

$(iii)$ For each vertex $u$ and each strong edge $vw$ *of* $G$, there exists a cycle containing the vertex $u$ and the edge $vw$ which is formed by two strongest strong $u - v$ paths or $u - w$ paths.

$(iv)$ For each pair of strong edges $xy$ and $uv$ of $G$ there exists a cycle containing the edges $xy$ and $uv$ which is formed by two strongest strong $x - u$ or $y - u$ paths.

$(v)$ For every three distinct vertices of $G$ there exist a strongest strong path joining any two of them not containing the third.

***Proof*** $(i) \Rightarrow (ii)$

Suppose that $G$ is a $p-$block. Consider a maximum spanning tree $T$ of $G$. By definition, every edge in an MST is strong. Also every $x - y$ path in $T$ is a strongest $x - y$ path in $G$. Thus between any two vertices of $G$ there exists a strongest strong path. Let $P$ be a strongest strong $x - y$ path in $G$. Assume that $P$ is the unique $x - y$ path in $G$. Then $P$ should belong to all maximum spanning trees. Also note that the length of $P$ is at least two since $G$ has no $p-$bridges. Thus all internal vertices of $P$ are internal vertices of every maximum spanning tree and by Theorem 2.1.14, they are all $p-$cut vertices contradicting the fact that $G$ is a $p-$block. Thus it follows that the strongest strong $x - y$ path $P$ does not belong to all maximum spanning trees. Hence there exists a maximum spanning tree say $T_1$ not containing $P$. Let $P_1$ be a strongest strong $x - y$ path in $T_1$. This strongest strong path $P_1$ together with $P$ gives a cycle in $G$ containing the vertices $x$ and $y$ as required. Note that $P$ and $P_1$ should be internally disjoint since otherwise the common vertices of $P$ and $P_1$ become $p-$cut vertices of $G$. ■

$(ii) \Rightarrow (iii)$

Let $u$ be a vertex and $vw$ a strong edge of $G$. Let $C_1$ be a cycle containing $u$ and $v$ satisfying the conditions in $(ii)$ and $C_2$ be a cycle containing $u$ and $w$ satisfying the conditions in $(ii)$. If $w$ is a neighbor of $v$ in $C_1$ or $v$ is a neighbor of $w$ in $C_2$, then we are done. So suppose that $vw$ is neither in $C_1$ nor in $C_2$. Let $P_1$ and $P_2$ be the strongest strong $u - v$ paths in $C_1$ and $Q$ be a strongest strong $u - w$ path in $C_2$. Let $z$ be the vertex in $Q$ before $w$ and nearest to it at which $Q$ meets $P_1$ or $P_2$. (Note that $z$ can be the vertex $u$ itself). Without loss of generality suppose that $P_2$ is the $u - v$ path which meets $Q$ at $z$. Let $P$ be the union of $w - z$ sub path of $Q$ and $z - u$ sub path of $P_2$. Then let $C = P_1 \bigcup vw \bigcup P$.

**Claim** $C$ is the required cycle.
Let $xy$ be an edge in $P_1$ such that strength of $P_1$ = strength of $P_2 = w(xy)$. Then three cases arise.

**Case-1** $w(vw) > w(xy)$

**Sub Claim-1** Strength of $P = w(xy)$

We have, strength of $P$ cannot exceed $w(xy)$, for otherwise $P \bigcup vw$ will become a strong $u - v$ path having strength more than the strongest $u - v$ path $P_1$, a contradiction. Therefore, strength of $P \leq w(xy)$. Since the strength of $P_2 = w(xy)$, the strength of the $u - z$ sub path of $P_2$ is greater than or equal to $w(xy)$. Hence if strength of $P < w(xy)$, then the strength of $z - w$ sub path of $Q < w(xy)$ and thus strength of $Q < w(xy)$. Thus we have $P_1 \bigcup (v, w)$ is a strong $u - w$ path which is stronger than the strongest $u - w$ path $Q$, a contradiction. Thus the only possibility is that strength of $P = w(xy)$ and hence Sub Claim-1 is proved.

Now we have two strongest strong paths between $u$ and $v$ namely $P_1$ and $P \bigcup vw$ whose union gives the required cycle containing the vertex $u$ and the edge $vw$.

**Case-2** $w(vw) = w(xy)$
If strength of $P < w(xy)$, then as in Case-1,the strength of $Q < w(xy)$ and hence $P_1 \bigcup vw$ become a strong $u - w$ path which is stronger than the strongest $u - w$ path $Q$, a contradiction. Thus, strength of $P \geq w(xy)$ and hence we have two strongest strong $u - v$ paths namely $P_1$ and $P \bigcup vw$ whose union gives the required cycle.

**Case-3** $w(vw) < w(xy)$

**Sub Claim-2** Strength of $P = w(vw)$

If strength of $P > w(vw)$, then all edges in $P$ and $P_1$ have strength more than $w(vw)$ and thus $vw$ becomes the unique weakest edge of the cycle $P_1 \bigcup vw \bigcup P$, contradicting our assumption that $vw$ is a strong edge. If strength of $P < w(vw)$, then the strength of $z - w$ sub path of $Q < w(vw)$ because, strength of the $u - z$ sub path of $P_2 \geq w(xy) > w(vw)$. Therefore strength of $Q < w(vw) < w(xy) =$ strength of $P_1$ and hence $P_1 \bigcup vw$ is a strong path having strength more than that of $Q$ which is a contradiction to the fact that $Q$ is a strongest strong $u - w$ path. Thus strength of $P = w(vw)$ and hence Sub Claim-2 is proved.

**Sub Claim-3** $P$ is a strongest $u - w$ path.

To prove Sub Claim-3 it is enough to prove that strength of $Q = w(vw)$ (since $Q$ is a strongest $u - w$ path). Clearly strength of $Q \geq w(vw)$. If strength of $Q < w(vw)$, $P$ will become a strong $u - w$ path which is stronger than the strongest $u - w$ path $Q$, a contradiction. If strength of $Q > w(vw)$, then because $w(vw) < w(xy)$, all edges in $P_1 \bigcup Q$ have strength more than $w(vw)$ and hence $vw$ becomes the unique weakest edge of the cycle $P_1 \bigcup vw \bigcup Q$ contradicting that $vw$ is a strong edge. Thus strength of $Q = w(vw)$ and the sub Claim-3 is proved.

Thus we have two strongest strong $u - w$ paths namely $P$ and $P_1 \bigcup vw$ whose union gives the required cycle containing the vertex $u$ and edge $vw$. Thus in all the three cases the claim is proved.

$(iii) \Rightarrow (iv)$ is similar to $(ii) \Rightarrow (iii)$.

$(iv) \Rightarrow (v)$

Let $x$, $u$ and $w$ be any three distinct vertices of $G$. Let $P$ be a strongest strong $x - u$ path in $G$ with strength $\alpha$ (say). Let $y$ be a strong neighbor of $x$ and $v$ a strong neighbor of $u$ in $P$. Then $xy$ and $uv$ are strong edges of $G$. By $(iv)$ there exists a cycle $C$ containing the edges $xy$ and $uv$ formed by two strongest strong $x - u$ paths or $y - u$ paths.

**Case-1** $C$ is the union of two strongest strong $x - u$ paths.
In this case since there exists two internally disjoint strongest strong $x - u$ paths at least one of them will not contain $w$ and the proof is complete.

**Case-2** $C$ is the union of two strongest strong $y - u$ paths.
Let $P_1$ be the $y - u$ strongest strong path containing the edge $uv$ and let $C - P_1$ be the other path containing edge $xy$. Since $P$ is a strongest strong $x - u$ path, containing $y$ with strength $\alpha$, we have $w(xy) \geq \alpha$. Also we have strength of $P_1 \geq \alpha$ for otherwise the $y - u$ sub path of $P$ will become a strong path stronger than the strongest. But strength of $P_1$ cannot exceed $\alpha$ because $C - P_1$ is also a strongest strong $y - u$ path which passes through the edge $xy$ with $w(xy) = \alpha$. Thus only possibility is, strength of $P_1 = \alpha$. In this case, strength of $P_1 \bigcup xy$ is also $\alpha$ and hence $P_1 \bigcup xy$ becomes a strongest strong $x - u$ path. Also strength of $C - \{P_1 \bigcup xy\}$ cannot exceed $\alpha$ for otherwise it is a contradiction to the fact that $P$ is a strongest strong $x - u$ path.

That is strength of $C - \{P_1 \bigcup xy\} \leq \alpha$.......(1). Also since $C - P_1$ is a strongest strong $y - u$ path, it follows that strength of $C - \{P_1 \bigcup xy\} \geq \alpha$.......(2). From (1) and (2), strength of $C - \{P_1 \bigcup xy\} = \alpha$.

Thus we have two internally disjoint strongest strong $x - u$ paths namely $P_1 \bigcup (x, y)$ and $C - \{P_1 \bigcup xy\}$ with at least one of them not containing the vertex $w$.

$(v) \Rightarrow (i)$
Assume $(v)$. Let $w$ be a vertex in $G$. By $(v)$, between any two vertices $u$ and $v$ other than $w$ there exists a strongest strong $u - v$ path not containing $w$. Thus $w$ is

not in all strongest paths between any pair of vertices and hence is not a partial cut vertex. Thus G is a $p-$block. ■

The remaining characterizations of blocks in graphs cannot be extended to $p-$blocks.

## 2.4 Partial Trees

In this section we present a new type of weighted graph, called a partial tree [4] which is not a tree, but at least one of its threshold subgraphs is a tree. Note that when $G$ is a weighted tree, its threshold subgraphs are trees or forests.

**Definition 2.4.1** A connected weighted graph $G(V, E)$ is called a **partial tree** if $G$ has a spanning subgraph $F(V, E')$ which is a tree, where for all edges $xy$ of $G$ which are not in $F$, we have $CONN_G(x, y) > w(xy)$.

When the graph $G$ is not connected and the condition is satisfied by all components of $G$, then $G$ is called a **partial forest**.

**Example 2.4.2** The following is an example of a partial tree (Fig. 2.7). By removing all the edges in the boundary we will get the spanning tree $F$.

Any weighted tree $T$ is a partial tree. But the converse is not be true as seen from this example.

Now we characterize partial trees.

**Theorem 2.4.3** *A connected weighted graph $G = (V, E, w)$ is a partial tree if and only if in any cycle $C$ of $G$, there exists an edge $e = xy$ such that $w(e) < CONN_{G-e}(x, y)$, where $G - e$ is the subgraph of $G$ obtained by deleting the edge $e$ from $G$.*

***Proof*** Let $G$ be a connected weighted graph. If there is no cycle, then $G$ is clearly a tree and hence is a partial tree. If there exists cycles in $G$, let $xy$ be an edge belonging to a cycle $C$ with the minimum weight in $G$. Delete the edge $xy$ from $G$.

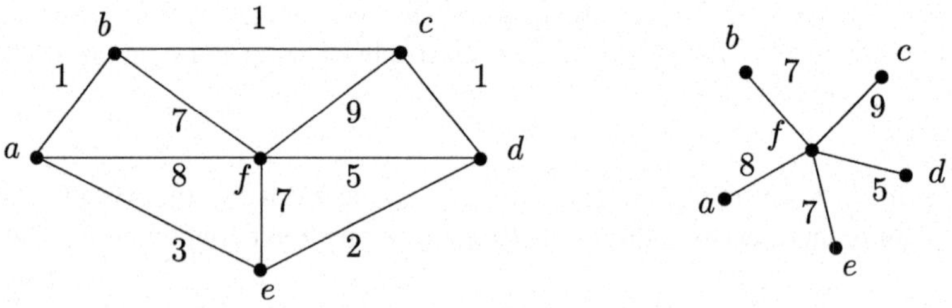


**Fig. 2.7** A partial tree and its unique MST

If there are still cycles in the graph, we can repeat the process. Now at each stage no previously deleted edge is stronger than the edge being currently deleted. When no cycle remains, the resulting subgraph is a tree say $F$. Let $xy$ not be an edge of $F$. Then $xy$ is one of the edges deleted in the process of constructing $F$. Since $F$ is a tree and the weight of $xy$ was minimum from the edges of a cycle in $G$, it follows that there exists a path from $x$ to $y$ stronger than $w(xy)$ and that does not involve $xy$ or any edges deleted prior to it. If this path involves edges that were deleted later, the path can be further diverted and so on. This process stabilizes with a path consisting entirely of edges of $F$. Thus $G$ is a partial tree.

Conversely, if $G$ is a partial tree and $P$ is a cycle, then some edge $e = xy$ of $P$ does not belong to $F$. Thus by definition we have $w(e) < CONN_{G-e}(x, y) \leq CONN_G(x, y)$. ■

Now we give a sufficient condition for a weighted graph to be a partial tree using the concept of strongest paths.

**Theorem 2.4.4** *If there exists at most one strongest path between any two vertices of G, then G must be a partial forest.*

***Proof*** Suppose $G$ is not a partial forest. Then by Theorem 2.1.12, there is a cycle $C$ in $G$ such that $w(xy) \geq CONN_G(x, y)$ for all edges $xy$ of the cycle $C$. Thus $xy$ is a strongest path from $x$ to $y$. If we choose $xy$ to be a weakest edge of $C$, then it follows that the rest of the cycle $C$ is also a strongest path from $x$ to $y$, a contradiction. ■

**Theorem 2.4.5** *If $z$ is a common vertex of at least two p-bridges, then $z$ is a p-cut vertex.*

***Proof*** Let $u_1z$ and $zu_2$ be two partial bridges. Then by Proposition 2.1.10, there exists $u, v$ such that $u_1z$ is on every strongest $u - v$ path. If $z$ is distinct from $u$ and $v$, then it follows that $z$ is a $p$-cut vertex. Now suppose that one of $v$ or $u$ is $w$ so that $u_1w$ is on every strongest $u - z$ path or $zu_2$ is on every strongest $w - v$ path. Suppose that $z$ is not a $p$-cut vertex. Then, between every two vertices there exist at least one strongest path not containing $z$. In particular, there exist at least one strongest path $P$ joining $u_1$ and $u_2$ not containing $z$. This path together with $u_1z$ and $zu_2$ forms a cycle.

We now consider two cases. First suppose that $u_1, z, u_2$ is a not a strongest path. Then one of $u_1z$, $zu_2$ or both becomes the weakest edges of the the cycle which contradicts that $u_1z$ and $zu_2$ are $p$-bridges.

Now suppose that $u_1, z, u_2$ is also a strongest path joining $u_1$ to $u_2$. Then $CONN_G(u_1, u_2) = \text{Min}\{w(u_1z), w(zu_2)\}$, the strength of $P$. Thus edges of $P$ are at least as strong as $w(u_1z)$ and $w(zu_2)$ which implies that $u_1z$ and $zu_2$ or both are the weakest edges of a cycle, which is again a contradiction. ■

The condition in the above theorem is not necessary as seen from the following example.

**Example 2.4.6** In Fig. 2.8, vertex $w$ is a cut vertex and hence is a $p$-cut vertex, but it is not a common vertex of two or more $p$-bridges.

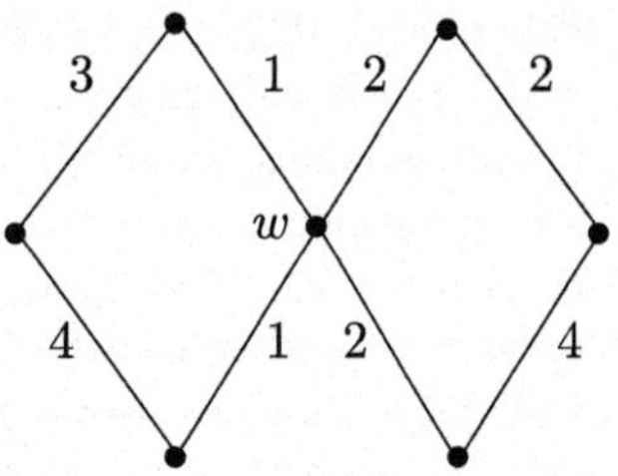


**Fig. 2.8** Graph with a cut vertex

**Theorem 2.4.7** *Let $G = (V, E, w)$ be a weighted graph. If $e = uv$ is a p-bridge in G, then $CONN_G(u, v) = w(e)$.*

***Proof*** Suppose that $uv$ is a $p$-bridge and that $CONN_G(u, v) > w(e)$. Then there exists a strongest $u - v$ path with strength greater than $w(uv)$ and all edges of this strongest path have strength greater than $w(uv)$. Now this path together with the edge $uv$ forms a cycle in which $uv$ is the weakest edge, contradicting that $uv$ is a $p$-bridge. ■

The converse of Theorem 2.4.7 is not true. The condition for the converse to be true is discussed later.

**Theorem 2.4.8** *If G is a partial tree and is not a tree, then there exist at least one edge $uv$ for which $w(uv) < CONN_G(u, v)$.*

***Proof*** If $G$ is a partial tree, then by definition there exists a spanning tree $F$ such that $w(uv) < CONN_G(u, v)$ for all edges $uv$ not in $F$. By hypothesis, there exist at least one such edge (since $G$ is not a tree) and the result follows. ■

Next theorem gives the information about partial cut vertices of a partial tree.

**Theorem 2.4.9** *If G is a partial tree and F, the spanning tree in the definition, then the edges of F are the partial bridges of G.*

***Proof*** Let $uv$ be an edge in $F$. Then this edge is the unique path between $u$ and $v$ in $F$. If there is no other path in $G$ from $u$ to $v$, then clearly $uv$ is a bridge of $G$ and hence is a partial bridge of $G$. If there exists a path say $P$ from $u$ to $v$ in $G$, then $P$ will definitely contain an edge $xy$ which is not in $F$ such that $CONN_G(x, y) > w(xy)$. Then $uv$ is not the weakest edge of any cycle in $G$ and hence by Theorem 2.1.12, $uv$ is a partial bridge. ■

**Theorem 2.4.10** *If G is a partial tree and F is the spanning tree in the definition, then the internal vertices of F are the partial cut vertices of G.*

***Proof*** Let $z$ be any vertex in $G$ which is not a pendent vertex of $F$. Then $z$ is the common vertex of two at least two edges in $F$, which are bridges of $G$. By Proposition 2.2.9, $z$ is a $p$-cut vertex. Also if $z$ is a pendant vertex of $F$, then $z$ is a not a $p$-cut vertex; else there would exist $u$, $v$ different from $z$ such that $z$ is on every strongest strongest $u - v$ path and one such path certainly lies in $F$. But since $z$ is a pendant vertex of $F$, this is impossible. ■

**Theorem 2.4.11** *$G = (V, E, w)$ is a partial tree if and only if the following are equivalent for all $u, v \in V$.*
$(i)$ $(u, v)$ is a $p$-bridge.
$(ii)$ $CONN_G(u, v) = w(uv)$.

***Proof*** Let $G$ be a partial tree and suppose that $e = uv$ is a $p$-bridge. Then by Theorem 2.4.7, $CONN_G(u, v) = w(e)$. Now let $uv$ be an edge in $G$ such that $CONN_G(u, v) = w(e)$. If the underlying graph (graphs obtained by removing weights) is a tree, then clearly $uv$ is a bridge and hence is a $p$-bridge. If the underlying graph is not a tree, then by Theorem 2.4.8, $uv$ is in $F$ and hence $uv$ is a $p$-bridge.

Conversely assume that $(i)$ and $(ii)$ are equivalent. Construct a maximum spanning tree $T$ for $G$. If $uv$ is in $T$, then $CONN_G(u, v) = w(uv)$ and hence $uv$ is a $p$-bridge. Now these are the only $p$-bridges of $G$; for if possible let $u'v'$ be a $p$-bridge of $G$ which is not in $T$. Consider a cycle $C$ consisting of $u'v'$ and the unique $u' - v'$ path in $T$. Now edges of this $u' - v'$ path are $p$-bridges and so they are not weakest edges of $C$ and thus $u'v'$ must be the weakest edge of $C$ and cannot be a bridge.

Moreover, for all edges $u'v'$ not in $T$, we have $w(u'v') < CONN_T(u', v')$; for if possible let $w(u'v') \geq CONN_T(u', v')$. But $w(u'v') < CONN_G(u', v')$ where strict inequality holds since $u'v'$ is not a $p$-bridge. Hence $CONNT(u', v') < CONN_G(u', v')$ which gives a contradiction since $CONN_T(u', v')$ is the strength of the unique $u' - v'$ path in $T$ and hence $CONN_G(u', v') = CONN_T(u', v')$. Thus $T$ is the required spanning subgraph $F$, which is a tree and hence $G$ is a partial tree. ■

For a partial tree the spanning subgraph $F$ is unique. It follows from the proof of Theorem 2.4.11 that $F$ is nothing but the maximum spanning tree of $G$. Thus we have the following theorem.

**Theorem 2.4.12** *A weighted graph is a partial tree if and only if it has a unique maximum spanning tree.*

## References


1. Mathew, S., Sunitha, M.S.: Some connectivity concepts in weighted graphs. Adv. Appl. Discret. Math. **6**(1), 45–54 (2010)
2. Mathew, S.: On cycle connectivity of graphs. J. Interconnect. Netw. **13**, 1250005(1–11) (2012)
3. Mathew, S., Sunitha, M.S.: Cycle connectivity in weighted graphs. Proyecciones J. Math. **30**(1), 1–17 (2001)
4. Mathew, S., Sunitha, M.S.: Partial trees in weighted graphs -1. Proyecciones J. Math. **30**(2), 163–174 (2011)
5. Rosenfeld, A.: Fuzzy Graphs. In: Zadeh, L.A., Fu, K.S., Shimura, M. (eds.) Fuzzy Sets and Their Applications, pp. 77–95. Academic, New York (1975)
6. Mathew, S., Sunitha, M.S.: A characterization of partial blocks in weighted graphs. Inf. Process. Lett. **112**(17–18), 706–710 (2012)

# Chapter 3
# More on Connectivity

This chapter discusses about some additional connectivity concepts in weighted graphs. A graph is said to be totally weighted if both its vertex set and edge set are weighted. Section 3.1 of this chapter introduces precisely weighted graphs and its features. Section 3.2 presents the weighted (generalized) version of Menger's theorem in classical graph theory. Section 3.3 deals with some generalized connectivity parameters named weighted vertex connectivity and weighted edge connectivity. The content of this chapter is based on [1].

## 3.1 Precisely Weighted Graphs

Precisely weighted graphs are introduced in this section. Strength of connectedness between any two vertices in a precisely weighted graph is obtained and its partial bridges are characterized. In any network, we note that flow through an edge cannot exceed the minimum of the capacities of its end vertices. The maximum flow through an edge is precisely the minimum of weights of its end vertices. Motivated by this concept, we have the following definition.

**Definition 3.1.1** A **precisely weighted graph** (PWG) is a complete graph $G(V, E)$ with weight functions $w_1 : V \to \Re^+$ and $w_2 : E \to \Re^+$ such that $w_2(xy) = w_1(x) \wedge w_1(y)$ for any pair of vertices $x, y \in V$. Here $\wedge$ denotes the minimum. It is denoted as $G(V, E, w_1, w_2)$.

**Example 3.1.2** Let $G(V, E, w_1, w_2)$ be a totally weighted graph with $V = \{u, v, w, x\}$, $w_1(u) = 22$, $w_1(v) = 7$, $w_1(w) = 10$, $w_1(x) = 33$, $w_2(xv) = w_2(uv) = w_2(vw) = 7$, $w_2(wx) = 10 = w_2(wu)$, $w_2(ux) = 22$ (See Fig. 3.1). $w_2$ value of any edge in $G$ is the minimum of the $w_1$ values of the vertices incident on it. Then $G$ is a $PWG$.



S. Mathew et al., *Weighted and Fuzzy Graph Theory*, Studies in Fuzziness and Soft Computing 429, https://doi.org/10.1007/978-3-031-39756-1_3

Fig. 3.1 A precisely weighted graph

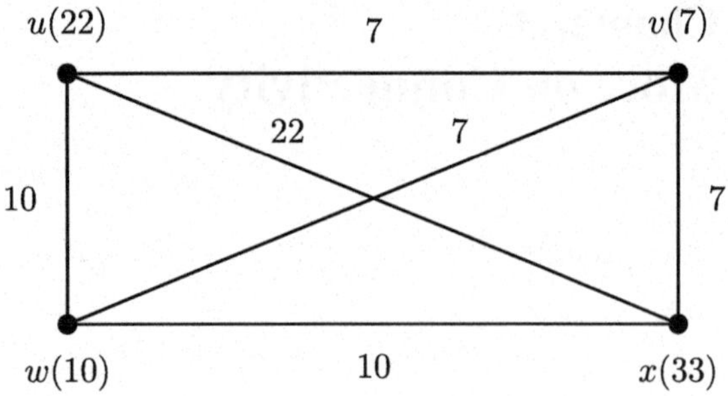


Example 3.1.2 illustrates Definition 3.1.1. Following proposition describes the evaluation of strength of connectedness between any two vertices of a $PWG$.

**Proposition 3.1.3** *If* $G = (V, E, w_1, w_2)$ *is a* $PWG$, *then* $CONN_G(u, v) = w_2(uv)$ *for any two vertices in* $V$.

***Proof*** Let $G = (V, E, w_1, w_2)$ be a $PWG$. Let $u, v \in V$. Consider all paths of length two from $u$ to $v$. We denote the maximum of strengths of all such paths from $u$ to $v$ by $w_2^2(u, v)$. Then, $w_2^2(u, v) = \vee_{w \in V}\{w_2(uw) \wedge w_2(wv)\} = \vee_{w \in V}\{w_1(u) \wedge w_1(w) \wedge w_1(v)\} \leq w_1(u) \wedge w_1(v) = w_2(u, v)$. Similarly the maximum of strengths of all $u - v$ paths of length 3, denoted by $w_2^3(u, v) \leq w_2(uv)$ and in the same way we can show that $w_2^k(u, v) \leq w_2(uv)$ for all positive integer $k$. By definition $CONN_G(u, v) = Sup\{s(P) : P \text{ is a } u - v \text{ path in } G\} = w_2(uv)$, which completes the proof. ■

In turn, an edge joining any two vertices $u$ and $v$ is a shortest strongest $u - v$ path in a $PWG$.

**Corollary 3.1.4** *A precisely weighted graph has no partial cut vertices.*

The concept of a partial bridge is given in Definition 2.1.6. A complete graph will not contain any bridges. Apart from this, a PWG can have a partial bridge as seen from the following theorem. The proof is by induction on $|V|$, the cardinality of $V$.

**Theorem 3.1.5** *A PWG can have at most one partial bridge.*

***Proof*** Let $G = (V, E, w_1, w_2)$ be a $PWG$, where $|V| = 3$. Then $G$ can have at most one $p$-bridge by above corollary. Assume that any precisely weighted graph with less than $k$ vertices possesses at most one $p$-bridge. Let $G_k$ denotes a PWG with $k$ vertices $v_1, v_2, \ldots, v_k$. Remove the vertex $v_k$ from $G_k$ and let the resulting weighted graph be $G_{k-1}$. By assumption $G_{k-1}$ can have at most one $p$-bridge. If $G_{k-1}$ has no $p$-bridges then the proof is over because all the $k - 1$ edges $v_1v_k, v_2v_k, \ldots, v_{k-1}v_k$ are adjacent and any two adjacent $p$-bridges will contribute a $p$-cut vertex (precisely the common vertex), which is not possible, by corollary above. If $G_{k-1}$ has a $p$-bridge say $v_rv_l$ where $1 \leq r, l \leq k - 1$, then none of the edges $v_1v_k, v_2v_k, \ldots, v_{k-1}v_k$ can be a $p$-bridge, for otherwise let $v_pv_k)$ where $1 \leq p \leq k - 1$ be a $p$-bridge. Now consider the four vertices $v_r$, $v_l$, $v_p$ and $v_k$. Arrange them so that their $w_1$ values are in ascending order and rename them as $u_1, u_2, u_3$ and $u_4$. Then $u_1u_2$, $u_1u_3$ and $u_1u_4$ are not $p$-bridges, they being the weakest edges of some cycle in the subgraph induced by $u_1, u_2, u_3$ and $u_4$ (a $p$-bridge cannot be the weakest edge of any cycle, by Theorem 2.1.10). Now the edges $u_2u_3$, $u_2u_4$ and $u_3u_4$ form a triangle say $C$. Since we have two $p$-bridges from among the combinations of $u_1, u_2, u_3$ and $u_4$, they must be adjacent edges in $C$, which is a contradiction since any two adjacent $p$-bridges contribute a $p$-cut vertex and a precisely weighted graph has no $p$-cut vertices. Thus it follows that $G_k$ has at most one $p$-bridge and the theorem is proved by induction. ■

Theorem 3.1.6 helps to locate the $p$-bridge of a $PWG$. More over, it is the characterization of $p$-bridges in a $PWG$.

**Theorem 3.1.6** *Let $G = (V, E, w_1, w_2)$ be a PWG with $|V| = n$. Then $G$ has a partial bridge if and only if there exists an increasing sequence $\{t_1, t_2, \ldots, t_n\}$ of positive real numbers such that $t_{n-2} < t_{n-1} \leq t_n$ where $t_i = w_1(u_i)$ for $i = 1, 2, \ldots, n$. Also the edge $u_{n-1}u_n$ is the $p$-bridge of $G$.*

***Proof*** Assume that $G = (V, E, w_1, w_2)$ is a PWG and that $G$ has a $p$-bridge $uv$. Now $w_2(uv) = w_1(u) \wedge w_1(v)$. Without loss of generality let $w_1(u) \leq w_1(v)$, so that $w_2(u, v) = w_1(u)$. Also note that $uv$ is not a weakest edge of any cycle in $G$. Now it is required to prove that $w_1(u) \geq w_1(w)$ for every $w \neq v$. On the contrary assume that there is at least one vertex $w \neq v$ such that $w_1(u) \leq w_1(w)$. Now consider the cycle $C : u, v, w, u$.

Then $w_2(uv) = w_2(uw) = w_1(u)$ and $w_2(vw) = w_1(v)$ if $w_1(u) = w_1(v)$ or $w_1(u) < w_1(v) \leq w_1(w)$ and $w_2(vw) = w_1(w)$ if $w_1(u) < w_1(w) < w_1(v)$.

In either case edge $uv$ becomes a weakest edge of a cycle which contradicts the fact that $uv$ is a $p$-bridge.

Conversely, let $t_1 \leq t_2 \leq .... \leq t_{n-2} \leq t_{n-1} \leq t_n$ and $t_i = w_1(u_i)$ for all $i$. We claim that edge $u_{n-1}u_n$ is the $p$-bridge of $G$. We have $w_2(u_{n-1}u_n) = w_1(u_{n-1}) \wedge w_1(u_n) = w_1(u_{n-1})$ and by hypothesis, all other edges of $G$ will have strength strictly less than $w_1(u_{n-1})$. Thus the edge $u_{n-1}u_n$ is not a weakest edge of any cycle in $G$ and hence is a $p$-bridge. ■

The concept of strong edges in weighted graphs was introduced in [2] by Mathew and Sunitha. Discussion on the classification of strong edges of weighted graphs can be seen in Chap. 2. This section discusses on different types of edges in a $PWG$. A PWG is peculiar that it has no $\delta$-edges and can have at most one $\alpha$-strong edge. Characterization of partial bridges in terms of $\alpha$-strong edges and the existence of a $\beta$-strong path between any two vertices of a PWG are also discussed here. Further, it is established that the concepts of strong path and strongest path coincide in a PWG without $\alpha$- strong edges. It is interesting that all edges in a PWG are strong. Although a complete graph never contain bridges, a PWG may contain $p$-bridges.

**Lemma 3.1.7** *A PWG has no $\delta$-edges.*

***Proof*** Let $G = (V, E, w_1, w_2)$ be a $PWG$. If possible, assume that $G$ contains a $\delta$-edge $uv$(say). Then,

$$w_2(uv) < CONN_{G-(u,v)}(u, v).$$

That is, there exists a stronger path $P$ other than the edge $uv$ from $u$ to $v$ in $G$. Let $w_2(uv) = p$ and the strength of the path $P$ be $q$. Then $p < q$. Let $w$ be the first vertex in $P$ after $u$. Then,

$$w_2(uw) > p.....(1)$$

Similarly let $x$ be the last vertex in $P$ before $v$. Then,

$$w_2(xv) > p.....(2)$$

Since $w_2(uv) = p$, at least one of $w_1(u)$ or $w_1(v)$ should be $p$. Now, $G$ being a $PWG$, (1) gives a contradiction if $w_1(u) = p$ and (2) gives a contradiction if $w_1(v) = p$; which completes the proof. ■

Theorem 2.2.6 states that an edge $e$ in a weighted graph G is a $p$-bridge if and only if $e$ is $\alpha$-strong. This result along with Theorem 3.1.5 leads to the fact that a PWG has at most one $p$-bridge. Hence we have the Lemma 3.1.8.

**Lemma 3.1.8** *There exists at most one $\alpha$-strong edge in a PWG.*

Following two theorems are the consequences of Lemmas 3.1.7 and 3.1.8. Here we use the notation $^nC_2$ to denote the number of combinations of $n$ things taken two at a time; $^nC_2 = \frac{n!}{2!(n-2)!}$.

**Theorem 3.1.9** *Let $G = (V, E, w_1, w_2)$ be a PWG with $\mid V \mid = n$. Then the number of $\beta$-strong edges in G is $^nC_2$ or $^nC_2 - 1$.*

**Theorem 3.1.10** *Let $G = (V, E, w_1, w_2)$ be a PWG. Then there exist $\beta$-strong paths between any two vertices of G.*

Next theorem shows the equivalence of certain strong paths and strongest paths in PWGs without $p$-bridges.

**Theorem 3.1.11** *Let G be a PWG without $\alpha$-strong edges. Suppose P is an $x - y$ path in G of maximum strength. Then P is a strong $x - y$ path if and only if P is a strongest $x - y$ path.*

***Proof*** Let $G$ be a PWG without $\alpha$-strong edges and let $P$ be any $x - y$ path in $G$. Assume that $P$ is a strong $x - y$ path. By Lemma 3.1.7, all edges of $G$ are $\beta$-strong. Then by definition of a $\beta$-strong edge,

$CONN_{G-xy}(x, y) = w_2(xy) =$ strength of $P$.....(1)
Now, since $G$ is a PWG, by Proposition 3.1.3,

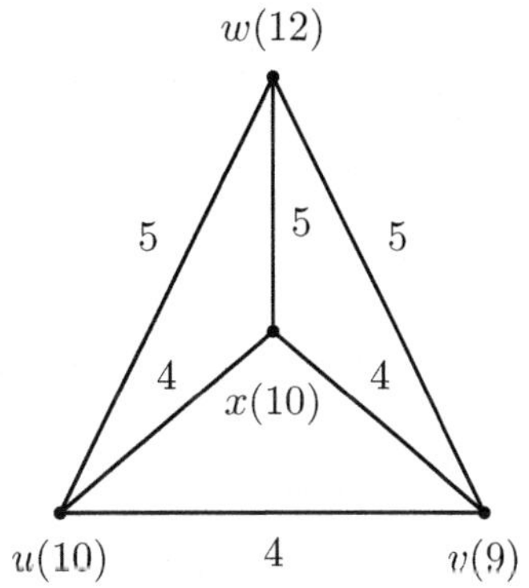


**Fig. 3.2** Illustration of Example 3.1.12

$$CONN_G(x, y) = w_2(xy)....(2)$$

From (1) and (2), $CONN_G(x, y)$ = strength of $P$, which implies that $P$ is a strongest $x - y$ path.

Now let $P$ be a strongest $x - y$ path in $G$. By Lemma 3.1.7, $P$ contains only $\beta$-strong edges and hence is a strong $x - y$ path, which completes the proof. ■

Converse of Theorem 3.1.11 does not hold in general as illustrated in Example 3.1.12.

**Example 3.1.12** Let $G(V, E, w_1, w_2)$ be a totally weighted graph with $V = \{u, v, w, x\}$ and $w_1(u) = 10$, $w_1(v) = 9$, $w_1(w) = 12$, $w_1(x) = 10$ and $w_2(uv) = w_2(vx) = w_2(ux) = 4$, $w_2(vw) = w_2(wx) = w_2(uw) = 5$ (Fig. 3.2). $G$ is not a PWG even if all strong paths are strongest paths and vice versa.

In an unweighted graph, the weight of each edge is equal to one. Hence all of the edges are trivially strong. In turn, the strong degree and usual degree of vertices coincide. Whereas in weighted graphs, both strong edges and non strong edges exist. Note that, in a weighted graph the strong degree and usual degree of vertices may not be equal always. Now we rewrite the usual definition of degree of a vertex in a weighted graph in terms of the weight function $w$ as follows.

The degree of a vertex $v \in V$ in a weighted graph $G(V, E, w)$ with weight function $w$ is defined as $d(v) = \sum_{u \neq v} w(uv)$. The minimum degree of $G$ is $\delta(G) = \wedge\{d(v) \mid v \in V\}$ and the maximum degree of $G$ is $\Delta(G) = \vee\{d(v) \mid v \in V\}$.

**Definition 3.1.13** Let $G = (V, E, w)$ be a weighted graph. The **strong degree of a vertex** $v \in V$ is defined as the sum of weights of all strong edges incident at $v$. It is denoted by $d_s(v)$.

If $N_s(v)$ denote the set of all strong neighbors of $v$, then $d_s(v) = \sum_{u \in N_s(v)} w(uv)$.

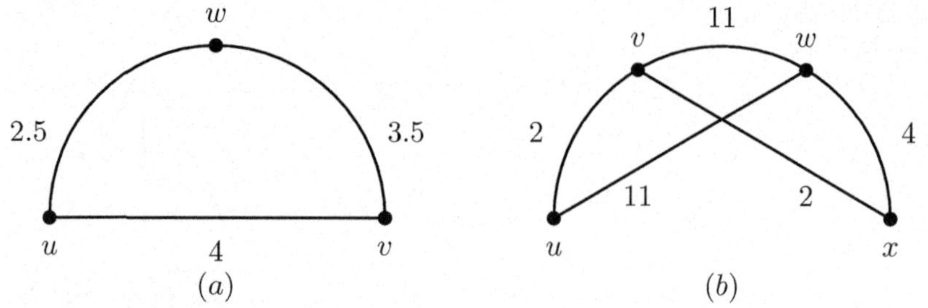


**Fig. 3.3** Graphs in Examples 3.1.14 and 3.1.15

**Example 3.1.14** Let $G(V, E, w)$ be a weighted graph with $V = \{u, v, w\}$, $w(uv) = 2.5$, $w(vw) = 3.5$ and $w(uv) = 4$ (Fig. 3.3a). Here edge $uv$ is not strong and hence $d_s(u) = 4$, $d_s(w) = 3.5$ and $d_s(v) = 7.5$.

Now the **minimum strong degree** of $G$ is $\delta_s(G) = \wedge\{d_s(v)|v \in V\}$ and **maximum strong degree** of $G$ is $\Delta_s(G) = \vee\{d_s(v), v \in V\}$.

**Example 3.1.15** Let $G(V, E, w)$ be with $V = \{u, v, w, x\}$ and $w(uv) = 2{=}w(vx)$, $w(vw) = 11 = w(uw)$, $w(wx) = 4$ (Fig. 3.3b). In $G$, all edges except $uv$ and $vx$ are strong. Thus $d_s(u) = 11 = d_s(v)$, $d_s(w) = 26$ and $d_s(x) = 4$. Hence $\delta_s(G) = 4$ and $\Delta_s(G) = 26$.

Following propositions are direct and can be derived easily.

**Proposition 3.1.16** ([2]) *The sum of strong degrees of all vertices in a weighted graph G is equal to twice the sum of weights of all strong edges of G.*

**Proposition 3.1.17** *In a precisely weighted graph there always exists at least one pair of vertices u and v such that* $d_s(u) = d_s(v)$.

The concept of strong degree has relevance in weighted graph applications, especially those problems related with network flow. This is because the flow through edges which are not strong can be redirected through a different strongest path. There exists at least a strong edge incident at each vertex of a nontrivial connected weighted graph and hence we have the following proposition.

**Proposition 3.1.18** *In a non trivial connected weighted graph* $G = (V, E, w)$, $0 < d_s(v) \leq d(v)$ *for every vertex* $v \in V$.

Apparently, for each vertex $v$ in an unweighted graph, $d(v) = d_s(v)$. Note that, every edge of a PWG $G = (V, E, w_1, w_2)$ is strong. Thus $d_s(v) = d(v)$ for all $v \in V$ and hence we have the expression $d_s(v) = \sum_{u \neq v} \wedge\{w_1(u), w_1(v)\}$ where $u \in V$.

Minimum and maximum strong degrees of a PWG $G = (V, E, w_1, w_2)$ can be evaluated in terms of the weights of its vertices. This is due to the property that $w_2(uv) = w_1(u) \wedge w_1(v)$ for any edge $uv \in E$. We move on to Lemma 3.1.19.

**Lemma 3.1.19** *Let $G = (V, E, w_1, w_2)$ be a PWG with $V = \{u_1, u_2, \ldots, u_n\}$ such that $w_1(u_1) \leq w_1(u_2) \leq w_1(u_3) \leq \cdots \leq w_1(u_n)$. Then $u_1u_j$ is an edge of minimum weight at $u_j$ for $2 \leq j \leq n$ and $u_iu_n$ is an edge of maximum weight at $u_i$ for $1 \leq i \leq n-1$. Also,*

$$d(u_1) = \delta_s(G) = (n-1)w_1(u_1) \text{ and}$$
$$d(u_n) = \Delta_s(G) = \sum_{i=1}^{n-1} w_1(u_i).$$

***Proof*** Throughout the proof, we suppose that $w_1(u_1) < w_1(u_2) \leq w_1(u_3) \leq \cdots \leq w_1(u_{n-1}) < w_1(u_n)$. If there are more than one vertex with minimum vertex strength or maximum vertex strength, the proof will be similar. First we prove that for $2 \leq j \leq n$, $u_1u_j$ is an edge of minimum weight at $u_j$. If possible, suppose that $u_1u_l$, $2 \leq l \leq n$ is not an edge of minimum weight at $u_l$. Also let $u_ku_l$, $2 \leq k \leq n$, $k \neq l$ be an edge of minimum weight at $u_l$. Being a PWG,

$w_2(u_1u_l) = w_1(u_1) \wedge w_1(u_l)$ and, $w_2(u_ku_l) = w_1(u_k) \wedge w_1(u_l)$.

Since $w_2(u_ku_l) < w_2(u_1u_l)$, we have, $w_1(u_k) \wedge w_1(u_l) < w_1(u_1) \wedge w_1(u_l) = w_1(u_1)$.

That is either $w_1(u_k) < w_1(u_1)$ or $w_1(u_l) < w_1(u_1)$. Since $l, k \neq 1$, this is a contradiction to our assumption that $w_1(u_1)$ is the unique minimum vertex degree. Thus for $2 \leq j \leq n$, $u_1u_j$ is an edge of minimum weight at $u_j$.

Next, we prove that $u_iu_n$ is an edge of maximum weight at $u_i$ for $1 \leq i \leq n-1$. On the contrary suppose that $u_ku_n$, $1 \leq k \leq n-1$ is not an edge of maximum weight at $u_k$ and let $u_ku_r$, $1 \leq r \leq n-1$, $k \neq r$ be an edge of maximum weight at $u_k$.

Then, $w_2(u_ku_r) > w_2(u_ku_n)$ and hence, $w_1(u_k) \wedge w_1(u_r) > w_1(u_k) \wedge w_1(u_n) = w_1(u_k)$, which implies that $w_1(u_r) > w_1(u_k)$. Therefore, $w_2(u_ku_r) = w_1(u_k) = w_2(u_ku_n)$, which is a contradiction to our assumption. Thus $u_ku_n$ is an edge of maximum weight at $u_k$.

Now we have,

$$d_s(u_1) = \sum_{i=2}^{n} w_2(u_1u_i) = \sum_{i=2}^{n}(w_1(u_1) \wedge w_1(u_i)) = \sum_{i=2}^{n} w_1(u_1) = (n-1)w_1(u_1).$$

If possible suppose that $d_s(u_1) \neq \delta_s(G)$ and let $u_k$, $k \neq 1$ be a vertex in $G$ with minimum strong degree.

Now $d_s(u_1) > d_s(u_k)$, implies $\sum_{i=2}^{n} w_2(u_1u_i) > \sum_{k \neq 1, j \neq k} w_2(u_ku_j)$.

That is, $\sum_{i=2}^{n}(w_1(u_1) \wedge w_1(u_i)) > \sum_{k \neq 1, j \neq k} (w_1(u_k) \wedge w_1(u_j))$.

Since $w_1(u_1) \wedge w_1(u_i) = w_1(u_1)$ for $i = 2, 3, \ldots, n$, $w_1(u_k) \wedge w_1(u_1) = w_1(u_1)$ and for all other indices $j$, $w_1(u_k) \wedge w_1(u_j) > w_1(u_1)$, hence it follows that

$$(n-1)w_1(u_1) > \sum_{k \neq 1, j \neq k} (w_1(u_k) \wedge w_1(u_j)) > (n-1)w_1(u_1).$$

That is, $d_s(u_1) > d_s(u_1)$, a contradiction. Thus $d_s(u_1) = \delta_s(G) = (n-1)w_1(u_1)$.

Finally we show that $d_s(u_n) = \Delta_s(G) = \sum_{i=1}^{n-1} w_1(u_i)$.

Since $w_1(u_n) > w_1(u_i)$ for $i = 1, 2, \ldots, n-1$ and $G$ is a PWG,

$$w_2(u_n u_i) = w_1(u_n) \wedge w_1(u_i) = w_1(u_i).$$

Therefore, $d_s(u_n) = \sum_{i=1}^{n-1} w_2(u_n, u_i) = \sum_{i=1}^{n-1} w_1(u_i)$.

Now if possible suppose that $d_s(u_n) \neq \Delta_s(G)$. Let $u_l$, $1 \leq l \leq n-1$ be a vertex in $G$ such that $d_s(u_l) = \Delta_s(G)$ and $d_s(u_n) < d_s(u_l)$.

$$\text{Now, } d_s(u_l) = \sum_{i=1}^{l-1} w_2(u_i u_l) + \sum_{i=l+1}^{n-1} w_2(u_i u_l) + w_2(u_n u_l).$$

$$\leq \sum_{i=1}^{l-1} w_1(u_i) + (n-l)w_1(u_l) + w_1(u_l) \leq \sum_{i=1}^{n-1} w_1(u_i) = d_s(u_n).$$

That is, $d_s(u_l) \leq d_s(u_n)$, a contradiction to our assumption. Thus the Lemma is proved. ■

We can associate a sequence of real numbers to any totally weighted graph, namely the vertex strength sequence as seen in the following definition.

**Definition 3.1.20** Let $G = (V, E, w_1, w_2)$ be a totally weighted graph with $|V| = n$. Then the **vertex-strength sequence** or **node strength sequence** (n-s sequence in short) of $G$ is defined to be $(p_1, p_2, \ldots, p_n)$ with $p_1 \leq p_2 \leq \cdots \leq p_n$ where $p_i > 0$ is the weight of vertex $i$ when vertices are arranged so that their weights are non decreasing. In particular $p_1$ is the smallest vertex weight and $p_n$, the largest vertex weight.

The following example illustrates this concept.

**Example 3.1.21** Let $G = (V, E, w)$ be with $V = \{a, b, c, d\}$ and $w(a) = w(c) = w(d) = 1.1$, $w(b) = 1.2$. Then the vertex-strength sequence of $G$ is $(1.1, 1.1, 1.1, 1.2)$ or $(1.1^3, 1.2)$.

The number of vertices of minimum strong degree and maximum strong degree in a PWG can be determined by observing the corresponding n-s sequence as shown in the next proposition.

**Proposition 3.1.22** *Let $G = (V, E, w_1, w_2)$ be a PWG with $|V| = n$. Then,*

*(i) If the n-s sequence of $G$ is of the form $(p_1^{n-1}, p_2)$, then $\delta_s(G) = \Delta_s(G) = (n-1)p_1 = d_s(u_i)$, $i = 1, 2, \ldots, n$.*
*(ii) If the n-s sequence of $G$ is of the form $(p_1^{r_1}, p_2^{n-r_1})$ with $0 < r_1 \leq n-2$, then there exist exactly $r_1$ vertices with degree $\delta_s(G)$ and $n - r_1$ vertices with degree $\Delta_s(G)$.*

*(iii) If the n-s sequence of $G$ is of the form $(p_1^{r_1}, p_2^{r_2}, \ldots, p_k^{r_k})$ with $r_k > 1$ and $k > 2$, then there exist exactly $r_1$ vertices with degree $\delta_s(G)$ and exactly $r_k$ vertices with degree $\Delta_s(G)$.*

*(iv) If the n-s sequence of $G$ is of the form $(p_1^{r_1}, p_2^{r_2}, \ldots, p_{k-1}^{r_{k-1}}, p_k)$ with $k > 2$, then there exist exactly $1 + r_{k-1}$ vertices with degree $\Delta_s(G)$.*

***Proof*** The proofs of $(i)$ and $(ii)$ are obvious. We present the proofs for $(iii)$ and $(iv)$.

$(iii)$ Let $v_i^{(j)}$, $j = 1, 2, \ldots, r_i$ be the set of vertices in $G$ with $d_s(v_i^{(j)}) = p_i$, $1 \leq i \leq k$. By Lemma 3.1.19, we have,

$$d_s(v_1^{(j)}) = \delta_s(G) = (n-1)p_1 \, for \, j = 1, 2, \ldots, r_1.$$

No vertex with weight more than $p_1$ can have degree $\delta_s(G)$ since,
$w_2(v_i^{(j)} v_{i+1}^{(l)}) = w_1(v_i^{(j)}) > p_1$ for $2 \leq i \leq k$, $j = 1, 2, \ldots, r_i, l = 1, 2, \ldots, r_{i+1}$.
Thus there exists exactly $r_1$ vertices with strong degree $\delta_s(G)$.

Next we prove that $d_s(v_k^t) = \Delta_s(G), t = 1, 2, \ldots, r_k$.

Since $w_1(v_k^t)$ is the maximum vertex weight, we have $w_2(v_k^t v_k^j) = p_k, t \neq j; t, j = 1, 2, \ldots, r_k$ and $w_2(v_k^t v_i^j) = w_1(v_k^t) \wedge w_1(v_i^j) = w_1(v_i^j)$ for $t = 1, 2, \ldots, r_k$; $j = 1, 2, \ldots, r_i$; $i = 1, 2, \ldots, k-1$.

Thus for $t = 1, 2, \ldots, r_k$,

$$\begin{aligned} d_s(v_k^t) &= \sum_{i=1}^{k-1} \sum_{j=1}^{r_i} w_1(v_i^j) + (r_k - 1)p_k \\ &= \sum_{i=1}^{n-1} w_1(u_i) = \Delta_s(G) \quad (By Lemma 3.1.19). \end{aligned}$$

Now if $u$ is a vertex such that $w_1(u) = p_{k-1}$, we have,

$$\begin{aligned} d_s(u) &= \sum_{i=1}^{k-2} \sum_{j=1}^{r_i} w_2(u, v_i^j) + (r_{k-1} - 1 + r_k)p_{k-1} \\ &= \sum_{i=1}^{k-2} \sum_{j=1}^{r_i} w_1(v_i^j) + \sum_{j=1}^{r_{k-1}} w_1(v_{k-1}^j) + (r_k - 1)p_{k-1}. \\ &< \sum_{i=1}^{k-2} \sum_{j=1}^{r_i} w_1(v_i^j) + \sum_{j=1}^{r_{k-1}} w_1(v_{k-1}^j) + (r_k - 1)p_k = \Delta_s(G). \end{aligned}$$

Thus there exists exactly $r_k$ vertices with degree $\Delta_s(G)$.

$(iv)$ Let $v_k^{(1)} = v_k$ be the vertex in $G$ such that $d_s(v_k) = p_k$.

Then by Lemma 3.1.19, $d_s(v_k) = \Delta_s(G) = \sum_{i=1}^{n-1} w_1(u_i)$.

Now let $v_{k-1}^t, t = 1, 2, \ldots, r_{k-1}$ be the vertices in $G$ with $d_s(v_{k-1}^t) = p_{k-1}$. Then for $t = 1, 2, \ldots, r_{k-1}$,

$$d_s(v_{k-1}^t) = \sum_{i=1}^{k-2} \sum_{j=1}^{r_i} w_2(v_i^j v_{k-1}^t) + \sum_{l \neq m} w_2(v_{k-1}^l v_{k-1}^m) + w_2(v_{k-1}^l, v_k).$$

But, $w_2(v_i^j, v_{k-1}^t) = w_1(v_i^j)$ for $i = 1, 2, \ldots, k-2$ and $j = 1, 2, \ldots, r_i$.

$w_2(v_{k-1}^l v_{k-1}^m) = p_{k-1}$ and $w_2(v_{k-1}^l v_k) = p_{k-1}$.

Thus,

$$\begin{aligned} d_s(v_{k-1}^t) &= \sum_{i=1}^{k-2} \sum_{j=1}^{r_i} w_1(v_i^j) + (r_{k-1} - 1)p_{k-1} + p_{k-1} \\ &= \sum_{i=1}^{k-2} \sum_{j=1}^{r_i} w_1(v_i^j) + r_{k-1} p_{k-1} \\ &= \sum_{i=1}^{n-1} w_1(u_i) = \Delta_s(G). \end{aligned}$$

Thus there exist $r_{k-1} + 1$ vertices with strong degree $\Delta_s(G)$.

Now if $u$ is a vertex such that $w_1(u) < p_{k-1}$, as in the proof of $(iii)$, we can show that $d_s(u) < \Delta_s(G)$. Thus there exist exactly $r_{k-1} + 1$ vertices with strong degree $\Delta_s(G)$ and the proof is complete. ■

## 3.2 Generalized Menger's Theorem

The contents of this section are based on [3]. Removal of single vertex or a collection of vertices may reduce the strength of flow between pairs of vertices. The same situation occurs on removal of single edge or a collection of edges. A close observation on this matter helps analyze the flow problems related to weighted graph networks. Here we define the concept of a strength reducing set of vertices.

**Definition 3.2.1** A **strength reducing set of vertices** in a weighted graph $G = (V, E, w)$ is a set of vertices $S \subseteq V$ with the property that either $CONN_{G-S}(u, v) < CONN_G(u, v)$ for some pair of vertices $u, v \in V - S$ or $G - S$ is trivial.

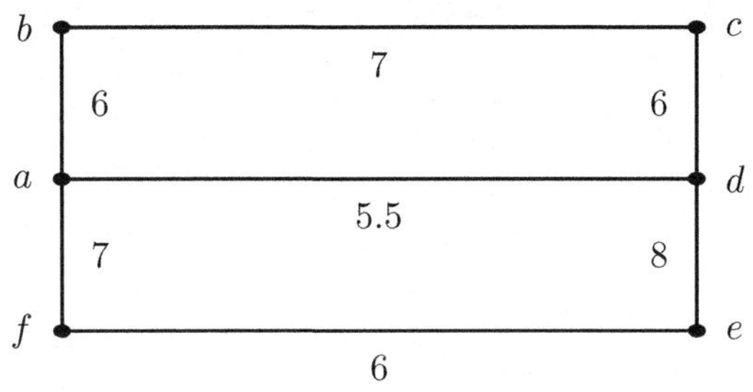


**Fig. 3.4** A weighted graph with strength reducing sets

If $S$ consists of a single vertex $w$, then $w$ is a partial cut vertex. Similar definition on edges holds and is termed as the strength reducing set of edges.

**Definition 3.2.2** A **strength reducing set of edges** in a weighted graph $G = (V, E, w)$ is a set of edges $F \subseteq E$ with the property that $CONN_{G-F}(u, v) < CONN_G(u, v)$ for some pair of vertices $u, v \in V$ with at least one of $u$ or $v$ different from the end vertices of edges in $F$, where $G - F$ is the graph obtained by deleting all edges in $F$ from $G$.

If $F$ contains a single edge $e$, then $e$ is a partial bond. Example 3.2.3 illustrates the former definitions.

**Example 3.2.3** Consider the weighted graph in Fig. 3.4.

Since $CONN_{G-S}(a, d) = 5.5 < 6 = CONN_G(a, d)$ for $S = \{b, f\}$, $S$ is a strength reducing set of vertices . Also $F = \{ab, ef\}$ is a strength reducing set of edges.

Now, let us consider a specific pair of vertices $u, v$ in $V$ and obtain sets of vertices or edges whose removal from $G$ reduces $CONN_G(u, v)$.

**Definition 3.2.4** Let $u, v$ be any two vertices of a weighted graph $G = (V, E, w)$. A $u - v$ **strength reducing set** of vertices in $G$ is a set of vertices $S \subseteq V$ such that $CONN_{G-S}(u, v) < CONN_G(u, v)$.

If $S$ contains a single vertex $w$, then $w$ will be a partial cut vertex. Note that, it is not necessary to have a $u - v$ strength reducing set of vertices always. In Example 3.2.3, no $b - c$ strength reducing set of vertices exists as the edge $bc$ itself is a strongest $b - c$ path.

**Definition 3.2.5** Let $u, v$ be any two vertices of a weighted graph $G = (V, E, w)$. A $u - v$ **strength reducing set of edges** in $G$ is a set of edges $F \subseteq E$ such that $CONN_{G-F}(u, v) < CONN_G(u, v)$. If $F$ contains a single edge $e$, then $e$ is a partial bridge.

**Definition 3.2.6** A $u - v$ strength reducing set of vertices(edges) with $n$ elements is said to be a **minimal** $u - v$ **strength reducing set** of vertices (edges) if there exists no $u - v$ strength reducing set of vertices (edges) with less than $n$ elements.

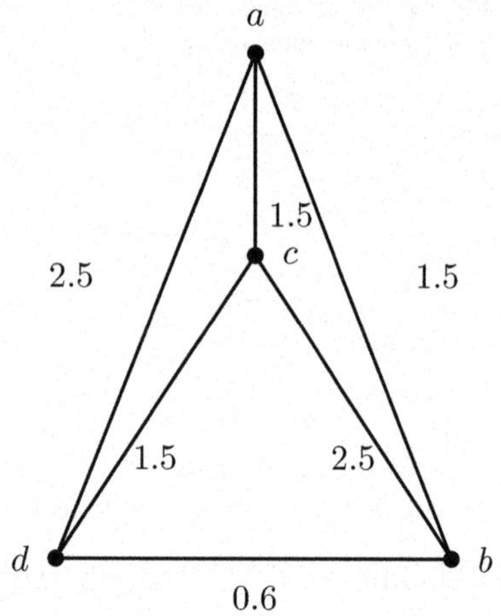


**Fig. 3.5** Minimum and maximum strong degrees

Next we have an important, but obvious result.

**Theorem 3.2.7** *Let $G = (V, E, w)$ be a connected weighted graph and let $u, v$ be two vertices in $G$ such that $uv$ is not a strong edge. A set $S \subseteq V$ is a $u - v$ strength reducing set in $G$ if and only if every strongest $u - v$ path in $G$ contains at least one vertex from $S$.*

A similar result for strength reducing set of edges is given below.

**Theorem 3.2.8** *Let $G = (V, E, w)$ be a connected weighted graph and let $u, v$ be two vertices in $G$. Then a set $F$ of edges in $G$ is a $u - v$ strength reducing set of edges if and only if every strongest path from $u$ to $v$ contains at least one edge of $F$.*

**Definition 3.2.9** The number of strong edges in a weighted graph $G$ is called the **strong size** of $G$ . It is denoted by $ss(G)$.

**Example 3.2.10** Consider the weighted graph in Fig. 3.5. In this graph, all edges except $bd$ are strong edges. Hence $ss(G) = 5$. So $d_s(b) = 4$ whereas $d_w(b) = 4.6$. Also $d_s(a) = d_s(c) = 5.5$, $d_s(d) = 4$, $d_w(d) = 4.6$. Thus $\delta_s(G) = 5.5$ and $\Delta_s(G) = 11.5$.

Now we move on to a lemma that is appropriately used to prove the generalization of the vertex version of Menger's Theorem.

**Lemma 3.2.11** *Let $G = (V, E, w)$ be a weighted graph. If $x$ is a vertex in $G$, with at least one $\alpha$-strong edge incident on $x$, then $ss(G - x) < ss(G)$.*

***Proof*** Let $G = (V, E, w)$ be a weighted graph with a vertex $x$ with properties as given in the statement. Let $wx$ be an $\alpha$-strong edge incident on $x$. We know that an edge $xy$ is strong if $CONN_G(x, y) \leq w(xy)$ and is $\alpha$-strong if

$CONN_{G-(x,y)}(x, y) \leq w(xy)$. Since $G - (x, y)$ is a subgraph of $G$, by Proposition 2.1.4, $CONN_{G-xy}(x, y) \leq CONN_G(x, y)$. Thus if $xy$ is $\alpha$-strong,

$$w(xy) \geq CONN_{G-(x,y)} \geq CONN_G(x, y)$$

which implies that $xy$ is strong. As the removal of the vertex $x$ from $G$ also removes the edge $xw$ from $G$, it follows that the number of strong edges of $G - x$ will be less than that of $G$, that is $ss(G - x) < ss(G)$. ■

Also note that if $E$ is a set of $n$ strong edges in $G$, the removal of $E$ from $G$ reduces the strong size of $G$ by $n$. That is $ss(G - E) = ss(G) - n$.

Theorem 3.2.12 is a generalization of one of the celebrated results in Graph Theory due to Karl Menger [4].

**Theorem 3.2.12** (Generalization of the vertex version of Menger's Theorem) *Let $G = (V, E, w)$ be a weighted graph. For any two vertices $u, v \in V$ such that $uv$ is not strong, the maximum number of internally disjoint strongest $u - v$ paths in G is equal to the number of vertices in a minimal $u - v$ strength reducing set.*

***Proof*** We prove the result by induction on strong size $ss(G)$ of $G$. When $ss(G) = 0$, the only possibility is that the graph $G$ is empty, so that between any pair of vertices $u$ and $v$, there does not exist a path in $G$ and both parameters given in the statement of the theorem reduces to zero and hence the result is trivially true for any pair of vertices $u, v \in V$.

Assume that the theorem is true for all weighted graphs $G$ with strong size less than $m$ where $m \geq 1$. Let $G$ be a weighted graph of strong size $m$. Let $u, v \in V$ such that $uv$ is not strong. If $u$ and $v$ are in different components of $G$, then the theorem is obviously true. So assume that $u$ and $v$ belongs to the same component of $G$. Then either $uv$ is not adjacent or $uv$ is a $\delta$-edge. In both cases $u - v$ strength reducing sets of vertices exists in $G$. (If $uv$ is strong, then reduction of any number of vertices will not reduce the strength of connectivity between $u$ and $v$ and hence no $u - v$ strength reducing set of vertices exists.)

Now suppose that $S_G(u, v)$ is a minimal strength reducing set of vertices in $G$ with $| S_G(u, v) |= k \geq 1$. By Theorem 3.2.7, each strongest $u - v$ path contain at least one member from $S_G(u, v)$. Hence a $u - v$ strength reducing set contains as many vertices as the number of internally disjoint strongest $u - v$ paths. In other words there exists at most $k$ internally disjoint strongest $u - v$ paths. We show that $G$ contains exactly $k$ internally disjoint strongest $u - v$ paths.

If $k = 1$, then $| S_G(u, v) |= 1$. Let $S_G(u, v) = \{z\}$. Then $CONN_{G-\{z\}}(u, v) < CONN_G(u, v)$. That is $z$ is a partial cut vertex of $G$. So every strongest $u - v$ path must pass through $z$. Hence the number of internally disjoint $u - v$ paths is one and the result is true. So assume that $k \geq 2$.

**Case I** $G$ has a minimal $u - v$ strength reducing set of vertices containing a vertex $x$ such that both $ux$ and $xv$ are $\alpha$-strong edges.
Let $S_G(u, v)$ be the minimal $u - v$ strength reducing set of vertices with the above mentioned property. Then $S_G(u, v) - \{x\}$ is a minimal $u - v$ strength reducing set

in $G - \{x\}$ having $k - 1$ vertices. Since both $ux$ and $xv$ are $\alpha$-strong, they are clearly strong and hence $ss(G - \{x\}) < ss(G)$. By induction, it follows that $G - \{x\}$ contains $k - 1$ internally disjoint strongest $u - v$ paths. Since $ux$ and $xv$ are $\alpha$-strong, $P : uxv$ is a strongest $u - v$ path. Thus we have $k$ internally disjoint strongest $u - v$ paths in $G$.

**Case II** For every minimal $u - v$ strength reducing set $S_G(u, v)$ in $G$, either every vertex in $S_G(u, v)$ is an $\alpha$- strong neighbor of $u$ (that is if $w$ is a vertex in $S_G(u, v)$, then, $uw$ is an edge which is the unique strongest $u - w$ path), but not of $v$ or every vertex in $S_G(u, v)$ is an $\alpha$-strong neighbor of $v$ but not of $u$.
Suppose that every vertex in $S_G(u, v)$ is an $\alpha$-strong neighbor of $u$ but not to $v$. Consider a strongest $u - v$ path $P$ in $G$. Let $x$ be the first vertex of $P$ which is in $S_G(u, v)$. Then $ux$ is $\alpha$-strong and since $xv$ is not $\alpha$-strong, there exist at least one vertex say $y$ other than $u$ and $v$ such that $xy$ is $\beta$-strong. Denote the edge $xy$ by $e$.

**Claim** Every $u - v$ strength reducing set in $G - \{e\}$ has exactly $k$ vertices.
If possible suppose that there exists a minimal $u - v$ strength reducing set in $G - \{e\}$ with $k - 1$ vertices say $Z = \{z_1, z_2, \ldots, z_{k-1}\}$. Then $Z \cup \{x\}$ is a minimal $u - v$ strength reducing set in $G$. Note that every $z_i, i = 1, 2, \ldots, k - 1$ and $x$ are $\alpha$-strong neighbors of $u$. Since $Z \cup \{y\}$ also is a minimal $u - v$ strength reducing set in $G$, it follows that $y$ is an $\alpha$-strong neighbor of $u$ contradicting the fact that edge $xy$ is $\beta$-strong (The edges $ux$, $uy$ and $xy$ forms a triangle with edge $xy$ as the weakest edge. The unique weakest edge of a cycle is a $\delta$-edge). Thus $k$ is the minimum number of edges in a $u - v$ strength reducing set in $G - \{e\}$. Since $ss(G - \{e\}) < ss(G)$, it follows by induction that there are $k$ internally disjoint $u - v$ paths in $G - \{e\}$ and hence in $G$.

**Case III** There exists a minimum $u - v$ strength reducing set $W$ in $G$ such that no member of $W$ is an $\alpha$-strong neighbor of both $u$ and $v$ and $W$ contains at least one vertex which is not an $\alpha$-strong neighbor of $u$ and at least one vertex which is not an $\alpha$-strong neighbor of $v$.

Let $W$ be a minimal $u - v$ strength reducing set with $k$ elements having the above properties. Let $W = \{w_1, w_2, \ldots, w_k\}$. Consider all strongest paths from $u$ to $v$. Since $W$ is minimal, $w_i, i = 1, 2, \ldots, k$ must belong to at least one such path. Let $G_u$ be the subgraph of $G$ consisting of all $u - w_i$ sub paths of all strongest $u - v$ paths in which $w_i \in W$ is the only vertex of the path belonging to $W$. Note that if $CONN_G(u, v) = t$, then $\mu(xy) \geq t$ for every edge $xy$ in these paths.

let $G'_u$ be the graph constructed from $G_u$ by adding a new vertex $v'$ and joining $v'$ to each vertex $w_i$ for $i = 1, 2, \ldots, k$. Let $w_1(v') = 1$ and $w_2(w_i, v') = w_1(w_i)$ for every $i = 1, 2, \ldots, k$. The graphs $G_v$ and $G'_v$ are defined similarly. (Note that if the graph is not totally weighted, then $w(w_i) = 1$ for every $i$).

Since $W$ contains a vertex that is not an $\alpha$-strong neighbor of $u$ and a vertex that is not an $\alpha$-strong neighbor of $v$ (Note that all newly introduced edges are strong), we have $ss(G'_u) < ss(G)$ and $ss(G'_v) < ss(G)$.

Clearly $S_{G'_u}(u, v') = k$ and $S_{G'_v}(u', v) = k$. So by induction $G'_u$ contains $k$ internally disjoint $u - v'$ paths say $A_i, i = 1, 2, \ldots, k$ where $A_i$ contains $w_i$. Also $G'_v$

contains $k$ internally disjoint $u' - v$ paths say $B_i$, $i = 1, 2, \ldots, k$ where $B_i$ contains $w_i$. Let $A_i'$ be the $u - w_i$ sub paths of $A_i$ and $B_i'$ be the $w_i - v$ sub path of $B_i$ for $1 \leq i \leq k$. Now k internally disjoint strongest $u - v$ paths can be constructed by joining $A_i$ and $B_i$ for $i = 1, 2, \ldots, k$ and the theorem is proved by induction. ■

For the sake of completion, we state the edge version of Theorem 3.2.12. Similar arguments as in Theorem 3.2.12 can be used to prove the following.

**Theorem 3.2.13** *Let $G = (V, E, w)$ be a connected weighted graph and let $u, v \in V$. Then the maximum number of edge disjoint strongest $u - v$ paths in $G$ is equal to the number of edges in a minimal $u - v$ strength reducing set.*

## 3.3 Generalized Connectivity Parameters

This section generalizes the vertex and edge connectivity parameters ($\kappa$ and $\kappa'$) of classical graph theory. In an unweighted graph, all weights are assumed to be one. Thus the new parameters coincide with the classical parameters. Most of the results are from [3].

**Definition 3.3.1** Let $X \subseteq V$ be a strength reducing set of vertices in a weighted graph $G = (V, E, w)$. The **strong weight** of $X$, denoted by $s(X)$ is defined as $s(X) = \sum_{x \in X} w(xy)$, where $w(xy)$ is the minimum of the weights of strong edges incident on $x$.

**Definition 3.3.2** The **weighted vertex connectivity** of a connected weighted graph $G = (V, E, w)$ is defined as the minimum strong weight of strength reducing set of vertices in $G$. It is denoted by $\kappa_w(G)$. For a disconnected or trivial graph $G$, $\kappa_w(G)$ is defined to be 0.

**Example 3.3.3** Consider the weighted graph in Fig. 3.6a. Clearly $X_1 = \{v\}$, $X_2 = \{w\}$, $X_3 = \{v, w\}$ and $X_4 = \{v, x\}$ are strength reducing sets. All edges except $ux$ are strong in $G$. Hence $s(X_1) = 10$, $s(X_2) = 7$, $s(X_3) = 17$ and $s(X_4) = 17$. Thus the weighted vertex connectivity of $G$ is $\kappa_w(G) = 7$.

Any $\delta$-edge can be replaced by a path having more strength than its edge weight. Any strength reducing set of edges containing at least a $\delta$-edge can be replaced by another strength reducing set having more number of strong edges. Thus we have,

**Definition 3.3.4** A strength reducing set of edges is said to be a **strong strength reducing set of edges (srs of edges)** if it contains no $\delta$-edges.

**Definition 3.3.5** The **strong weight** of a strength reducing set of edges $F$ in a weighted graph $G = (V, E, w)$ is defined as $s'(F) = \sum_{e_i \in E} w(e_i)$ where $e_i$ is a strong edge in $F$.

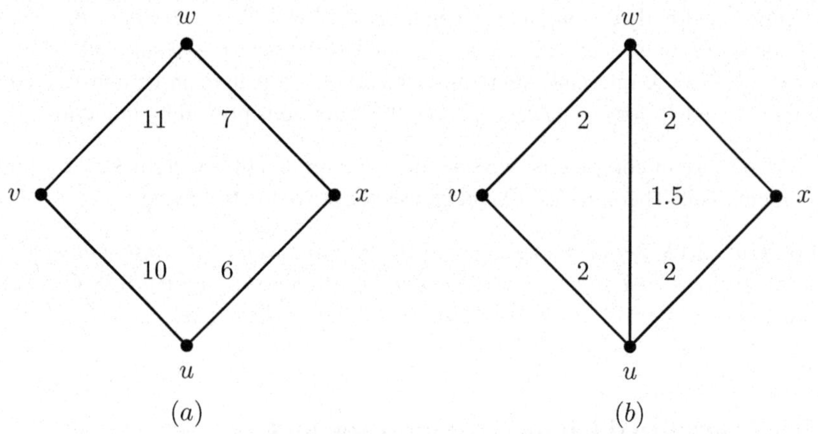


**Fig. 3.6** Graphs in Examples 3.3.3 and 3.3.7

**Definition 3.3.6** The **weighted edge connectivity** $\kappa'_w(G)$ of a connected weighted graph $G = (V, E, w)$ is defined as the minimum strong weight of strength reducing set of edges in $G$. $\kappa'_w(G)$ of a disconnected or trivial weighted graph is defined to be 0.

**Example 3.3.7** Consider the graph in Fig. 3.6b. The reduction of a single edge will not reduce the strength of connectedness between any pairs of vertices. Hence any strength reducing set of edges will contain at least two edges. Among them, $F = \{uv, wx\}$ is the strength reducing set with a minimum strong weight of 4. Hence $\kappa'_w(G) = 4$.

**Theorem 3.3.8** *In a weighted tree $T$, $\kappa_w(G) = \kappa'_w(G) =$ minimum edge weight in $T$.*

***Proof*** Let $T$ be a weighted tree. Any internal vertex of $T$ is a cut vertex and hence is a partial cut vertex. Thus any such vertex forms a strength reducing set and any strong edge in $T$ must have at least one such vertex as an end vertex. Hence weighted vertex connectivity $\kappa_w(T)$ of $T$ is the minimum weight of strong edges in $T$. Since each edge in $T$ is strong, $\kappa_w(T)$ is the minimum weight of edges in $T$.

Also each edge in a tree is a bridge and hence is a partial bridge. Thus each edge is a strength reducing set. Since each edge of $T$ is strong, it follows that the weighted edge connectivity of a weighted tree is the minimum weight of a strong edge in $T$. Since each edge of $T$ is strong, it follows that the weighted edge connectivity of a weighed tree is the minimum weight of edges in $T$. ■

Theorem 3.3.9 is the weighted analogue of a distinguished result relating vertex connectivity, edge connectivity and minimum degree of a graph due to Hassler Whitney.

**Theorem 3.3.9** *In a connected weighted graph* $G = (V, E, w)$, $\kappa_w(G) \leq \kappa'_w(G) \leq \delta_s(G)$.

***Proof*** First we prove the second inequality. Let $G = (V, E, w)$ be a connected weighted graph. Let $v$ be a vertex in $G$ such that $d_s(v) = \delta_s(G)$. Let $F$ be the set of strong edges incident at $v$. If these are the only edges incident at $v$, then $G - F$ is disconnected. If not, let $vu$ be an edge which is not strong, incident at $v$. Then $u$ is a vertex different from the end vertices of the edges in $F$. By definition of a strong edge, $w(uv) < CONN_G(u, v)$ which implies that there exists a strongest $u - v$ path say $P$ in $G$ which should definitely pass through one of the strong edges at $v$. Thus the removal of $F$ from $G$ reduce the strength of connectedness between $v$ and $u$. Thus in both cases, $F$ is a strength reducing set of edges. The strong weight of this srs is $\delta_s(G)$. Hence it follows that $\kappa'_w(G) \leq \delta_s(G)$.

Next to prove $\kappa_w(G) \leq \kappa'_w(G)$, let $F$ be an srs of edges with strong weight $\kappa'_w(G)$. We have the following cases.

**Case 1** Every edge in $F$ has one vertex $v$ (say) in common.
In this case let $F = \{e_i = vv_i, i = 1, 2, \ldots, n\}$.

Let $X = \{v_1, v_2, \ldots, v_n\}$. Then clearly $X$ is an srs of vertices. Now,
$min_{u \in V} w(v_i u) \leq w(vv_i)$.
Therefore,
$\sum_i (min_{u \in V} w(v_i u) \leq w(v_1 v) + w(v_2 v) + \cdots + w(v_n v)$.
Thus $\kappa_w(G) \leq \kappa'_w(G)$.

**Case 2** Not all edges in $E$ have a vertex in common.
Let $F = \{e_i = u_i v_i, i = 1, 2, \ldots, n\}$ for some $n$. Let $X_1 = \{u_1, u_2, \ldots, u_n\}$ and $X_2 = \{v_1, v_2, \ldots, v_n\}$. By assumption, $CONN_{G-F}(x, y) < CONN_G(x, y)$ for some pair of vertices $x, y \in V$ with at least one of $x$ or $y$ is different from both $u_i$ and $v_i$ for $i = 1, 2, \ldots, n$.

**Subcase 1** $x$ and $y$ are not members of $X_1 \bigcup X_2$.
In this case, take $X = X_1$ or $X = X_2$. Then clearly $X$ is an srs of vertices, since its deletion from $G$ reduces the strength of connectedness between $x$ and $y$ and, $\kappa_w(G) \leq$ strong weight of $X \leq$ strong weight of $F = \kappa'_w(G)$.

**Subcase 2** Either $x$ or $y$ is in $X_1 \bigcup X_2$.
Without loss of generality suppose that $x$ is in $X_1 \bigcup X_2$. Let $x \in X_1$. Then take $X = X_2$. Clearly $X$ is an srs of vertices for; the deletion of $X$ from $G$ will reduce the strength of connectedness between $x$ and $y$. Thus,

$\kappa_w(G) \leq$ strong weight of $X \leq$ strong weight of $F = \kappa'_w(G)$.

Thus in all cases, $\kappa_w(G) \leq \kappa'_w(G) \leq \delta_s(G)$. ■

## References

1. Mathew, S., On totally weighted interconnection networks, J. Interconnection Networks, 14 (2013) 1350004(1- 16)
2. Mathew, S., Sunitha, M.S.: Some connectivity concepts in weighted graphs. Advances and Applications in Discrete Mathematics **6**(1), 45–54 (2010)
3. Mathew, S., Sunitha, M.S.: A generalization of classical connectivity parameters for graphs. World Applied Sciences Journal **22**, 18–22 (2013)
4. Menger, K.: Zur allgemeinen Kurventheorie. Fund Math **10**, 96–115 (1927)

# Chapter 4
# Cycle Connectivity

Connectivity concepts in terms of weighted cycles are presented in this chapter. A kind of weighted cycle named strongest strong cycle is introduced. Partial blocks in weighted graphs are characterized using strongest paths. A new class of weighted graphs named $\theta$—weighted graphs are introduced in Sect. 4.2 and partial blocks in $\theta$—weighted graphs are characterized. In Sects. 4.3 and 4.4, we introduce a connectivity parameter called cycle connectivity which is very much useful for vulnerability analysis of weighted networks. Section 4.5 discusses cyclically balanced weighted graphs and an analogous result on Whitney's theorem can be seen towards the end.

## 4.1 Strongest Strong Cycles

We have discussed in Chap. 1 about cycles in weighted graphs. We can classify cycles on the basis of the nature of edges present in them. Here we introduce a special category of cycles in weighted graphs called strongest strong cycles.

**Definition 4.1.1** A cycle $C$ in a weighted graph $G$ is called a **strongest strong cycle (SSC)** if $C$ is the union of two strongest strong $u - v$ paths for every pair of vertices $u$ and $v$ in $C$ except when $uv$ is a $p$-bridge of $G$ in $C$.

According to Definition 4.1.1, an SSC may contain $p$-bridges. The relaxation on the condition that $C$ is the union of two strongest strong $u - v$ paths is only given to those vertices which are the end vertices of $p$-bridges of $G$ in $C$. Also, $CONN_G(x, y) = CONN_C(x, y)$ for all vertices $x, y$ in $C$.



S. Mathew et al., *Weighted and Fuzzy Graph Theory*, Studies in Fuzziness and Soft Computing 429, https://doi.org/10.1007/978-3-031-39756-1_4

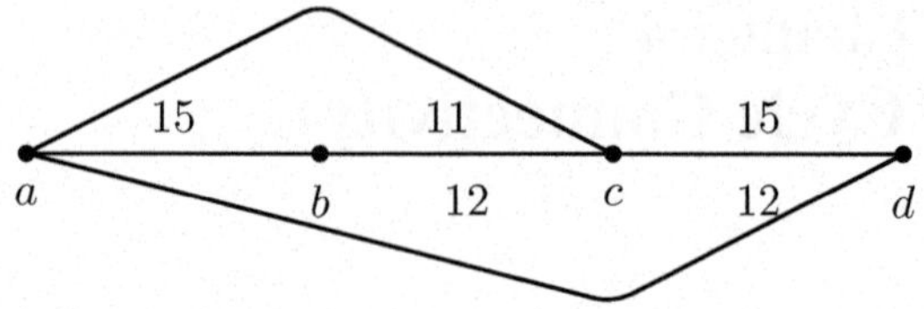


**Fig. 4.1** A strongest strong cycle

**Example 4.1.2** Let $G(V, E, w)$ be a weighted graph with $V = \{a, b, c, d\}$, $w(ab) = 15 = w(cd)$, $w(bc) = 12$, $w(ac) = 11$ and $w(ad) = 12$ (in Fig. 4.1).

In this graph, $C = abcda$ is a strongest strong cycle. There are two $p$-bridges ($ab$ and $cd$) of $G$ in $C$. We can find two strongest paths joining any other pair of vertices in $C$.

A locamin cycle in a weighted graph $G$ need not be an SSC and vice versa. Both the concepts (locamin cycle and SSC) coincide when $G$ is a cycle as seen in the next theorem.

**Theorem 4.1.3** *Let $G$ be a weighted cycle. Then the following are equivalent.*

*$(i)$ $G$ is a p-block.*

*$(ii)$ $G$ is an SSC.*

*$(iii)$ $G$ is a locamin cycle.*

***Proof*** $(i) \Rightarrow (ii)$

First assume that $G$ is a $p$-block. Then by Theorem 2.3.3, there exist two internally disjoint strongest $u - v$ paths for every pair of vertices $u, v$ in $G$ such that $uv$ is not a $p$-bridge. Clearly all edges belonging to these paths are strong; for otherwise if $xy$ is not strong (i.e. a $\delta$-edge), then $G - xy$ will be the only strongest $x - y$ path in $G$, getting a contradiction.

$(ii) \Rightarrow (iii)$

Suppose that $G$ is an SSC. If possible suppose that $G$ is not locamin. Then there exists some vertex $w$ such that $w$ is not on a weakest edge of $G$. Let $uw$ and $wv$ be the two edges incident on $w$, which are not weakest edges. This implies that the path $uwv$ is the unique strongest $u - v$ path in $G$, contradiction to the assumption that $G$ is an SSC.

$(iii) \Rightarrow (i)$

Let $G$ be a locamin cycle. If possible suppose that $G$ has a $p$-cut vertex say $w$. Then for some pair of vertices $u, v$ in $G$, $CONN_{G-w}(u, v) < CONN_G(u, v)$. This implies that there exists a unique strongest path between $u$ and $v$ in $G$, which contradicts the assumption that $G$ is an SSC. ■

**Theorem 4.1.4** *If any two vertices of a weighted graph G lie on a common SSC, then G is a p-block.*

***Proof*** Let $G$ be a weighted graph satisfying the condition of the Theorem. Clearly $G$ is connected. Let $w$ be a vertex of $G$. For any two vertices $x$ and $y$ such that $x \neq w \neq y$, there exists an SSC containing $x$ and $y$. That is there exist two internally disjoint strongest $x - y$ paths in $G$. At most one of these paths can contain the vertex $w$ and hence $w$ cannot be a $p$-cut vertex of $G$. Since $w$ is arbitrary, it follows that $G$ is a block. ■

## 4.2 Cycle Connectivity

Apart from graphs, cycles of different strengths pass through the same pair of vertices in weighted graphs. As the strength of each edge is one in a classical graph, each cycle in it is of strength one. Based on this fact, we introduce two new connectivity concepts in weighted graphs, namely **$\theta$-evaluation** and **cycle connectivity $C^G_{u,v}$**. Using which a new type of weighted graph called $\theta$—**weighted graph** is defined. This section characterizes $p$-blocks in $\theta$—weighted graphs.

**Definition 4.2.1** Let $G = (V, E, w)$ be a weighted graph. Then for any two vertices $u$ and $v$ in $V$, there associated a set say $\theta(u, v)$ called the $\theta$—**evaluation** of $u$ and $v$ and is defined as $\theta(u, v) = \{\alpha$: where $\alpha$ is the strength of a strong cycle passing through both $u$ and $v\}$.

If there is no strong cycle containing both $u$ and $v$, then define $\theta(u, v) = \phi$.

**Definition 4.2.2** $Max\{\alpha : \alpha \in \theta(u, v); u, v \in V\}$ is defined as the cycle connectivity between $u$ and $v$ in $G$ and is denoted by $C^G_{u,v}$. If $\theta(u, v) = \phi$ for some pair of vertices $u$ and $v$, define the cycle connectivity between $u$ and $v$ to be 0.

**Example 4.2.3** Consider the weighted graph in Fig. 4.2.

In Fig. 4.2, there are 5 strong cycles passing through $a$ and $c$. They are $C_1 = abca$, $C_2 = acda$, $C_3 = abcda$, $C_4 = abeca$ and $C_5 = abecda$. Here $s(C_1) = s(C_3) = 13$, $s(C_2) = 15$ and $s(C_4) = s(C_5) = 11$. Thus $\theta(a, c) = \{11, 13, 15\}$ hence $C^G_{a,c} = max\{11, 13, 15\} = 15$.

Cycle connectivity is a measure of connectedness of a weighted graph and it is always less than or equal to the strength of connectedness between any two vertices $u$ and $v$. More over, non- zero cycle connectivity between $u$ and $v$ assures two disjoint strongest paths between them. In a graph without weights, the cycle connectivity of any two vertices is 1 if they belongs to a common cycle and 0 otherwise.

**Definition 4.2.4** Let $G$ be a weighted graph. $G$ is said to be a $\theta$—**weighted graph** if $\theta$- evaluation of each pair of vertices in $G$ is either empty or a singleton set. In

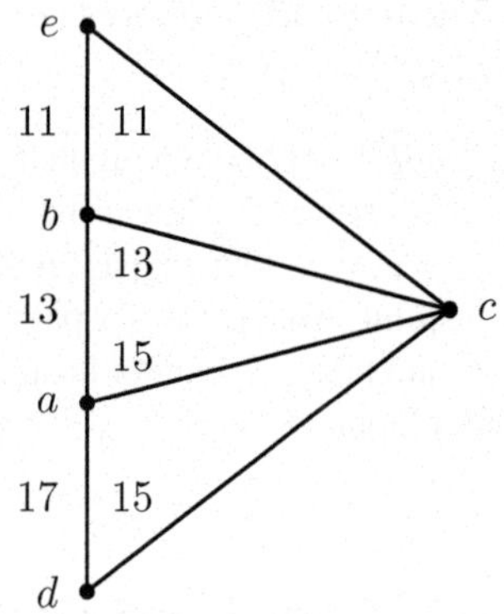


**Fig. 4.2** Cycle connectivity-illustration

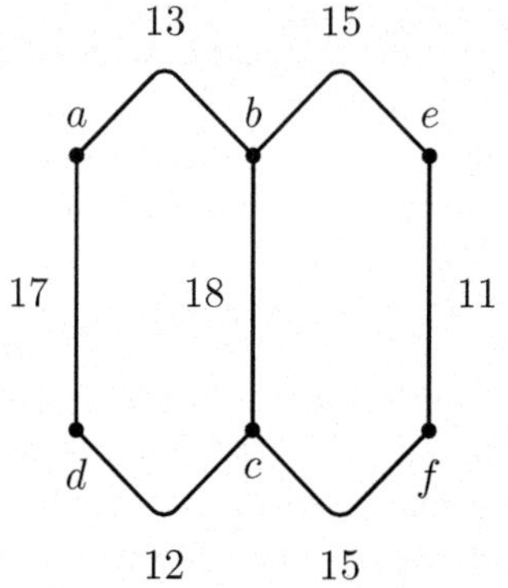


**Fig. 4.3** A trivial $\theta$-weighted graph

other words $G$ is called a $\theta$—weighted graph if for each pair of vertices $u$ and $v$, either there is no strong cycle passing through $u$ and $v$ or all strong cycles passing through $u$ and $v$ have the same strength.

**Example 4.2.5** The graph in Fig. 4.3 is a $\theta$-weighted graph. In $G$, the edges $dc$ and $fe$ are not strong. Clearly $G$ is a trivial $\theta$—weighted graph as it has no strong cycles.

**Example 4.2.6** A non trivial $\theta$-weighted graph is given in Fig. 4.4. It is also clear from this example that any edge disjoint weighted graphs are $\theta$-weighted graphs.

In this graph, $\theta(u, v) = 12$ for any pair of vertices $u, v \in A$ where $A = \{a, c, d, b, f, e\}$. $\theta(u, v) = 0$ if $u \in A$ and $v \in B$ where $B = v^* \setminus A$. Also $\theta(u, v) = 17$ if $u, v \in B$. Hence $G$ is a $\theta$—weighted graph.

Consider a $\theta$—weighted graph $G$ which is a $p$-block. We now show that, the concepts of strong path and strongest path coincide in $G$ and as a result, the concepts of strong cycle and strongest strong cycle also coincide. Further, the necessary and sufficient conditions hold for blocks can be generalized to $p$-blocks in weighted graphs.

**Fig. 4.4** A non trivial $\theta$-weighted graph

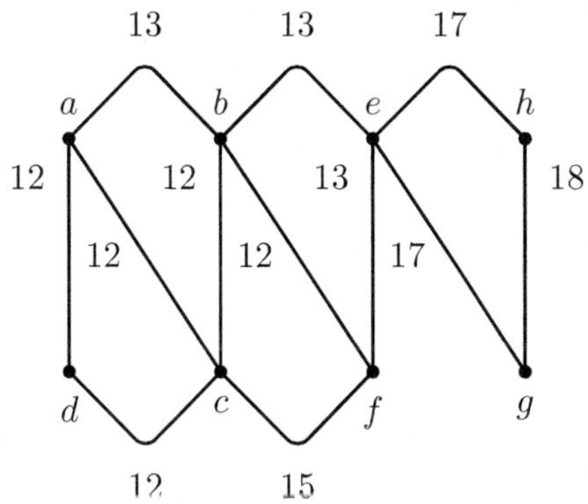


**Lemma 4.2.7** *Let $G$ be a $\theta$—weighted graph which is a block. Then any strong $u - v$ path such that $uv$ is not a $p$-bridge is a strongest $u - v$ path and hence any strong cycle in $G$ is a strongest strong cycle.*

***Proof*** Let $G = (V, E, w)$ be a $\theta$—weighted graph which is a $p$-block. Clearly $G$ is connected. Let $u, v \in V$ be such that $uv$ is not a $p$-bridge. Let $P$ be a strong $u - v$ path in $G$. If $P$ is not a strongest $u - v$ path, then because $G$ is a $p$-block, there exist two internally disjoint strongest strong $u - v$ paths say $P_1$ and $P_2$. Then $P_1 \bigcup P$ is a strong cycle with strength less than that of the cycle $P_1 \bigcup P_2$. Both these cycles pass through $u$ and $v$ and hence $\theta(u, v)$ is not a singleton or empty set, which is a contradiction to the fact that $G$ is a $\theta$-weighted graph. Thus $P$ must be a strongest strong $u - v$ path.

To prove the second assertion of the lemma, let $C$ be a strong cycle in $G$. Let $u, v$ be two vertices in $C$ such that $uv$ is not a $p$-bridge in $C$. Then by first part both these $u - v$ paths in $C$ are strongest $u - v$ paths. Thus $G$ is a strongest strong cycle. ■

**Theorem 4.2.8** *Let $G$ be a $\theta$-weighted graph. Then the following statements are equivalent.*

*$(i)$ $G$ is a p-block.*

*$(ii)$ Every two vertices of $G$ lie on a common strongest strong cycle.*

*$(iii)$ Each vertex and a strong edge of $G$ lie on a common strongest strong cycle.*

*$(iv)$ Any two strong edges of $G$ lie on a common strongest strong cycle.*

*$(v)$ For any two given vertices $u$ and $v$ such that $uv$ is not a $p$-bridge and a strong edge $xy$ in $G$, there exists a strongest strong $u - v$ path containing the edge $xy$.*

*$(vi)$ For every three distinct vertices $u_i, i = 1, 2, 3$ of $G$ such that $(u_i, u_j), j = 1, 2, 3$ and $i \neq j$, is not a $p$-bridge, there exist strongest strong paths joining any two of them containing the third.*

*$(vii)$ For every three distinct vertices $u_i, i = 1, 2, 3$ of $G$ such that $(u_i, u_j), j = 1, 2, 3$ and $i \neq j$, is not a $p$-bridge, there exist strongest strong paths joining any two of them not containing the third.*

***Proof*** Theorem 2.3.4 and Lemma 4.2.7 together give all the required implications except $(vii) \Rightarrow (i)$.

To prove $(vii) \Rightarrow (i)$, note that for any vertex $w$ of $G$ and for every pair of vertices $x, y$ other than $w$, there exists a strongest $x - y$ path not containing the vertex $w$. Clearly $G$ is connected. Thus vertex $w$ is not in every strongest $x - y$ path for all pair of vertices $x$ and $y$ and hence $w$ is not a $p$-cut vertex. Since $w$ is arbitrary, it follows that $G$ is a $p$-block. ■

## 4.3 Cycle Connectivity of Weighted Graphs

In the previous section we discussed cycle connectivity between pairs of vertices in a weighted graph. Now we define and discuss the cycle connectivity of weighted graphs. Also we will find out the cycle connectivity of partial trees, cycles and precisely weighted graphs.

**Definition 4.3.1** Let $G = (V, E, w)$ be a weighted graph. **Cycle connectivity of** $G$ is defined as

$$CC(G) = Max\{C^G_{u,v} : u, v \in V\}$$

That is cycle connectivity of a weighted graph $G$ is defined as the maximum of cycle connectivity of different pairs of vertices of $G$.

Note that in a graph $G$ without weights, $CC(G)$ is one if $G$ contains at least a cycle and it is zero if $G$ is a tree.

**Example 4.3.2** Consider the graph in Fig. 4.5 on 4 vertices. Here $C^G_{a,d} = 13$, $C^G_{b,c} = 11 = C^G_{b,d}$. Thus $CC(G) = 13$.

Clearly the cycle connectivity of a weighted graph is the maximum value among the strengths of all strong cycles in $G$.

**Theorem 4.3.3** *A weighted graph $G$ is a partial tree if and only if $CC(G) = 0$.*

Fig. 4.5 A weighted graph with cycle connectivity 13

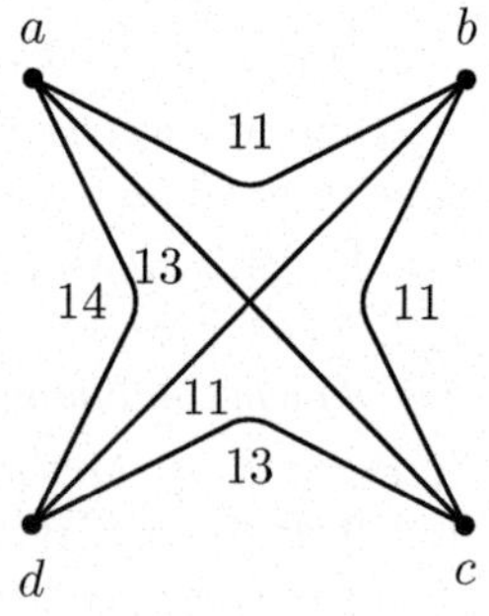

***Proof*** If $G$ is a partial tree, then $C^G_{u,v} = 0$ for every pair of vertices $u$ and $v$ in $G$. That is, $G$ has no strong cycles passing through $u$ and $v$. Since it is true for all $u, v \in V$, we have, $CC(G) = 0$.

Conversely suppose that $CC(G) = 0$. That is, $C^G_{u,v} = 0$ for every pair of vertices in $G$. That is $G$ has no strong cycles. Hence it follows that $G$ is a $p$-tree. ■

Also it follows from the definition that the cycle connectivity of a weighted strong cycle $G$ is the strength of $G$ itself. Now we find the cycle connectivity of a precisely weighted graph.

**Theorem 4.3.4** *Let $G = (V, E, w)$ be a precisely weighted graph with vertices $v_1, v_2, \ldots, v_n$ such that $\sigma(v_i) = t_i$ and $t_1 \leq t_2 \leq \cdots \leq t_{n-2} \leq t_{n-1} \leq t_n$. Then $CC(G) = t_{n-2}$.*

***Proof*** Assume the conditions of the Theorem. Since any three vertices of $G$ are adjacent, any three vertices are on a 3-cycle. Also by Lemma 3.1.7, all edges in a PWG are strong. Thus to find the maximum strength of cycles in $G$, it is sufficient to find the maximum strength of all 3-cycles in $G$. Let us consider a 4-cycle $C = abcda$ in $G$ (case of $n$-cycle is similar). Since $G$ is complete, there exists parts of two 3-cycles in $C$, namely $C_1 = abca$ and $C_2 = acda$. Let the strength $s(C) = t$. For all edges $xy$ in $C$, $w(xy) \geq t$. In particular $w(ab) \geq t$ and $w(bc) \geq t$. Since $G$ is a precisely weighted graph, it has no $\delta$-edges. Thus $w(ac) \geq \min\{w(ab), w(bc)\} \geq t$. That is $w(ac) \geq t$.

Suppose $w(ac) = t$, then $s(C_1) = s(C_2) = s(C) = t$.

Suppose $w(ac) > t$, then since $s(C) = t$, at least one of $C_1$ or $C_2$ will have strength equal to $t$.

In either case, $s(C) = \min\{s(C_1), s(C_2)\}$

Thus the strength of a 4-cycle is nothing but the strength of a 3-cycle in $G$. Among all 3-cycles, the 3-cycle formed by three vertices with maximum vertex strength will have the maximum strength. Thus the cycle $C = v_{n-2}v_{n-1}v_nv_{n-2}$ is a cycle with maximum strength in $G$. Also strength of $C = t_{n-2} \wedge t_{n-1} \wedge t_n = t_{n-2}$ where $\wedge$ stands for the minimum. Thus $CC(G) = t_{n-2}$. ■

## 4.4 Cyclic Cut Vertices and Cyclic Bridges

In this section we introduce two concepts in weighted graphs namely cyclic cut vertex and cyclic bridge. Cut vertices and bridges affect connectivity of a weighted graph on their removal from the graph. Analogously, cyclic cut vertices and cyclic bridges affect the cycle connectivity of a weighted graph on their removal.

**Definition 4.4.1** A vertex $w$ in a weighted graph is called a **cyclic cut vertex** if $CC(G - w) < CC(G)$.

**Definition 4.4.2** An edge $uv$ of a weighted graph is called a **cyclic bridge** if $CC[G - (u, v)] < CC(G)$.

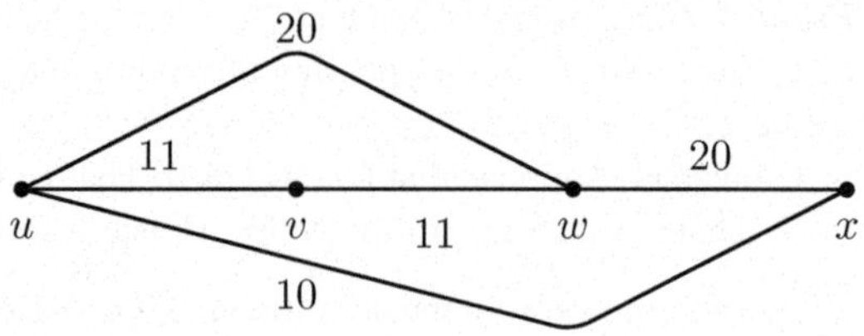


**Fig. 4.6** A weighed graph having cyclic cut vertices and cyclic bridges

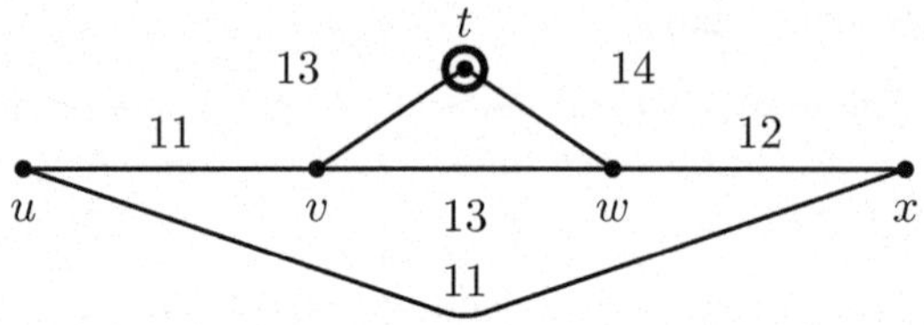


**Fig. 4.7** Example for a cyclic cut vertex which is not a cut vertex

From the above definitions, it follow that a vertex in a non-weighted connected graph $G$ containing cycles, is a cyclic cut vertex if $G - w$ is a forest and $uv$ is a cyclic bridge if $G - uv$ is a tree. As an example, consider a polygon in $n$ vertices. The number of cyclic cut vertices and cyclic bridges in it is $n$ each. Also an $n$-connected graph with $n \geq 3$ will have no cyclic bridges.

**Example 4.4.3** Figure 4.6 contains both cyclic cut vertices and cyclic bridges. $u$, $v$ and $w$ are the cyclic cut vertices. Whereas $uv$, $vw$ and $uw$ are the cyclic bridges.

**Remark** In a weighted graph, a cyclic cut vertex need not be a cut vertex and a cyclic bridge need not be a bridge as seen from Fig. 4.7.

Here $t$ is not a cut vertex. But it is a cyclic cut vertex since $CC(G - t) = 11 < CC(G) = 13$. Also edges $vt, tw$ and $vw$ are cyclic bridges, but not bridges of $G$.

**Definition 4.4.4** A weighted graph $G$ is said to be **cyclically balanced** if $G$ has no cyclic cut vertices and cyclic bridges.

A weighted graph containing two disjoint strong cycles of equal strength is cyclically balanced. All edge disjoint graphs need not be cyclically balanced; if there is a common vertex for them, then it will be a cyclic cut vertex.

**Proposition 4.4.5** *In a weighted graph $G$, if edge $uv$ is a cyclic bridge, then both $u$ and $v$ are cyclic cut vertices.*

***Proof*** Let $G(V, E, w)$ be a weighted graph and $uv$ be a cyclic bridge in $G$. Then $CC(G - (u, v)) < CC(G)$.

Hence $CC(G - u) \leq CC(G - (u, v)) < CC(G)$ and $CC(G - v) \leq CC(G - (u, v)) < CC(G)$. Thus $u$ and $v$ are cyclic cut vertices. ■

**Proposition 4.4.6** *Let G be a weighted cycle. Then,*

*(a) G has no cyclic cut vertices or cyclic bridges if G is a partial tree.*

*(b) All edges in G are cyclic bridges and all vertices in G are cyclic cut vertices if G is a strong cycle.*

***Proof*** $(a)$ Follows from the fact that a partial tree has no strong cycles.

$(b)$ If $G$ is a strong cycle, then $CC(G) =$ strength of $G$. The removal of any edge or vertex will reduce its cycle connectivity 0. ■

**Remark** Part $(b)$ of the last proposition says that every partial bridge in a weighted strong cycle is a cyclic bridge. In addition to this, each partial cut vertex of a weighted strong cycle is a cyclic cut vertex. Proposition 4.4.7 generalizes this result.

**Proposition 4.4.7** *Let G be a weighted graph containing at most one cycle. Then every partial bridge of G is a cyclic bridge and every partial cut vertex of G is a cyclic cut vertex. The converse is not true.*

**Proposition 4.4.8** *Let G be a weighted graph such that G has a unique weighted cycle C such that strength of* $C = CC(G)$. *Then every partial bridge in C is a cyclic bridge and every partial cut vertex in C is a cyclic cut vertex.*

It can be inferred from Proposition 4.4.6 $(b)$ that every vertex is a cyclic cut vertex in a PWG with three vertices and that all its edges are cyclic bridges. But when a CFG contains more than three vertices the condition may not be true as seen from the following theorem.

**Theorem 4.4.9** *Let* $G = (V, E, w_1, w_2)$ *be a PWG with* $|V| \geq 4$. *Let* $v_1, v_2, \ldots, v_n \in V$ *and* $\sigma(v_i) = c_i$ *for* $i = 1, 2, \ldots, n$ *and* $c_1 \leq c_2 \leq \cdots \leq c_n$. *Then G has a cyclic cut vertex ( or a cyclic bridge) if and only if* $c_{n-3} < c_{n-2}$. *Further there exist three cyclic cut vertices (or cyclic bridges) in a PWG (if exists).*

***Proof*** Let $v_1, v_2, \ldots, v_n \in V$ and $\sigma(v_i) = c_i$ for $i = 1, 2, \ldots, n$ and $c_1 \leq c_2 \leq \cdots \leq c_n$. Suppose that $G$ has a cyclic cut vertex $u$ (say). Then $CC(G-u) < CC(G)$. That is $u$ belongs to a unique cycle $C$ with $\alpha =$ strength of $C >$ strength of $C'$ for any other cycle $C'$ in $G$. Since $c_1 \leq c_2 \leq \cdots \leq c_n$, it follows that the strength of the cycle $v_{n-2}v_{n-1}v_n$ is $\alpha$. Hence $u \in \{v_{n-2}, v_{n-1}, v_n\}$.

To prove $c_{n-3} < c_{n-2}$. Suppose not. That is $c_{n-3} = c_{n-2}$. Then $C_1 = v_n v_{n-1} v_{n-2}$ and $C_2 = v_n v_{n-1} v_{n-3}$ have the same strength and hence the removal of $v_{n-2}, v_{n-1}$ or $v_n$ will not reduce $CC(G)$ which is a contradiction to the fact that $u \in \{v_{n-2}, v_{n-1}, v_n\}$t. Hence $c_{n-3} < c_{n-2}$.

Conversely suppose that $c_{n-3} < c_{n-2}$. To prove $G$ has a cyclic cut vertex. Since $c_n \geq c_{n-1} \geq c_{n-2}$ and $c_{n-2} > c_{n-3}$, all cycles of $G$ have strength less than that of strength of $v_n v_{n-1} v_{n-2}$. Hence the deletion of any of the three vertices in $v_n, v_{n-1}, v_{n-2}$ will reduce the cycle connectivity of $G$. Hence $v_n, v_{n-1}$ and $v_{n-2}$ are cyclic cut vertices of $G$. Also from the proof it follows that there exist three cyclic cut vertices if exists. The case of cyclic bridges is similar. ■

**Fig. 4.8** Illustration to Theorem 4.4.9

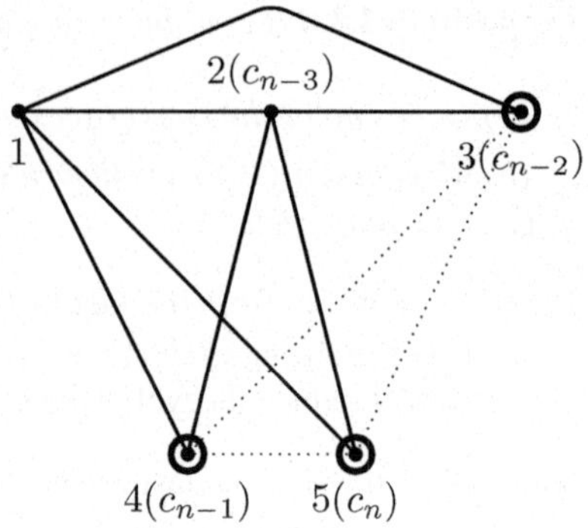


The above result may be verified with Fig. 4.8.

In this example, $2 = c_{n-3} < c_{n-2} = 3$. $G$ has three cyclic cut vertices namely the vertices with weights $c_n$, $c_{n-1}$ and $c_{n-2}$ (Given in circles). Also $c_n c_{n-1}$, $c_{n-1} c_{n-2}$ and $c_{n-2} c_n$ (as labels, shown in dotted lines) are cyclic bridges.

Now we state a theorem without proof related with edge disjoint weighted graphs.

**Theorem 4.4.10** *Let $G$ be an edge disjoint weighted graph which is not a tree. A cut vertex of $G$ is a cyclic cut vertex if it is the common vertex of all cycles of $G$ or it is a vertex of a cycle with maximum strength, which is unique.*

## 4.5 Cyclically Balanced Graphs

This section introduces some specific sets named cyclic vertex cut and cyclic edge cut that has influence on the cyclic connectivity of a weighted graph. cyclic vertex connectivity and cyclic edge connectivity are parameters which are important in the analysis of cycle connectivity of graphs and weighted graphs.

**Definition 4.5.1** Let $G = (V, E, w)$ be a weighted graph. A **cyclic vertex cut** $(CVC)$ of $G$ is a set of vertices $X \subseteq V$ such that $CC(G - X) < CC(G)$, provided $CC(G) > 0$, where $CC(G)$ is the cycle connectivity of $G$.

**Example 4.5.2** Let $G$ be a connected weighted graph with vertex set $V = \{a, b, c, d\}$ (Fig. 4.9) such that $w(ab) = 15 = w(da)$, $w(bc) = w(cd) = w(ca) = w(bd) = 12$. Here $abcda$ and $acda$ are cycles of strength 12. Also $CC(G) = 12$. Vertex set $X$= $\{\{a, c\}\}$ is a $2 - CVC$ of $G$.

**Definition 4.5.3** Let $X$ be cyclic vertex cut of $G$. The **cyclic strong weight** of $X$ is defined as $S_c(X) = \sum\limits_{x \in X} w(xy)$, where $w(xy)$ is the minimum of weights of strong edges incident on $x$.

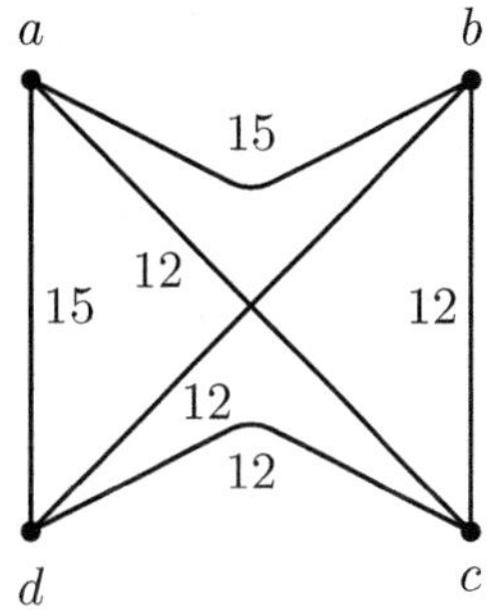


**Fig. 4.9** Cyclic vertex cuts and cyclic bridges

**Definition 4.5.4** **Cyclic vertex connectivity** of a weighted graph $G$, denoted by $\kappa_c(G)$, is the minimum of the cyclic strong weights of cyclic vertex cuts in $G$.

In example 4.5.2, $C_1 = \{a, c\}$, $C_2=\{a, d\}$, $C_3 = \{b, c\}$, $C_4 = \{b, d\}$, $C_5 = \{a, b\}$ and $C_6 = \{c, d\}$ are $2 - CVCs$ with $S(C_1) = S(C_2) = S(C_3) = S(C_4) = S(C_5) = S(C_6) = 14$. {d} is a $1 - CVC$ in $G$. Thus the cyclic vertex connectivity is 12.

**Definition 4.5.5** Let $G = (V, E, w)$ be a weighted graph. A **cyclic edge cut** $(CEC)$ of a weighted graph $G$ is a set of edges $Y \subseteq E$ such that $CC(G - Y) < CC(G)$, provided $CC(G) > 0$, where $CC(G)$ is the cyclic connectivity of $G$.

**Example 4.5.6** In Fig. 4.9, $\{ab, cd\}$ and $\{db, cd, cb\}$ are $2 - CEC$ and $3 - CEC$ of $G$ respectively.

**Definition 4.5.7** Let $G = (V, E, w)$ be a weighted graph of $G$. The **strong weight** of a cyclic edge cut $Y$ of $G$ is defined as $S_c^{'}(Y) = \sum_{e_i \in E} w(e_i)$, where $e_i$ is a strong edge of $Y$.

**Definition 4.5.8** **Cyclic edge connectivity** of a weighted graph $G$ is denoted by $\kappa_c^{'}(G)$ is the minimum of the cyclic strong weights of cyclic edge cuts in $G$.

In Fig. 4.9, all edges other than $bd$ are strong. $\{cd\}$ and $\{ad\}$ are $1 - CECs$. $\{cd, ac\}$, $\{cd, bc\}$, $\{ad, ab\}$ are some of the $2 - CECs$ of $G$. Thus cyclic edge connectivity of the graph is 12.

Next we have a relation between the cyclic vertex connectivity and vertex connectivity of a precisely weighted graph.

**Theorem 4.5.9** *For a precisely weighted graph $G$, $\kappa_c(G) \leq \kappa_w(G)$.*

***Proof*** Let $G$ be a precisely weighted graph. Label the vertices of $G$ as $v_1, v_2, \ldots, v_n$ such that $d(v_1) \leq d(v_2) \leq \cdots \leq d(v_n)$. Let $v_1$ be a vertex such that $d(v_1) = \delta_s(G)$ (Fig. 4.10).

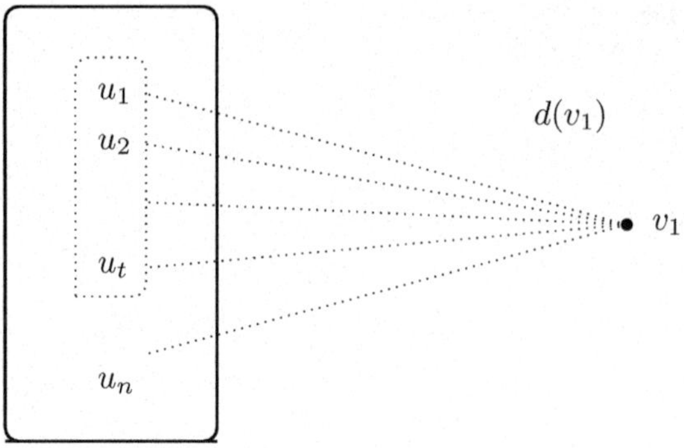


**Fig. 4.10** Minimum strong degree of a vertex

**Case 1.** $v_1$ is a cyclic cut vertex.

In this case, $\{v_1\}$ is a cyclic cut set. Then,

$S(v_1) = \wedge_{v_i \in V} w(v_1 v_i) \leq \sum_{v_i v_1 \in E} w(v_1 v_i) = \delta_s(G)$

Now, since $\kappa_c(G) = min\{S(V)\}$, where $V$ is a cyclic cut set of $G$, we have, $\kappa_c(G) \leq S(v_1) = \delta_s(G) = \kappa_w(G)$.

**Case 2.** $\{v_1\}$ is not a cyclic cut vertex.

Let $F = \{u_1, u_2, \ldots, u_t\}$ be a cyclic cut set such that $S(F) = \kappa_c(G)$, Now, $\kappa_c(G) = S(F) = \sum_{i=1}^{t} \min\{w(u_i u_j)\}, \forall u_i u_j \in E$, (for some $j \neq i, j = 1, 2, \ldots, n$) $= \sum_{i=1}^{t} \min\{w(u_i v_1)\} \leq d(v_1) = \delta_s(G) = \kappa_w(G)$. ■

Absence of strong cycles in a partial tree leads to the statement that $\kappa_c(G) = 0$ for a partial tree $G$.

**Corollary 4.5.10** *For a partial tree $G$, $\kappa_c(G) \leq \kappa_w(G)$.*

**Theorem 4.5.11** *A vertex in a weighted graph is a cyclic cut vertex if and only if it is a common vertex of all strong cycles with maximum strength.*

***Proof*** Let $G$ be a weighted graph. Let $w$ be a cyclic cut vertex of $G$. Then $CC(G - w) < CC(G)$. i.e.; Max $\{S(C)$, where $C$ is a strong cycle in $G - w\}$ < Max $\{S(C')$, where $C'$ is a strong cycle in $G\}$. All strong cycles in $G$ with maximum strength will be removed by the deletion of $w$. Hence $w$ is common to all strong cycles with maximum strength.

Conversely, Let $w$ be a common vertex of every strong cycle with maximum strength. The removal of $w$ results in the deletion of all strong cycles with maximum strength. Hence it will result in the reduction of cycle connectivity of $G$. Thus $w$ is a cyclic cut vertex of $G$. ■

**Definition 4.5.12** A vertex $w \in V$ of a weighted graph $G = (V, E, w)$ is said to be a **cyclic end vertex** , if it lies on a strong cycle but which is not a cyclic cut vertex.

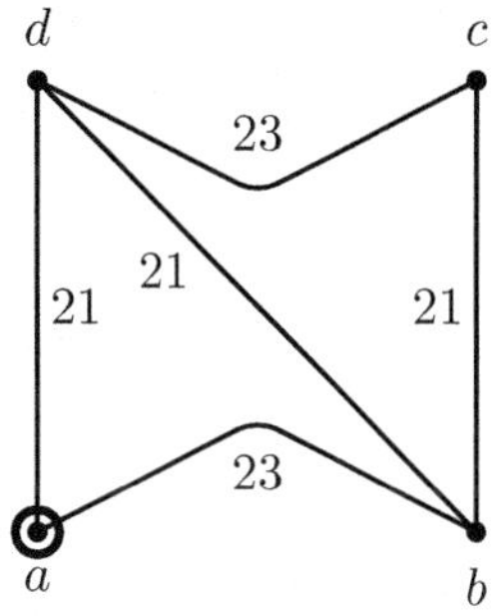


**Fig. 4.11** Cyclic end vertex of a graph

**Example 4.5.13** Let $G$ be a connected weighted graph with vertex set $V = \{a, b, c, d\}$ (Fig. 4.11) such that $w(ab) = 23$, $w(bc) = 21$, $w(cd) = 23$, $w(da) = 21$ and $w(bd) = 21$. Here $abcda$, $adb$ and $cdb$ are strong cycles and each of which has strength 21. Thus $CC(G) = 21$. $a$ and $c$ are the cyclic end vertices of $G$.

**Theorem 4.5.14** *Let $G$ be a weighted graph. Then no cyclic cut vertex is an end vertex of $G$.*

***Proof*** Let $G$ be a weighted graph. Let $w$ be a cyclic cut vertex of $G$. Then $w$ lies on a strong cycle with maximum strength in $G$. Clearly $w$ has at least two strong neighbors in $G$. Hence $w$ cannot be an end vertex of $G$.

Conversely, If $w$ is an end vertex of $G$ with $|N(w)| = 1$ then $w$ cannot lie on a strong cycle in $G$ which implies that $w$ is not a cyclic cut vertex. ∎

**Corollary 4.5.15** *No cyclic cut vertex is a cyclic end vertex of $G$.*

Next we characterize cyclically balanced graphs which are precisely weighted.

**Theorem 4.5.16** *Let $G = (V, E, w_1, w_2)$ be a PWG with $|V| \geq 4$. Let $v_1, v_2, \ldots, v_n \in V$ and $w_1(v_i) = c_i$ for $i = 1, 2, \ldots, n$ and $c_1 \leq c_2 \leq \cdots \leq c_n$. Then $G$ is cyclically balanced if and only if $c_{n-3} = c_{n-2}$.*

***Proof*** Let $v_1, v_2, \ldots, v_n \in V$ and $\sigma(v_i) = c_i$ for $i = 1, 2, \ldots, n$ and $c_1 \leq c_2 \leq \cdots \leq c_n$. Suppose $G$ is cyclically balanced. To prove that $c_{n-3} = c_{n-2}$. Suppose not, that is $c_{n-3} < c_{n-2}$. Since $c_{n-2} \leq c_{n-1} \leq c_n$ and $c_{n-3} < c_{n-2}$, all cycles of $G$ have strength less than that of strength of $v_n v_{n-1} v_{n-2} v_n$. Hence the deletion of any of the three vertices $v_n, v_{n-1}, v_{n-2}$ reduce the cycle connectivity of $G$. Hence $v_n, v_{n-1}$ and $v_{n-2}$ are cyclic cut vertices of $G$, which is a contradiction to the fact that $G$ is cyclically balanced.

Conversely, suppose that $c_{n-3} = c_{n-2}$. Then $C_1 = v_n v_{n-1} v_{n-2}$ and $C_2 = v_n v_{n-1} v_{n-3}$ have the same strength and hence the removal of $v_{n-2}, v_{n-1}$ and $v_n$ will not reduce the cyclic connectivity of $G$. That is, there does not exist any cyclic cut vertex in $G$. Hence $G$ is cyclically balanced. ∎

**Theorem 4.5.17** *Let G be a PWG. Then G is cyclically balanced if there exists a* $K_4$ *as a sub graph of G in which every cycle is of equal maximum strength.*

***Proof*** Let $G = (V, E, w_1, w_2)$ be a precisely weighted graph with $|V| \geq 4$. Let $v_1, v_2, \ldots, v_n \in V$ and $\sigma(v_i) = c_i$ for $i = 1, 2, \ldots, n$ and $c_1 \leq c_2 \leq \cdots \leq c_n$. Let $K_4$ be a subgraph of $G$ with vertex set $\{v_{n-3}, v_{n-2}, v_{n-1}, v_n\}$ such that $c_{n-3} \leq c_{n-2} \leq c_{n-1} \leq c_n$. Suppose all the strong cycles in $K_4$ are of equal maximum strength. This happens only when $c_{n-3} = c_{n-2}$. Then by Theorem 4.5.15, $G$ is cyclically balanced. ■

**Theorem 4.5.18** *Let* $G = (V, E, w_1, w_2)$ *be a PWG and* $v \in V$ *such that* $d_s(v) = \Delta_s(G)$. *If v lies on a strong cycle C, then* $S(C) = CC(G)$.

***Proof*** Let $G = (V, E, w_1, w_2)$ be a precisely weighted graph and $v \in V$ be a vertex such that $d_s(v) = \Delta_s(G)$. Let $v_1, v_2, \ldots, v_n \in V$ and $\sigma(v_i) = c_i$ for $i = 1, 2, \ldots, n$ and $c_1 \leq c_2 \leq \cdots \leq c_n$. Since $c_{n-2} \leq c_{n-1} \leq c_n$ and $c_{n-3} < c_{n-2}$, all cycles of $G$ have strength less than strength of $v_n v_{n-1} v_{n-2} v_n$. First to prove that for all $v_i$, $i = 1, \ldots, n-2$, $d_s(v_i) < d_s(v_n)$.

$$\begin{aligned} d_s(v_i) &= \sum_{j=1, j \neq i}^{n} min(c_i, c_j) \\ &= c_1 + c_2 + \cdots + (n-i)c_i \\ &= \sum_{i=1}^{n-1} c_i \\ &= d_s(v_n) \end{aligned}$$

Also

$$\begin{aligned} d_s(v_n) &= (c_{n-1} \wedge c_n) + \sum_{j=1}^{n-2} (c_n \wedge c_i) \\ &= (c_{n-1} \wedge c_n) + \sum_{j=1}^{n-2} (c_{n-1} \wedge c_i) \\ &= d_s(v_{n-1}) = \sum_{i=1}^{n-1} c_i = \Delta_s(G) \end{aligned}$$

Therefore, $v_n$ belongs to the strong cycle $c_n c_{n-1} c_{n-2} c_n$, where strength of this cycle is equal to the cycle connectivity of graph $G$. ■

Following inequality on weighted graphs is analogous to the Whitney's theorem for unweighted graphs.

**Theorem 4.5.19** *For a PWG,* $\kappa_c(G) \leq \kappa_c'(G) \leq \Delta_s(G)$.

***Proof*** Consider all cycles, having strength equal to cycle connectivity of $G$. Let $X = \{e_1, e_2, \ldots, e_n\}$, where $e_i = u_i v_i$ be one of the edges in each such cycles. Then $X$ form a cyclic edge cut of $G$. Let $S_c(X)$ be the cyclic strong weight of $X$. Then by the definition of cyclic edge connectivity,

$$\kappa_c^{'}(G) \leq S_c(X)\ldots\ldots\ldots\ldots..(1)$$

Let $Y = \{v_1, v_2, \ldots\}$ be the collection of one of the end vertices of each edge in the cyclic edge cut $X$ of $G$. Then $Y$ form a cyclic vertex cut of $G$. Let $S_c(Y)$ be the cyclic strong weight of $Y$. Then by the definition of cyclic vertex connectivity of $G$, we have

$$S_c(Y) \leq \kappa_c^{'}(G)\ldots\ldots\ldots\ldots..(2)$$

Thus,

$$\kappa_c(G) \leq S_c(Y)\ldots\ldots\ldots\ldots\ldots..(3)$$

From equations (1), (2) and (3)

$$\kappa_c(G) \leq S_c(Y) \leq \kappa_c^{'}(G) \leq \Delta_s(G).$$

Hence the theorem is proved. ■

**Theorem 4.5.20** *A weighted graph $G$ with $n \geq 6$ is cyclically balanced if there exist two disjoint cycles $C_1$ and $C_2$ such that $S(C_1) = S(C_2) = CC(G)$.*

***Proof*** Let $G$ be a weighted graph with $n \geq 6$ and cycle connectivity of $G$ is equal to $CC(G)$. Let $C_1$ and $C_2$ be two disjoint cycles in $G$ such that $S(C_1) = S(C_2) = CC(G)$.

If a vertex say $u$ not in $V(C_1 \cup C_2)$ is deleted, the cycle connectivity of the graph remains the same. If the vertex $u$ in the cycle $C_1$ is deleted, the cycle connectivity remains the same. Since there exists another cycle $C_2$, with strength of $C_2$ is equal to the cycle connectivity of the graph. Similarly if $u$ is in $C_2$, $u$ is not a cyclic cut vertex (Fig. 4.12).

Suppose $uv$ is an edge outside the two cycles, removal of which will not reduce the cycle connectivity of the graph. If $uv$ is an edge either on $C_1$ or on $C_2$ then the removal of $uv$ from any one of these cycles will not affect the cycle connectivity of $G$. Hence $uv$ is not a cyclic bridge. ■

A graph with 4 or 5 vertices is cyclically balanced if and only if there exist at least four strong cycles having equal maximum strength as seen in Example 4.12.

Next we construct a cyclically balanced graphs on more than 5 vertices.

**Theorem 4.5.21** *For $n \geq 6$, There is a connected cyclically balanced graph.*

***Proof*** For $n = 4$ and 5, we have cyclically balanced graphs from the above example (Fig. 4.5). For $n \geq 6$ We prove this by induction. For $n = 6$, let $v_1, v_2, \ldots, v_6$ be the 6 vertices, Construct two disjoint cycles (say) $C_1 = v_1v_2v_3$ and $C_2 = v_4v_5v_6$ with maximum strength.

Join each pair of vertices from the two cycles and make the graph complete. Then the removal of an edge or vertex will not reduce the cycle connectivity of $G$. So the newly obtained weighted graph is cyclically balanced.

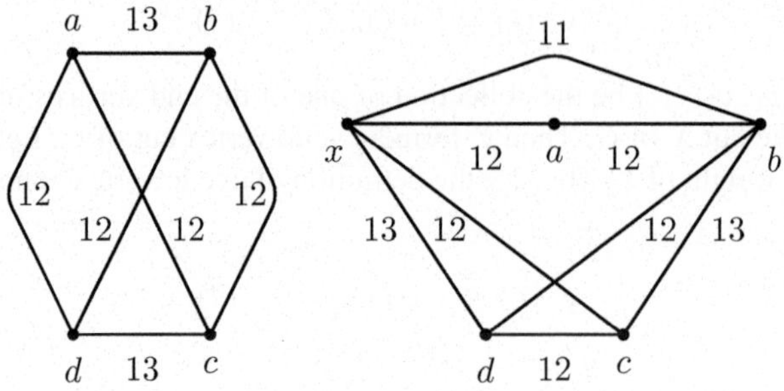


**Fig. 4.12** Cyclically balanced graphs on 4 and 5 vertices

Assume that the result is true for $n = k$. Let $G_k$ be a cyclically balanced graph with $k$ vertices. Then there exist two disjoint cycles of maximum strength in $G_k$.

Let $G_{k+1}$ be the graph obtained from $G_k$ by adding one more vertex $u$ and make the graph a complete with connecting all vertices of $G_k$ with $u$. Also assign a weight to all newly joined edges, which is less than or equal to the cycle connectivity of $G_k$.

In this case, if we remove the vertex $u$, then the cycle connectivity of $G_k$ remains the same. In a similar way the removal of any edge incident on $k + 1$th vertex $u$ will not change the cycle connectivity of $G$. Therefore cycle connectivity of $G_{k+1}$ remains the same. Hence the graph is cyclically balanced. ■

# Chapter 5
# Distance and Convexity

## 5.1 Weighted Distance

In classical graph theory, we can see studies on different types of distances and their applications. The concept of distance is defined and studied in weighted graphs also. The distance between two vertices in a weighted graph is defined as the minimum of the sum of weights over all geodesics connecting them. This section defines a new kind of distance in 2-connected weighted graphs named as the weighted distance.

**Definition 5.1.1** Let $G = (V, E, w)$ be a weighted graph. Then the weighted distance between two vertices $u$ and $v$ in $G$ is defined and denoted by $d_w(u, v) = \wedge_P\{l(P) * S(P)/P$ is a $u - v$ path, $l(P)$ is the length and $S(P)$ is the strength of the path $P\}$. $\wedge$ represents the minimum and $*$ represents ordinary product.

A $u - v$ path $P$ is called a **weighted** $u - v$ **geodesic** if $d_w(u, v) = l(P) * S(P)$.

This means a $u - v$ path $P$ is called a weighted $u - v$ geodesic, if the weighted distance between $u$ and $v$ is calculated along $P$.

**Example 5.1.2** Consider the weighted graph in Fig. 5.1 on four vertices $\{a, b, c, d\}$. In this graph, $d_w(a, b) = \wedge\{1 * 23, 2 * 23, 3 * 22\} = 23$, $d_w(a, c) = \wedge\{2 * 23, 3 * 22, 2 * 22, 3 * 23\} = 44$, $d_w(a, d) = \wedge\{1 * 23, 2 * 23, 3 * 22\} = 23$, $d_w(b, c) = \wedge\{1 * 25, 2 * 22, 3 * 22\} = 25$, $d_w(b, d) = \wedge\{1 * 24, 2 * 22, 2 * 23\} = 24$ and $d_w(c, d) = \wedge\{1 * 22, 2 * 24, 3 * 23\} = 22$.

**Remark 5.1.3** The weighted distance $d_w$ satisfies the following metric properties [1].

1. $d_w(u, v) \geq 0$ for all $u, v \in V$
2. $d_w(u, v) = 0$ if and only if $u = v$
3. $d_w(u, v) = d_w(v, u)$ for all $u, v \in V(G)$ (symmetry).



S. Mathew et al., *Weighted and Fuzzy Graph Theory*, Studies in Fuzziness and Soft Computing 429, https://doi.org/10.1007/978-3-031-39756-1_5

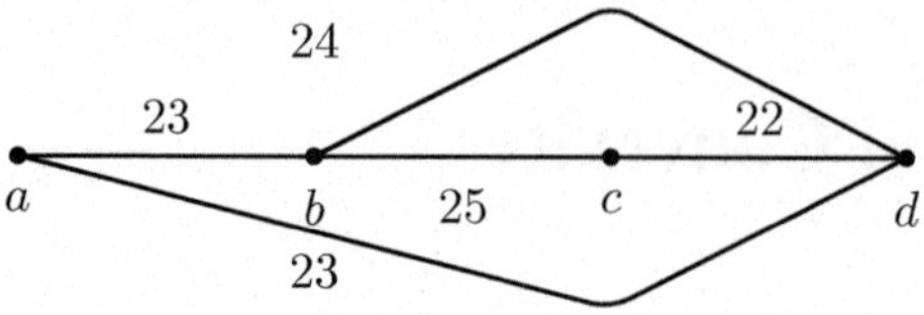


**Fig. 5.1** Distance in a weighted graph

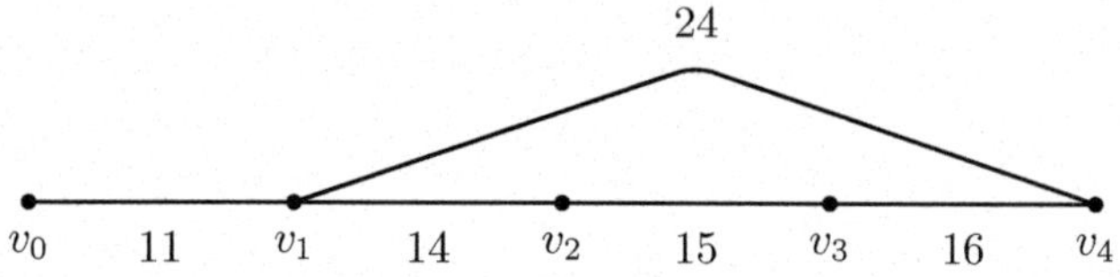


**Fig. 5.2** Weighted distance is not a metric—example

**Remark 5.1.4** Consider an arbitrary weighted graph $G = (V, E, w)$ and the metric property called triangle inequality, $d_w(u, v) \leq d_w(u, w) + d_w(w, v)$ for all $u, v, w \in V(G)$. We can not always expect $d_w$ to satisfy the triangle inequality. This is clear from the following example.

**Example 5.1.5** Consider the weighted graph in Fig. 5.2.

Consider the vertices $v_0$, $v_2$ and $v_4$. $d_w(v_2, v_4) = 2 * 14 = 28$, $d_w(v_2, v_0) = 2 * 11 = 22$ and $d_w(v_0, v_4) = 2 * 11 = 22$. Now we can see that, $d_w(v_2, v_4) > d_w(v_2, v_0) + d_w(v_0, v_4)$. This means that the distance $d_w$ is not a metric as it violates the triangle inequality.

If we restrict the underlying unweighted graph $G^*$ to be $k-$ connected, $k \geq 2$, then the triangle inequality holds good [1], and the weighted distance $d_w$ will be a metric. The detailed proof is shown in the following theorem.

**Theorem 5.1.6** *Let $G = (V, E, w)$ be a weighted graph such that the underlying unweighted graph $G^*$ is $k-$ connected, $k \geq 2$. Then for any three vertices $u$, $v$ and $w$ in $G$, we have, $d_w(u, v) \leq d_w(u, w) + d_w(w, v)$.*

***Proof*** Let $G = (V, E, w)$ be a weighted graph such that the underlying unweighted graph $G^*$ is $k-$ connected, where $k \geq 2$. We want to prove that for any three vertices $u$, $v$ and $w$ in $G$, $d_w(u, v) \leq d_w(u, w) + d_w(w, v)$. For convenience, let $k = 2$. Then between every pair of vertices, there exists two internally disjoint paths. This means every pair of vertices in $G$ lie on a common cycle. Let $C$ be a cycle containing the vertices $u$ and $v$. We prove the inequality in the following 3 different cases.

**Case 1.** The third vertex $w$ lies on a weighted $u - v$ geodesic $P_g$ confined to $C$ (Fig. 5.3).

Let the weighted $u - v$ geodesic $P_g = P \cup Q$, where $P$ is the $u - w$ sub path and $Q$ is the $w - v$ sub path of $P_g$. Therefore, $l(P \cup Q) * S(P \cup Q) = d_\omega(u, v)$. Let $P'$

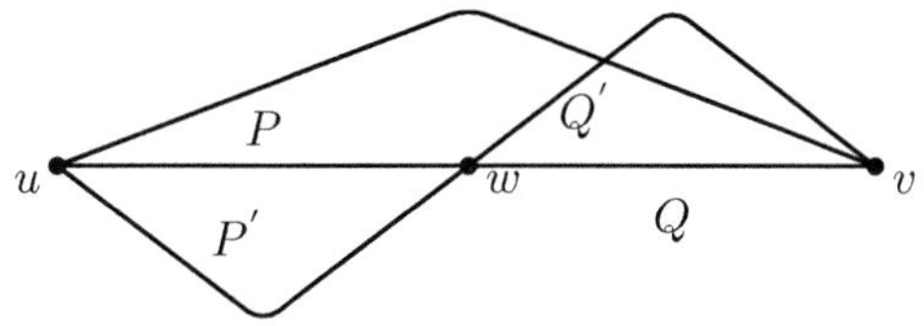


**Fig. 5.3** Illustration—Case 1

be another $u - w$ path and $Q'$ be another $w - v$ path in $G$, which are not belonging to $C$.

**Subcase 1.1.** Both $P$ and $Q$ are weighted geodesics.

Let $P$ be the weighted $u - w$ geodesic and $Q$ be the weighted $w - v$ geodesic. Then $l(P) * S(P) = d_w(u, w)$ and $l(Q) * S(Q) = d_w(w, v)$. We want to prove $d_w(u, v) \leq d_w(u, w) + d_w(w, v)$.

We have $l(P \cup Q) = l(P) + l(Q)$ and $S(P \cup Q) = min\{S(P), S(Q)\} = S(P)$, say. So $d_w(u, v) = l(P \cup Q) * S(P \cup Q) = (l(P) + l(Q)) * S(P) = l(P) * S(P) + l(Q) * S(P) \leq l(P) * S(P) + l(Q) * S(Q) = d_w(u, w) + d_w(w, v)$.

Thus $d_w(u, v) \leq d_w(u, w) + d_w(w, v)$ is true.

When $min\{S(P), S(Q)\} = S(Q)$, the case is similar.

**Subcase 1.2.** Neither $P$ nor $Q$ is a weighted geodesic.

Let $P'$ be a weighted $u - w$ geodesic and $Q'$ be a weighted $w - v$ geodesic. Then $l(P') * S(P') = d_w(u, w)$ and $l(Q') * S(Q') = d_w(w, v)$. We want to prove $d_w(u, v) \leq d_w(u, w) + d_w(w, v)$.

If possible, suppose the contrary. Let $d_w(u, v) > d_w(u, w) + d_w(w, v)$. That is $l(P \cup Q) * S(P \cup Q) > l(P') * S(P') + l(Q') * S(Q')$. Let $min\{S(P'), S(Q')\} = S(P')$, say. Then the above inequality becomes $l(P \cup Q) * S(P \cup Q) > l(P') * S(P') + l(Q') * S(P') = (l(P') + l(Q')) * S(P') = l(P' \cup Q') * S(P') = l(P' \cup Q') * S(P' \cup Q')$, which means the path $P \cup Q$ is not a weighted $u - v$ geodesic, and we may consider $P' \cup Q'$ as a weighted $u - v$ geodesic. This is a contradiction to our main assumption. Hence the triangle inequality $d_w(u, v) \leq d_w(u, w) + d_w(w, v)$ holds good.

**Subcase 1.3.** Either $P$ or $Q$ is a weighted geodesic.

Let $P$ be the weighted $u - w$ geodesic and $Q'$ be the weighted $w - v$ geodesic. Then $l(P) * S(P) = d_w(u, w)$ and $l(Q') * S(Q') = d_w(w, v)$. Similar to the subcase 1.2, we get $P \cup Q'$ is a weighted $u - v$ geodesic and $P \cup Q$ is not, a contradiction, and the triangle inequality will become true. The proof is same when $P'$ is a weighted $u - w$ geodesic and $Q$ is a weighted $w - v$ geodesic.

**Case 2.** The third vertex $w$ lies on a cycle $C$ containing $u$ and $v$ and not on a weighted $u - v$ geodesic (Fig. 5.4).

Let $P$ be a weighted $u - v$ geodesic. Then $l(P) * S(P) = d_w(u, v)$.

Let $P'$ be a $u - w$ sub path and $P''$ be a $w - v$ sub path of $C$. Let $Q'$ and $Q''$ be any other $u - w$ and $w - v$ paths in $G$ not belonging to $C$.

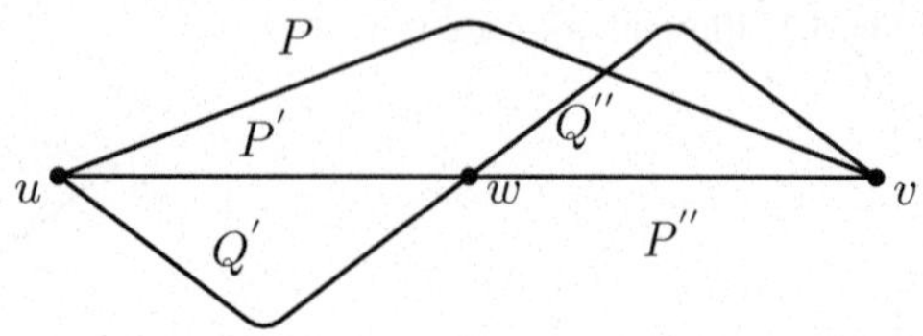

Fig. 5.4 Illustration—Case 2

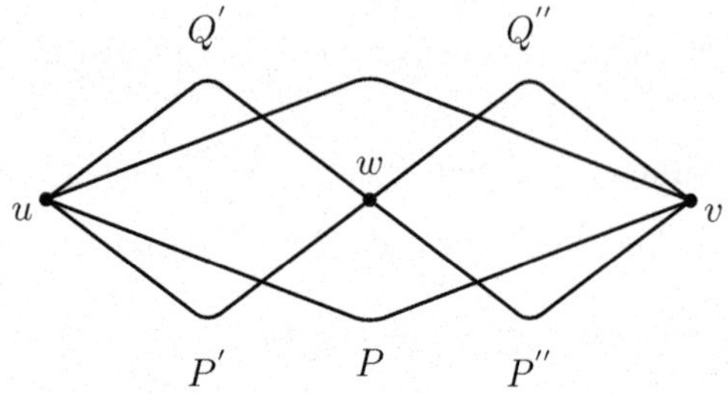

Fig. 5.5 Illustration—Case 3

The proof of this case is the same as that of the subcase 1.2. In all the subcases, $P$ is not a weighted $u-v$ geodesic, which is a contradiction.

**Case 3.** The third vertex $w$ does not lie on a cycle containing $u$ and $v$ (Fig. 5.5).

The proof of this case is also similar to that of Case 2. In all the subcases, we get contradictions. Thus the proof of theorem is complete. ■

In short, vertex set of a $k-$ connected weighted graph for $k \geq 2$ is a metric space with respect to the weighted distance metric $d_w$. Remark 5.1.7 is a statement on this.

**Remark 5.1.7** Consider a weighted graph $G = (V, E, w)$ such that its underlying unweighted graph $G^*$ is $k-$ connected with $k \geq 2$. Let the weighted distance between two vertices $u$ and $v$ in $G$ is as in Definition 5.1.1; $d_w(u, v) = \wedge_P\{l(P) * S(P)/P$ is a $u-v$ path, $l(P)$ is the length and $S(P)$ is the strength of the path $P\}$. Then $d_w$ satisfies all the four properties of a metric and hence $(V, d_w)$ is a metric space.

1. $d_w(u, v) \geq 0$ for all $u, v \in V$
2. $d_w(u, v) = 0$ for all $u = v$.
3. $d_w(u, v) = d_\omega(v, u)$ for all $u, v \in V$.
4. $d_w(u, v) \leq d_w(u, w^{'}) + d_w(w^{'}, v)$ for all $u$, $v$ and $w^{'} \in V(G)$.

We can generalize distance related parameters of unweighted graphs to weighted graphs. Following are the formal definitions of weighted eccentricity, weighted radius, weighted diameter, etc. with respect to the weighted distance.

We intentionally recall them for the better understanding of the development of this chapter.

Let $G = (V, E, w)$ be a weighted graph. Then the **weighted eccentricity** of a vertex $u \in V$ is defined and denoted by $e_w(u) = \vee_{v \in V}\{d_w(u, v)\}$, where $\vee$ represents the maximum. The minimum of the weighted eccentricities of all vertices is called the

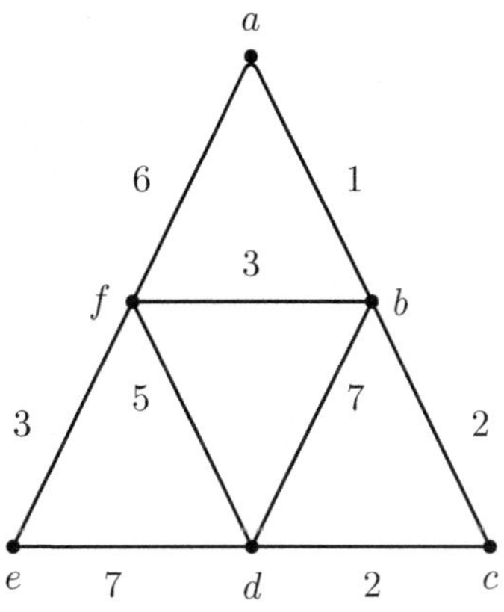


**Fig. 5.6** Weighted eccentricity and weighted center

**weighted radius** of the graph $G$. It is denoted as $r_w(G)$. Thus $r_w(G) = \wedge_{u \in V}\{e_w(u)\}$. The maximum of the weighted eccentricities of all the vertices is called the **weighted diameter** of the graph $G$. It is denoted as $d_w(G)$. That is, $d_w(G) = \vee_{u \in V}\{e_w(u)\}$.

Note that in Example 5.1.2, $e_w(a) = 44$, $e_w(b) = 25$, $e_w(c) = 44$, $e_w(d) = 24$. The weighted radius and weighted diameter of the graph $G$ are 24 and 44 respectively.

**Remark 5.1.8** For a weighted graph $G = (V, E, w)$ having weights $\geq 1$, $d(u, v) \leq d_w(u, v)$, where $d(u, v)$ is the distance between $u$ and $v$ in the underlying unweighted graph $G^*$ of $G$.

A vertex $v \in V$ is called a **weighted eccentric vertex** of another vertex $u$ in $V$ if $e_w(u) = d_w(u, v)$. The set of all weighted eccentric vertices of a vertex $u$ is denoted by $u_w^*$. In set theoretic notation, $u_w^* = \{v \in V \mid d_w(u, v) = e_w(u)\}$. Vertices with minimum weighted eccentricity are called weighted central vertices or **weighted radial vertices**, and vertices with maximum weighted eccentricity are called **weighted diametral vertices** or weighted peripheral vertices. Consider the illustration of the above definitions in Example 5.1.9.

**Example 5.1.9** Consider the weighted graph in Fig. 5.6. weighted distance between all different pairs of vertices are given below.
$d_w(a, b) = 1$, $d_w(a, c) = 2$, $d_w(a, d) = 2$, $d_w(a, e) = 3$, $d_w(a, f) = 2$, $d_w(b, c) = 2$, $d_w(b, d) = 3$, $d_w(b, e) = 3$, $d_w(b, f) = 2$, $d_w(c, d) = 2$, $d_w(c, e) = 4$, $d_w(c, f) = 3$, $d_w(d, e) = 4$, $d_w(d, f) = 3$ and $d_w(e, f) = 3$.

Now we compute the weighted eccentricities of each vertex. $e_w(a) = e_w(b) = e_w(f) = 3$, $e_w(c) = e_w(d) = e_w(e) = 4$. The weighted radius $r_w(G)$ and the weighted diameter $d_w(G)$ of the graph are 3 and 4 respectively.

Table 5.1 shows eccentric vertex set of each vertex in Fig. 5.6. Note that $a$, $b$ and $f$ are not weighted eccentric vertices of any other vertices in $G$. Also, these three

**Table 5.1** Eccentric vertex sets in Example 5.1.9

| vertex (u) | $a$ | $b$ | $c$ | $d$ | $e$ | $f$ |
|---|---|---|---|---|---|---|
| $u_w^*$ | $\{e\}$ | $\{d, e\}$ | $\{e\}$ | $\{e\}$ | $\{d, e\}$ | $\{c, d, e\}$ |

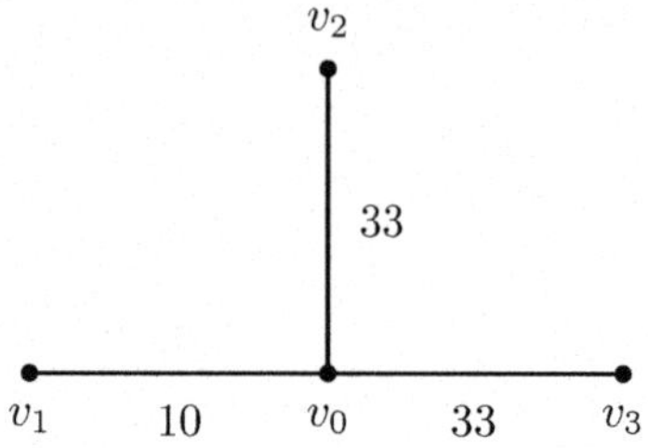


**Fig. 5.7** Violation of the radius and diameter relation

vertices are the weighted central vertices in $G$. The weighted diametral vertices in $G$ are $c, d$ and $e$.

In Example 5.1.2, $r_w(G) = 24, d_w(G) = 44$. Thus the weighted central vertex and the weighted diametral vertices in $G$ are $d$, $a$ and $c$ respectively.

Similar to the classical geodesic distance in weighted graphs, we have the following inequality with respect to weighted distance.

**Theorem 5.1.10** *For any k-connected weighted graph G, where $k \geq 2$, we have, $r_w(G) \leq d_w(G) \leq 2r_w(G)$.*

***Proof*** The first inequality follows from the definition itself. To prove the other, let $u$ and $v$ be any two vertices such that $d_w(u, v) = d_w(G)$. Let $w$ be any weighted central vertex of $G$. Then by triangle inequality, $d_w(u, v) \leq d_w(u, w) + d_w(w, v)$. But $d_w(u, w) \leq r_w(G)$ and $d_w(w, v) \leq r_w(G)$.
Thus $d_w(u, v) \leq r_w(G) + r_w(G) = 2r_w(G)$.

**Remark 5.1.11** Theorem 5.1.10 does not hold good for all the weighted graphs in general. It can be seen from the following Example 5.1.12.

**Example 5.1.12** Figure 5.7 is a weighted graph on four vertices.

In this weighted graph, $d_w(v_0, v_1) = 10, d_w(v_0, v_2) = 33, d_w(v_0, v_3) = 33$, $d_w(v_1, v_2) = 20, d_w(v_1, v_3) = 20$, and $d_w(v_2, v_3) = 66$.

The weighted eccentricities are $e_w(v_0) = 33, e_w(v_1) = 20, e_w(v_2) = 66$ and $e_w(v_3) = 66$. Here the weighted radius is $r_w = 20$. Also the weighted diameter is $d_w = 66$. Clearly $r_w(G) \leq d_w(G) \leq 2r_w(G)$ is not true.

The subgraph induced by the set of all weighted central vertices is called the **weighted center** of the graph $G$ and the subgraph induced by the set of all weighted diametral vertices is called the **weighted periphery** of the graph $G$.

The weighted center of a weighted graph need not be the same as the center of its underlying unweighted graph. Following example illustrates this.

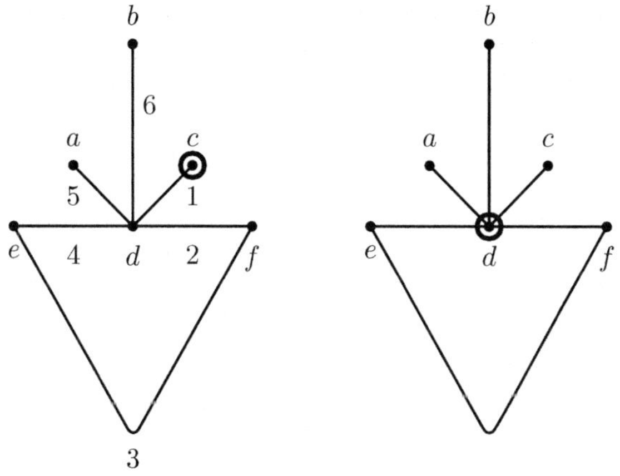


**Fig. 5.8** The weighted center is non isomorphic with center—Example

**Example 5.1.13** . For the weighted graph in Fig. 5.8, the weighted distances between each pair of vertices are computed below.

$$\begin{aligned} d_w(a,b) &= 10, \ d_w(a,c) = 2, \ d_w(a,d) = 5, \\ d_w(a,e) &= 6, \ d_w(a,f) = 4, \ d_w(b,c) = 2, \\ d_w(b,d) &= 6, \ d_w(b,e) = 6, \ d_w(b,f) = 4, \\ d_w(c,d) &= 1, \ d_w(c,e) = 2, \ d_w(c,f) = 2, \\ d_w(d,e) &= 4, \ d_w(d,f) = 2, \ d_w(e,f) = 3. \end{aligned}$$

The weighted eccentricities of each vertex is as follows. $e_w(a) = e_w(b) = 10$, $e_w(c) = 2$, $e_w(d) = e_w(e) = 6$ and $e_w(f) = 4$. Thus the weighted center consists of the isolated vertex $c$. Now for the underlying unweighted graph, the distances between each pair of vertices are,

$$\begin{aligned} &d(a,b) = 2, \ d(a,c) = 2, \ d(a,d) = 1, \ d(a,e) = 2, \ d(a,f) = 2, \\ &d(b,c) = 2, \ d(b,d) = 1, \ d(b,e) = 2, \ d(b,f) = 2, \ d(c,d) = 1, \\ &d(c,e) = 2, \ d(c,f) = 2, \ d(d,e) = 1, \ d(d,f) = 1, \ d(e,f) = 1. \end{aligned}$$

The eccentricities of the vertices are $e(a) = e(b) = e(c) = e(e) = e(f) = 2$, and $e(d) = 1$. This makes the center of the unweighted graph as an isolated vertex $d$.

## 5.2 Self Centered Weighted Graphs

This section introduces weighted self centered graphs which is a concept analogous to self centered graphs in classical graph theory. we shall discuss the properties of self centered graphs with respect to the weighted distance. A weighted graph $G = (V, E, w)$ is called **weighted self centered** , if it is isomorphic with its weighted center.

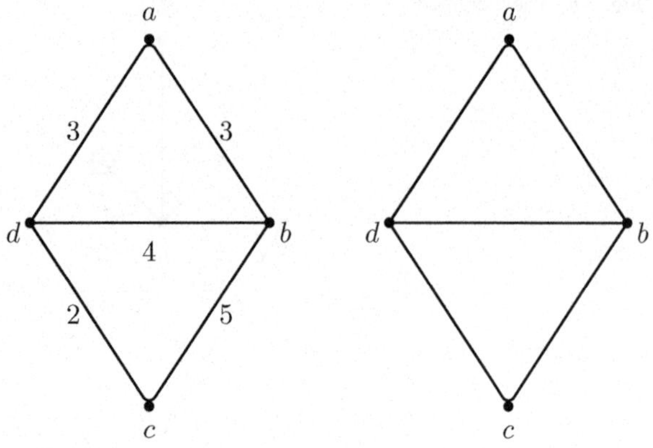


**Fig. 5.9** A weighted graph and its underlying unweighted graph

**Example 5.2.1** A weighted graph and its underlying unweighted graph are shown in Fig. 5.9. Here, $d_w(a, b) = 3, d_w(a, c) = 4, d_w(a, d) = 3, d_w(b, c) = 4, d_w(b, d) = 4, d_w(c, d) = 2$. Also note that $e_w(a) = e_w(b) = e_w(c) = ew_w(d) = 4$.

For the underlying unweighted graph, $d(a, b) = 1, d(a, c) = 2, d(a, d) = 1, d(b, c) = 1, d(b, d) = 1, d(c, d) = 1$. Also note that $e(a) = 2, e(b) = 1, e(c) = 2, e(d) = 1$.

The weighted graph is weighted self centered but the underlying unweighted graph is not self centered. Following theorem presents a necessary condition for any weighted graph $G$ to be a weighted self centered graph.

**Theorem 5.2.2** *If a weighted graph $G = (V, E, w)$ is weighted self centered, then each vertex of $G$ is weighted eccentric.*

***Proof*** Assume that the weighted graph $G$ is weighted self centered. Let $u$ be any vertex of $G$ and let $v \in u^*_w$. By the definition of a weighted eccentric vertex, $e_w(u) = d_w(u, v)$. But since $G$ is weighted self centered, $e_w(u) = e_w(v)$ and hence $e_w(v) = d_w(u, v) = d_w(v, u)$. Thus $u$ is a weighted eccentric vertex of $v$. Hence every vertex of $G$ is weighted eccentric.

The next theorem is also a necessary condition for a weighted graph to be weighted self centered (Fig. 5.10).

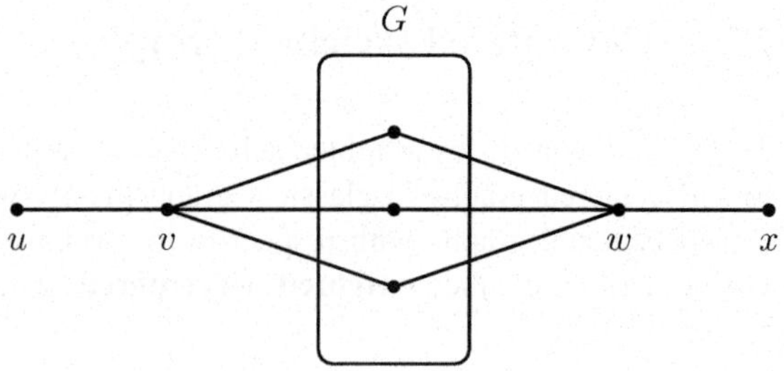


**Fig. 5.10** Graph in Theorem 5.2.4

**Theorem 5.2.3** *If a connected weighted graph $G = (V, E, w)$ is weighted self centered, then for every pair of vertices $u, v$ in $G$ such that whenever $u$ is a weighted eccentric vertex of $v$ then $v$ should be one of the weighted eccentric vertices of $u$.*

***Proof*** Assume that $G$ is weighted self centered. Also assume that $u$ is a weighted eccentric vertex of $v$. This means, $e_w(v) = d_w(v, u)$. Since $G$ is weighted self centered, all vertices will be having the same weighted eccentricity. Therefore $e_w(v) = e_w(u)$.

From the above two equations, $e_w(u) = d_w(v, u) = d_w(u, v)$. Thus $e_w(u) = d_w(u, v)$. That is, $v$ is a weighted eccentric vertex of $u$.

Theorem 5.2.4 discusses about the construction of a weighted graph from a given weighted graph in such a way that the given graph is the weighted center of the constructed graph.

**Theorem 5.2.4** *Every weighted graph $G = (V, E, w)$ is the weighted center of some connected weighted graph $H$.*

***Proof*** Construct $H$ from $G$ by inserting 4 more vertices, say, $u, v, w$, and $x$. Construct $H$ as the sequential join $u + v + G + w + x$. Assign the least weight among the weights of edges of $G$ to all the new edges of $H$. Let that minimum weight be $k$. Thus, $e_w(u) = e_w(x) = 4k$, $e_w(v) = e_w(w) = 3k$, $e_w(y) = 2k$ for every $y \in V$.

Hence each vertex in $G$ is a weighted central vertex of $H$ and hence $G$ is the weighted center of $H$.

## 5.3 The Distance Matrix

In this section, we present an easy check for a weighted graph $G$ to find whether it is weighted self centered or not using its weighted distance matrix.

**Definition 5.3.1** Let $G = (V, E, w)$ be a connected weighted graph with $V = \{v_1, v_2, v_3, \ldots, v_n\}$. The **weighted distance matrix** $d_w = (d_{i,j})$ is a square matrix of order $n$ and is defined by $(d_{i,j}) = d_w(v_i, v_j)$. Note that the weighted distance matrix is a symmetric matrix.

**Definition 5.3.2** The **max-max composition** of a square matrix with itself is again a square matrix of the same order whose $(i, j)^{th}$ entry is given by $d_{i,j} = \max\{\max(d_{i1}, d_{1j}), \max(d_{i2}, d_{2j}), \ldots, \max(d_{in}, d_{nj})\}$

**Example 5.3.3** In this example we find the weighted distance matrix and its max-max composition.

The weighted distance matrix of the weighted graph in Fig. 5.11 and its max-max composition are given below.

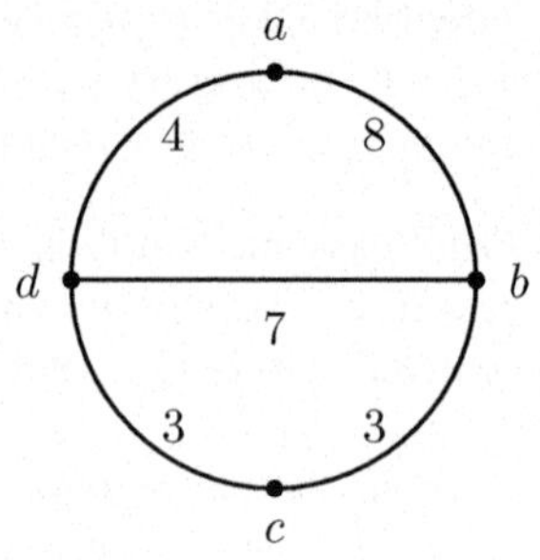


**Fig. 5.11** Example 5.3.3—Distance matrix

$$d_\omega = \begin{pmatrix} 0\,8\,6\,4 \\ 8\,0\,3\,6 \\ 6\,3\,0\,3 \\ 4\,6\,3\,0 \end{pmatrix}, \; d_\omega o d_\omega = \begin{pmatrix} 8\,8\,8\,8 \\ 8\,8\,8\,8 \\ 8\,8\,6\,6 \\ 8\,8\,6\,6 \end{pmatrix}.$$

Next we have a theorem related with the weighted eccentricities of vertices using the max-max composition of the weighted distance matrices.

**Theorem 5.3.4** *Let $G = (V, E, w)$ be a connected weighted graph. The diagonal elements of the max-max composition of the weighted distance matrix of G with itself are the weighted eccentricities of the vertices.*

***Proof*** Let $d_w = (d_{i,j})$ be the weighted distance matrix of $G$. Then $(d_{i,j}){=}d_w(v_i, v_j)$. In the max-max composition, $d_w o d_w$, the $i^{th}$ diagonal entry, $d_{i,i} = \max\{\max(d_{i,1}, d_{1,i}), \max(d_{i,2}, d_{2,i}), \max(d_{i,3}, d_{3,i}), \ldots, \max(d_{i,n}, d_{n,i})\} = \max\{d_{i,1}, d_{i,2}, d_{i,3}, \ldots, d_{i,n}\} = \max\{d_w(v_i, v_1), d_w(v_i, v_2), d_w(v_i, v_3), \ldots, d_w(v_i, v_n)\} = e_w(v_i)$. This completes the proof of the theorem.

**Theorem 5.3.5** *A connected weighted graph $G = (V, E, w)$ is weighted self centered if and only if all the entries in the principal diagonal of the max-max composition of the weighted distance matrix with itself are the same.*

***Proof*** As proved in Theorem 5.3.4, the principal diagonal entries in the max–max composition of the weighted distance matrix with itself are the weighted eccentricities of the vertices. If they are same, this means $e_w(u)$ is the same for all $u$ in $G$. Then $G$ is weighted self centered.

Examples 5.3.6 and 5.3.7 together illustrate Theorem 5.3.5

**Example 5.3.6** Consider the weighted graph in Fig. 5.12a The weighted distance matrix and the corresponding max-max composition are shown below. It is evident that the graph is weighted self centered.

$$d_\omega = \begin{pmatrix} 0\,4\,8\,5 \\ 4\,0\,5\,8 \\ 8\,5\,0\,4 \\ 5\,8\,4\,0 \end{pmatrix}, \; d_\omega o d_\omega = \begin{pmatrix} 8\,8\,8\,8 \\ 8\,8\,8\,8 \\ 8\,8\,8\,8 \\ 8\,8\,8\,8 \end{pmatrix}$$

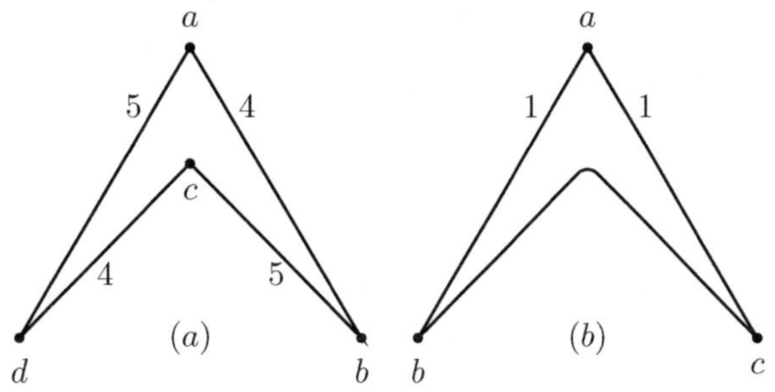


**Fig. 5.12** **a** A weighted self centered graph. **b** A weighted non-self centered graph

**Example 5.3.7** Consider the weighted graph shown in Fig. 5.12b. The weighted distance matrix and the corresponding max-max composition are given below.

$$d_\omega = \begin{pmatrix} 0 & 1 & 1 \\ 1 & 0 & 2 \\ 1 & 2 & 0 \end{pmatrix}, \; d_\omega o d_\omega = \begin{pmatrix} 1 & 2 & 2 \\ 2 & 2 & 2 \\ 2 & 2 & 2 \end{pmatrix}$$

It is clear that the graph is not weighted self centered since the diagonal elements in the composition are equal.

Next section discusses the central properties of weighted trees. In general, the weighted distance does not satisfy all the properties of a metric in weighted trees.

## 5.4 Weighted Center of Trees

This section makes a comparison between the weighted center of a weighted tree and the center of its underlying unweighted tree. Here we use the term unweighted tree to represent trees in which unit weight is assigned with each edge.

**Remark 5.4.1** In unweighted trees, the center is either $K_1$ or $K_2$ (Theorem 1.3.20). Whereas in weighted trees, the weighted center need not be $K_1$ or $K_2$.

**Example 5.4.2** Consider the following tree in Fig. 5.13.

In this tree, $d_w(a, b) = 1 = d_w(e, f), d_w(a, c) = d_w(b, c) = d_w(c, f) = 2, d_w(a, d) = d_w(a, e) = d_w(b, f) = d_w(c, d) = d_w(d, f) = 3, d_w(a, f) = d_w(b, d) = d_w(b, e) = d_w(c, e) = 4, d_w(d, e) = 6$.

**Fig. 5.13** A weighted tree

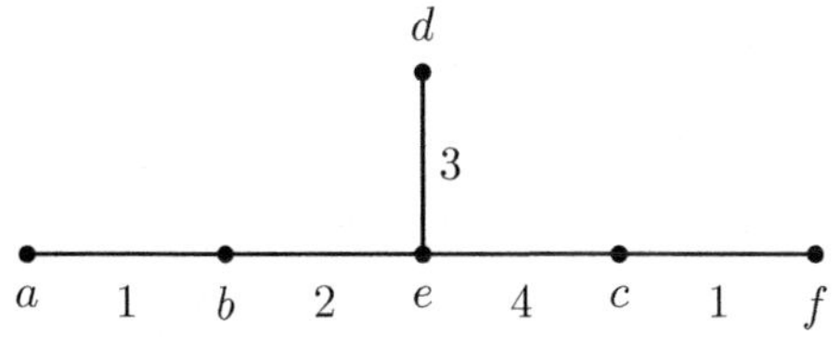

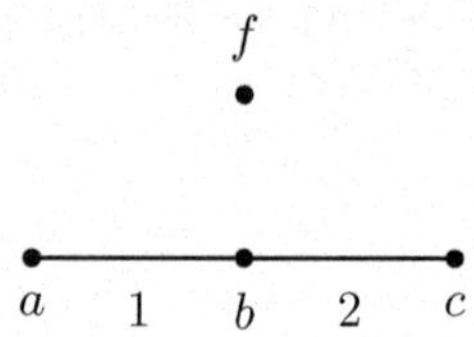


**Fig. 5.14** Weighted center of the weighted tree in Fig. 5.14

The weighted eccentricities are, $e_w(a) = e_w(b) = e_w(c) = e_w(f) = 4$, $e_w(d) = e_w(e) = 6$.

**Example 5.4.3** Figure 5.14 shows the weighted center of the weighted tree in Fig. 5.15. It is neither $K_1$ nor $K_2$.

**Remark 5.4.4** The weighted center of a tree can be disconnected.

**Remark 5.4.5** In a tree with number of vertices greater than or equal to 3, an end vertex cannot be a central vertex. But in a weighted tree, an end vertex can be a central vertex.

**Remark 5.4.6** Let the minimum weighted edge (of weight $k$) in a weighted tree be attached to a pendent vertex $v$. Then $e_w(v) = nk$, where $n$ is the length of the longest path from $v$ in the weighted tree.

## 5.5 Complement of a Weighted Graph

This section introduces the complement of a weighted graph. Discussion on its central properties are also given.

**Definition 5.5.1** Let $G = (V, E, w)$ be a simple weighted graph with largest edge weight $k$. Then the **weighted complement** of $G$ is the weighted graph $G_c = (V, E_c, w_c)$, where $E_c$ is the set of all edges $e$ in $E$ with positive weights $k - w(e)$. If a pair of non adjacent vertices exists in $G$, then they will be adjacent in $G_c$ by an edge with weight $k$. If a pair of vertices in $G$ are adjacent by an edge with weight $k$, then they will not be adjacent in $G_c$.

**Example 5.5.2** Consider the weighted graph and its complement in Fig. 5.15. Corresponding weighted distance matrix and composition are given below.

$$d_w(G) = \begin{pmatrix} 0\,3\,4\,3 \\ 3\,0\,4\,4 \\ 4\,4\,0\,2 \\ 3\,4\,2\,0 \end{pmatrix} \quad d_w o d_w = \begin{pmatrix} 4\,4\,4\,4 \\ 4\,4\,4\,4 \\ 4\,4\,4\,4 \\ 4\,4\,4\,4 \end{pmatrix}.$$

$$d_w(G_c) = \begin{pmatrix} 0\,2\,3\,2 \\ 2\,0\,2\,1 \\ 3\,2\,0\,3 \\ 2\,1\,3\,0 \end{pmatrix} \quad d_w o d_w = \begin{pmatrix} 3\,3\,3\,3 \\ 3\,2\,3\,3 \\ 3\,3\,3\,3 \\ 3\,3\,3\,3 \end{pmatrix}.$$

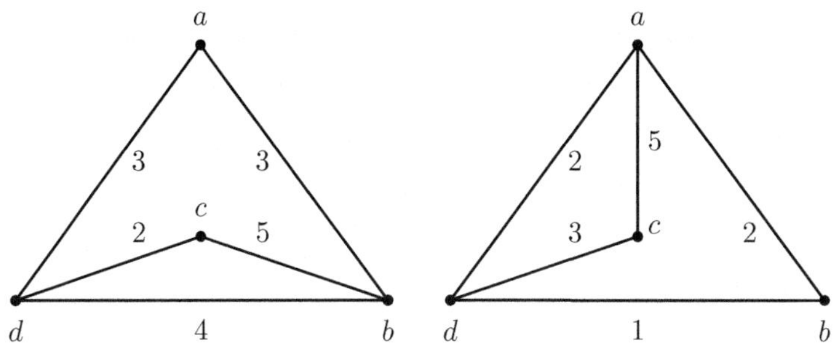


**Fig. 5.15** A weighted graph and its complement

**Remark 5.5.3** If $n$ is the number of vertices in $G$ and $k$ is the highest edge weight in $G$ then, sum of the weights of all edges in $G$ and $G_c$ is equal to $nC_2k$. It is clear from the example in Fig. 5.15.

**Remark 5.5.4** The complement of a weighted self centered graph may not have the weighted self centered property. In Example 5.5.2, $G$ is weighted self centered and the weighted eccentricity is 4 for all the four vertices. Whereas in $G_c$, $e_w(a) = e_w(c) = e_w(d) = 3$ and $e_w(b) = 2$. This shows that $G_c$ is not weighted self centered.

## 5.6 Geodetic Convex Sets

The concept of geodetic convexity and related discussions are included in this section. Geodetic convexity is based on the weighted distance. A subset of the vertex set of a weighted graph is said to be convex (with respect to a distance) if the induced weighted subgraph of that convex subset is capable to identify all the graph parameters of the subgraph. It is the relevance of this study in application point of view. A subset $W$ of $V$ is called weighted geodetic convex if the weighted geodetic closure of $W$ is $W$ itself.

**Definition 5.6.1** Let $G = (V, E, w)$ be a connected weighted graph without loops and multiple edges. Let $u$, $v$ be any two vertices of $G$. A $u - v$ path $P$ is called a **weighted** $u - v$ **geodesic** if $d_w(u, v) = S(P) * l(P)$. In other words, a $u - v$ path $P$ is called a weighted $u - v$ geodesic if the weighted distance between $u$ and $v$ is calculated along the path $P$.

**Definition 5.6.2** For any two vertices $u$ and $v$ of $G$, the **weighted geodetic closed interval** $I_w[u, v]$ is the set of all vertices in all weighted $u - v$ geodesics including $u$ and $v$.

**Definition 5.6.3** Let $G = (V, E, w)$ be a connected weighted graph without loops and multiple edges and let $S \subseteq V$. The union of all geodetic closed intervals $I_w[u, v]$

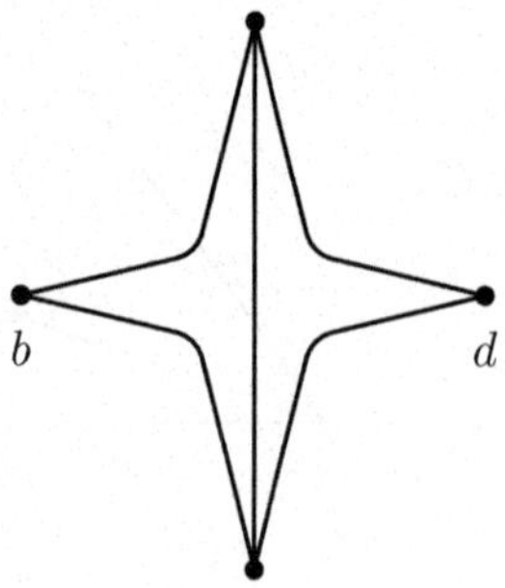


**Fig. 5.16** Graph in Example 5.6.5

over all pairs $u, v \in S$ is called the **weighted geodetic closure** of $S$. It is denoted by $I_w(S)$.

**Definition 5.6.4** Let $G = (V, E, w)$ be a connected weighted graph without loops and multiple edges. Any subset $S$ of $V$ is called **weighted geodetic convex** if $I_w(S) = S$.

In Example 5.6.5, we illustrate these concepts.

**Example 5.6.5** Consider the weighted graph in Fig. 5.16.

Corresponding weighted distance matrix for the graph (Fig. 5.16) is given below.

$$\begin{pmatrix} 0 & 2 & 1 & 2 \\ 2 & 0 & 2 & 3 \\ 1 & 2 & 0 & 2 \\ 2 & 3 & 2 & 0 \end{pmatrix}$$

In this figure, the path $a - c - b$ is a weighted $a - b$ geodesic. More over, it is the unique $a - b$ geodesic. The edge $ac$ is the unique weighted $a - c$ geodesic. It can be seen that $I_w[a, d] = \{a, c, d\}$, $I_w[a, c] = \{a, c\}$ and $I_w[a, b] = \{a, c, b\}$. If $S = \{a, b\}$, then $I_w[S] = \{a, b, c\} \neq S$; hence $S$ is not weighted geodetic convex. If $S = \{a, c\}$, then $I_w[S] = S$. This shows that $S = \{a, c\}$ is weighted geodetic convex.

The following proposition is obvious. The proof is left to the reader.

**Proposition 5.6.6** *Let $G = (V, E, w)$ be a connected weighted graph. Then the empty set $\Phi$, the vertex set $V$ and all singletons in the power set of $V$ are weighted geodetic convex.*

**Remark 5.6.7** A weighted geodetic convex set $S$ with $2 \leq |S| < |V|$ is termed as a nontrivial weighted geodetic convex set.

In Theorem 5.6.8, we show that the arbitrary intersection of weighted geodetic convex sets is again weighted geodetic convex.

**Theorem 5.6.8** *The intersection of two weighted geodetic convex sets is again weighted geodetic convex.*

***Proof*** Let $G = (V, E, w)$ be a connected weighted graph. Let $S$ and $T$ be any two weighted geodetic convex sets of $V$. We have to prove that $S \cap T$ is weighted geodetic convex. Let $u$ and $v$ be any two vertices of $S \cap T$. That is, $u$ and $v$ are two vertices in both $S$ and $T$, which are weighted geodetic convex. Therefore all vertices in all weighted $u - v$ geodesics are both in $S$ and $T$, and hence in $S \cap T$. This shows that $S \cap T$ is weighted geodetic convex.

In general, the union of two weighted geodetic convex sets is not weighted geodetic convex. Consider $S = \{a\}$ and $T = \{b\}$ in Fig. 5.16. Here $S \cup T = \{a, b\}$, which is not weighted geodetic convex.

## 5.7 Geodetic Blocks

This section introduces the concept of weighted geodetic blocks. A characterization of weighted geodetic blocks in terms of weighted geodetic convex sets can be seen. Some necessary conditions are also included. If the underlying unweighted graph $G^*$ of a weighted graph $G$ is complete, then $G$ is referred as a complete weighted graph.

**Definition 5.7.1** Let $G = (V, E, w)$ be a connected weighted graph and let $H$ be a complete weighted subgraph of $G$. Then $H$ is called a **weighted geodetic block** of $G$ if there exists no edge $e = uv$ in $H$ such that $w(e) \geq d_w(u, v)$.

**Remark 5.7.2** From the definition of weighted geodetic blocks of a weighted graph $G$, it is clear that all edges $e = uv$ in a weighted geodetic block are the only weighted $u - v$ geodesics in the weighted graph.

**Example 5.7.3** A weighted graph and its geodetic blocks are given in Figs. 5.17 and 5.18.

In Fig. 5.18, $H_1$ and $H_4$ are weighted geodetic blocks of $G$, but $H_2$ and $H_3$ are not. Even though all the edges in $H_2$ are weighted geodesics between their end vertices, its underlying graph $H_2^*$ is not complete. In $H_3$, the edge $cd$ is not a weighted $c - d$ geodesic.

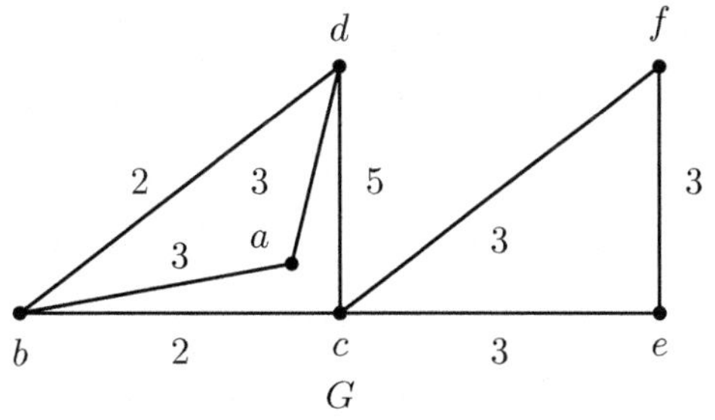


**Fig. 5.17** The weighted graph in Example 5.7.3

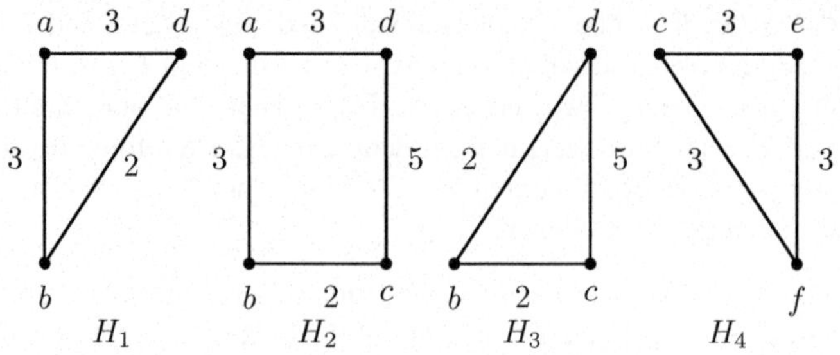


**Fig. 5.18** Weighted geodetic blocks of Fig. 5.17

The following theorem is a necessary condition for a complete weighted subgraph to be a weighted geodetic block.

**Theorem 5.7.4** *Let $G = (V, E, w)$ be a connected weighted graph. Let $H$ be a complete weighted subgraph of $G$. If $H$ is a weighted geodetic block of $G$, then $V(H)$ is a weighted geodetic convex set of $G$.*

***Proof*** Let $G = (V, E, w)$ be a connected weighted graph and $H$ be a complete weighted subgraph of $G$. Suppose that $H$ is a weighted geodetic block of $G$. We have to prove that $V(H)$ is a weighted geodetic convex set of $V$. Since $H$ is a weighted geodetic block of $G$, there exists no edge $e = uv$ in $H$ such that $w(e) \geq d_w(u, v)$. This means every edge $e = uv$ in $H$ is the only weighted $u - v$ geodesic in $H$. Therefore the vertices in all weighted $u - v$ geodesics are exactly $u$ and $v$. This is true for every pair of vertices in $V(H)$. Hence $V(H)$ is a weighted geodetic convex set of $V$.

Converse of the above theorem do hold good always. In other words, if $H$ is a complete weighted subgraph of $G$ such that $V(H)$ is a weighted geodetic convex subset of $V$, then $H$ need not be a weighted geodetic block of $G$. This is explained in Example 5.7.5.

**Example 5.7.5** Consider the graph in Fig. 5.19.

Here, the vertex induced sub graph $H$ induced by $\{d, e, f\}$ is not a weighted geodetic block, even though the set $\{d, e, f\}$ is a weighted geodetic convex set of $V$. Note that the edges $ef$ and $df$ are not weighted geodesics between their end points.

The sequential deletion of vertices from a weighted geodetic block results in a nested sequence of weighted geodetic blocks. This fact is explained in Theorem 5.7.6.

**Theorem 5.7.6** *Every vertex deleted subgraph of a weighted geodetic block is again a weighted geodetic block.*

***Proof*** Let $G = (V, E, w)$ be a connected weighted graph and $H$ be a weighted geodetic block of $G$. Let $u$ be any vertex of $H$. Consider the vertex deleted subgraph

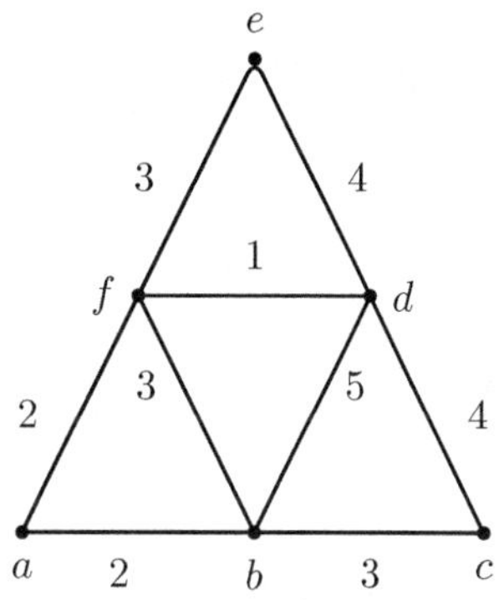


**Fig. 5.19** Weighted graph in Example 5.7.5

$H - u$. We have to prove that $H - u$ is again a weighted geodetic block of $G$. Let $e = xy$ be any edge of $H - u$. Clearly $x \neq u \neq y$, and since $xy$ is an edge of $H$, which is weighted geodetic convex, $e$ is the only weighted $x - y$ geodesic in $G$. This is true for any edge in $H - u$. Thus $H - u$ is a weighted geodetic block of $G$.

By Theorem 5.7.6, it is clear that, if we have a weighted geodetic block of order $p$, then there exist at least a weighted geodetic block of order $1, 2, 3, \ldots, (p - 1)$ each.

Next theorem characterizes weighted geodetic blocks of a connected weighted graph $G$.

**Theorem 5.7.7** *Let $G = (V, E, w)$ be a connected weighted graph and let $H$ be a complete weighted subgraph of $G$. Then $H$ is a weighted geodetic block of $G$ if and only if $\{u, v\}$ is weighted geodetic convex for every two vertices $u$, $v$ in $H$.*

***Proof*** Let $G = (V, E, w)$ be a connected weighted graph and let $H$ be a complete weighted subgraph of $G$. Suppose that $H$ is a weighted geodetic block of $G$. Let $u, v$ be any two vertices of $H$. We have to prove that $\{u, v\}$ is weighted geodetic convex. Since $H$ is a weighted geodetic block, it is complete. Therefore every pair of vertices of $H$ are adjacent and every edge is a unique weighted geodesic between its end vertices. In particular the vertices $u$ and $v$ are adjacent and $uv$ is the only weighted $u - v$ geodesic. This proves that $\{u, v\}$ weighted geodetic convex.

Conversely suppose that for any two vertices $u, v$ in $H$, $\{u, v\}$ is weighted geodetic convex. We have to prove that $H$ is a weighted geodetic block of $G$. That is to prove every edge in $H$ is the only weighted geodesic between its end vertices.

By the assumption and since $H$ is complete, each edge $e = uv$ is a weighted geodesic between its end vertices. Now it remains to prove the uniqueness of the geodesic. Suppose the contrary. Let there be another weighted $u - v$ geodesic in $H$, say, $P$. Since we consider only simple weighted graphs throughout this section, the path $P$ has at least one vertex, say, $w$ other than $u$ and $v$. Since $P$ is a weighted $u - v$ geodesic, we have $I_w[u, v] = \{u, v, w\} \neq \{u, v\}$. This means $\{u, v\}$ is not weighted geodetic convex, a contradiction. Hence our assumption is wrong. So all the edges in

$H$ are the only weighted geodesics between their end vertices. Thus $H$ is a weighted geodetic block.

**Corollary 5.7.8** *A complete connected weighted graph $G = (V, E, w)$ is a weighted geodetic block if and only if it has $(2^{|V|} - (| V | +2))$ nontrivial weighted geodetic convex sets.*

***Proof*** Let $G = (V, E, w)$ be a complete connected weighted graph. Suppose that $G$ is a weighted geodetic block. Then by Theorem 5.7.7, for any two vertices $u$ and $v$, $\{u, v\}$ is weighted geodetic convex. If we extend this set by adding any number of vertices, the resultant set is again a weighted geodetic convex set. Thus there are $2^{|V|}$ weighted geodetic convex. Therefore the number of nontrivial weighted geodetic convex sets is $(2^{|V|} - (| V | +2))$.

Conversely if there are $(2^{|V|} - (| V | +2))$ number of nontrivial weighted geodetic convex sets of $G$, all the two element subsets of $V(G)$ are weighted geodetic convex. By the above Theorem 5.7.7, $G$ is weighted geodetic convex.

In Theorem 5.7.9, we see that the vertex set of any weighted tree can be considered as the nested union of weighted geodetic convex subsets.

**Theorem 5.7.9** *Let $G = (V, E, w)$ be a weighted tree, then there exists a sequence of sets $V = V_n \supset V_{n-1} \supset, \ldots, \supset V_1$, where for each $i$, $V_i$ is weighted geodetic convex and $| V_i |= n_i$.*

***Proof*** Let $G = (V, E, w)$ be a weighted tree with $n$ vertices. Then between any two vertices of $G$, there exists only one path, and hence it is the unique weighted geodesic between them. So all vertices in all weighted geodesics between any two vertices are again in $V(G)$. Therefore $V(G) = V_n$ is weighted geodetic convex and $| V_n |= n$. Let $v_1$ be a pendent vertex of $G$ and set $V_{n-1} = V_n - \{v_1\}$. Then clearly the vertex induced subgraph of $G$ by the set $V_{n-1}$ is again a weighted tree. By the above same argument, we see that $V_{n-1}$ is weighted geodetic convex and $| V_{n-1} |= (n - 1)$. Let $v_2$ be a pendent vertex of $G - v_1$, set $V_{n-2} = V_{n-1} - v_2$. We can easily prove $V_{n-2}$ is weighted geodetic convex, and $| V_{n-2} |= (n - 2)$. Continue the above procedure of deleting pendent vertices until we get a singleton set. So finally we get the nested sequence $V = V_n \supset V_{n-1} \supset, \ldots, \supset V_1$.

**Remark 5.7.10** Even if we can find a nested sequence of vertex sets satisfying the condition mentioned in Theorem 5.7.9, the graph may not be a weighted tree.

## 5.8 Geodetic Boundary and Interior

In this section, we find the boundary and interior vertices of a weighted geodetic convex set. Some of their properties are discussed.

**Definition 5.8.1** Let $G = (V, E, w)$ be a connected weighted graph and $S$ be a weighted geodetic convex set of $G$. A vertex $u \in S$ is called a **weighted geodetic boundary vertex** of $S$ if and only if $S \setminus \{u\}$ is weighted geodetic convex.

**Definition 5.8.2** Let $G = (V, E, w)$ be a connected weighted graph and $S$ be a weighted geodetic convex set of $G$. A vertex $u \in S$ is called a **weighted geodetic interior vertex** of $S$ if and only if $S \setminus \{u\}$ is not weighted geodetic convex.

**Example 5.8.3** Consider the weighted graph in Fig. 5.20.

In Fig. 5.20, consider the set $S = \{a, b, c\}$. This is a weighted geodetic convex set. Here, vertex $c$ is a weighted geodetic interior vertex of $S$ because $S \setminus \{c\}$ is not a weighted geodetic convex set. Since $S \setminus \{b\}$ is weighted geodetic convex, $b$ is a weighted geodetic boundary vertex of $S$.

In the following two theorems, we characterize the boundary and interior vertices of a weighted geodetic convex set.

**Theorem 5.8.4** *Let $G = (V, E, w)$ be a connected weighted graph, and $S$ be a weighted geodetic convex set of $G$. Then a vertex $u \in S$ is a weighted geodetic boundary vertex of $S$ if and only if $u$ does not lie on any weighted $v - w$ geodesic for all $v, w \in S \setminus \{u\}$.*

***Proof*** Let $G = (V, E, w)$ be a connected weighted graph, and $S$ be a weighted geodetic convex set of $G$. Let $u \in S$.

Suppose that $u$ is a weighted geodetic boundary vertex of $S$. We have to prove that $u$ does not lie on any weighted $v - w$ geodesic for all $v, w \in S \setminus \{u\}$. Since $u$ is a weighted geodetic boundary vertex of $S$, $S \setminus \{u\}$ is weighted geodetic convex. Let $v, w \in S \setminus \{u\}$. Since $S \setminus \{u\}$ is weighted geodetic convex, all the vertices in all weighted $v - w$ geodesics are in $S \setminus \{u\}$ itself. This means that all the weighted $v - w$ geodesics are independent of $u$. Hence $u$ does not lie in any weighted $v - w$ geodesic for all $v, w \in S \setminus \{u\}$.

Conversely suppose that $u \in S$ and $u$ does not lie in any weighted $v - w$ geodesic for all $v, w \in S \setminus \{u\}$. We have to prove that $u$ is a weighted geodetic boundary

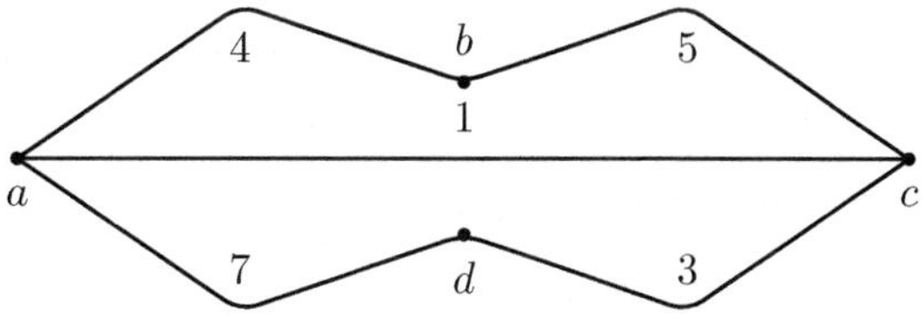


Fig. 5.20 Boundary and interior vertices

vertex of $S$. It is enough, if we prove that $S \setminus \{u\}$ weighted geodetic convex. Let $v, w \in S \setminus \{u\}$. By the assumption, all the weighted $v - w$ geodesics are free from $u$. Also $(S \setminus \{u\}) \cup \{u\} = S$, which is weighted geodetic convex. Hence all the vertices in all weighted geodesics between any two vertices in $S \setminus \{u\}$ lie in $S \setminus \{u\}$ itself, which proves $S \setminus \{u\}$ is weighted geodetic convex. So $u$ is a weighted geodetic boundary vertex of $S$.

**Theorem 5.8.5** *Let $G = (V, E, w)$ be a connected weighted graph, and $S$ be a weighted geodetic convex set of $G$. Then a vertex $u \in S$ is a weighted geodetic interior vertex of $S$ if and only if $u$ lies in at least one weighted $v - w$ geodesic for any $v, w \in S \setminus \{u\}$.*

***Proof*** Let $G = (V, E, w)$ be a connected weighted graph, and $S$ be a weighted geodetic convex set of $G$. Let $u \in S$.

Suppose that $u$ is a weighted geodetic interior vertex of $S$. We have to prove that $u$ lies in at lest one weighted $v - w$ geodesic for any $v, w \in S \setminus \{u\}$. Since $u$ is a weighted geodetic interior vertex of $S$, $S \setminus \{u\}$ is not weighted geodetic convex. Let $v, w \in S \setminus \{u\}$. Since $S \setminus \{u\}$ is not weighted geodetic convex, at least one vertex in any one of the weighted $v - w$ geodesic does not belong to $S \setminus \{u\}$. At the same time $(S \setminus \{u\}) \cup \{u\} = S$, which is weighed geodetic convex. This means at least one of the weighted $v - w$ geodesic contains the vertex $u$.

Conversely suppose that $u \in S$ and $u$ lies in any weighted $v - w$ geodesic for some $v, w \in S \setminus \{u\}$. We have to prove that $u$ is a weighted geodetic interior vertex of $S$. It is enough, if we prove that $S \setminus \{u\}$ is not weighted geodetic convex. Let $v, w \in S \setminus \{u\}$. By the assumption, there exist at least one weighted $v - w$ geodesic which contains $u$. Also $(S \setminus \{u\}) \cup \{u\} = S$, which is weighted geodetic convex. Hence $S \setminus \{u\}$ is not weighted geodetic convex. So $u$ is a weighted geodetic interior vertex of $S$.

In Theorem 5.8.6, we characterize the boundary and interior vertices of a weighted geodetic block.

**Theorem 5.8.6** *Let $G = (V, E, w)$ be a connected weighted graph, and let $H$ be a weighted geodetic block of $G$. Then every vertex of $H$ is a weighted geodetic boundary vertex of $V(H)$.*

***Proof*** Let $G = (V, E, w)$ be a connected weighted graph and let $H$ be a weighted geodetic block of $G$. Let $u$ be any vertex of $H$. We have to prove that $u$ is a weighted geodetic boundary vertex of $V(H)$. It is enough, if we prove $V(H) \setminus \{u\}$ is weighted geodetic convex. Since a weighted geodetic block is a complete weighted graph and each of its edges are unique weighted geodesics between their end vertices, by Theorems 5.7.4 and 5.7.6, $V(H) \setminus \{u\}$ is weighted geodetic convex. This proves that $u$ is a weighted geodetic boundary vertex of $V(H)$. Since $u$ is arbitrary, the proof is completed.

**Corollary 5.8.7** *Let $G = (V, E, w)$ be a connected weighted graph and let $H$ be a weighted geodetic block of $G$. Then no vertex of $H$ is a weighted geodetic interior vertex of $V(H)$.*

## 5.9 Monophonic Convex Sets

The concept of weighted monophonic convexity is explained in this section. Here we focus on weighted graphs without chords in it. A subset $W$ of $V$ is called weighted monophonic convex if the weighted monophonic closure of $W$ coincides with $W$. Let us look at some elementary definitions of weighted monophonic convexity.

**Definition 5.9.1** Let $G = (V, E, w)$ be a connected, weighted graph without loops and multiple edges. Let $P = v_0e_1v_1e_2v_2 \ldots e_nv_n$ be a $v_0 - v_n$ path. A **chord** of $P$ is an edge $e = v_iv_j$, $j \geq i + 2$ (that is; $e$ joins two nonconsecutive vertices of the path), such that $S(P)) * l(P) > S(P') * l(P')$, where $P' = v_0e_1v_1 \ldots v_iev_j \ldots e_nv_n$.

**Definition 5.9.2** Let $G = (V, E, w)$ be a connected, simple weighted graph. Let $u$, $v \in V$. A $u - v$ path $P$ is called a **weighted monophonic** $u - v$ **path** if it has no chords.

**Definition 5.9.3** The **weighted monophonic closed interval** $J_w[u, v]$ is the set of all vertices in all weighted monophonic $u - v$ paths including $u$ and $v$.

**Definition 5.9.4** Let $G = (V, E, w)$ be a connected weighted graph and $S \subseteq V$. The the union of all monophonic closed intervals $J_w[u, v]$ over all pairs $u, v \in S$ is called the **weighted monophonic closure** of $S$. It is denoted by $J_w[S]$.

**Definition 5.9.5** A subset $S$ of $V(G)$ is called **weighted monophonic convex** if $J_w[S] = S$.

**Definition 5.9.6** A subset $S$ of $V(G)$ is called **weighted monophonic** if $J_w[S] = V(G)$.

All the above definitions are illustrated in Example 5.9.7.

**Example 5.9.7** Consider the graph in Fig. 5.21.

For the vertices $a$ and $e$, the weighted monophonic interval $J_w[a, e]= \{a, b, e\}$. Consider the path $a - f - e$ of strength $6(= 2 * 3)$. The direct edge $ae$ is a chord of this path. For the vertices $e$ and $c$, the direct edge $ec$ is the only weighted $e - c$ monophonic path. Since $ec$ is a chord of the path $e - d - c$, it is not a weighted $e - c$ monophonic path.

In this graph, $\{a, b\}$, $\{e, c\}$, $\{f, e\}$, $\{a, b, e\}$, $\{c, d, e\}$, etc. are weighted monophonic convex; but $\{a, e\}$, $\{a, f, e\}$, etc. are not.

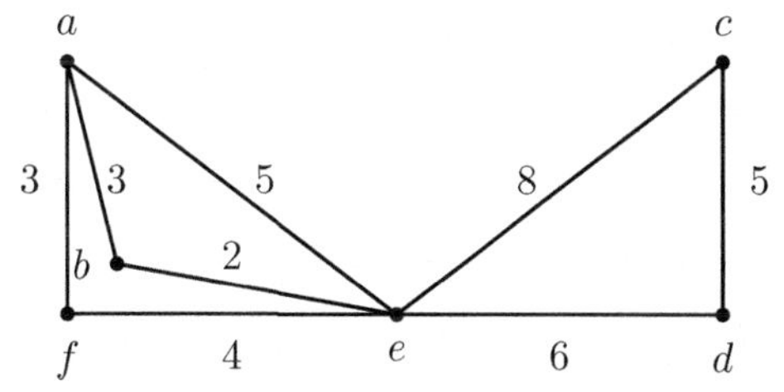


**Fig. 5.21** A weighted graph—monophonic and geodetic paths

**Remark 5.9.8** In a connected weighted graph $G$, the weighted $u - v$ geodesics are the paths along which we compute the weighted distance between vertices $u$ and $v$. Certainly, a weighted $u - v$ geodesic is a weighted monophonic path. This is because of the absence of chords in it.

Following proposition says that a weighted $u - v$ geodesic is always a weighted monophonic $u - v$ path.

**Proposition 5.9.9** *Let $G = (V, E, w)$ be a connected weighted graph and $u$, $v$ be any two vertices of $G$. Then any weighted $u - v$ geodesic is a weighted monophonic $u - v$ path in $G$.*

***Proof*** Let $G = (V, E, w)$ be a connected weighted graph and $u, v$ be any two vertices of $G$. Let $P = ue_1v_1e_2v_2 \dots e_nv$ be a weighted $u - v$ geodesic in $G$. We have to prove that $P$ is a weighted monophonic $u - v$ path in $G$. It is enough if we prove, $P$ has no chords. Suppose the contrary. Let the edge $v_iv_j$ be a chord of $P$. Let $P'$ be the $u - v$ path along this edge. This means $P' = ue_1v_1e_2v_2 \dots e_iv_ie_jv_j \dots e_nv$, where $j \geq i + 2$. Since $v_iv_j$ is a chord of $P$, we have $S(P) * l(P) > S(P') * l(P')$. This means $P'$ is a weighted $u - v$ geodesic and $P$ is not, which is a contradiction to our assumption. Hence our assumption is wrong. So $P$ is a weighted $u - v$ monophonic $u - v$ path, where $P$ is arbitrary. Hence all weighted $u - v$ geodesics are weighted monophonic paths also.

**Remark 5.9.10** Converse of Proposition 5.9.9 is not true all the time. In the weighted graph in Fig. 5.21, $a - b - e$ and $a - f - e$ are two weighted monophonic $a - e$ paths. But they are not weighted $a - e$ geodesics. Infact, edge $ae$ is the only weighted $a - e$ geodesic.

In the immediate proposition, we see that all edges in a connected simple weighted graph are weighted monophonic paths between the respective end vertices.

**Proposition 5.9.11** *Let $G = (V, E, w)$ be a connected simple weighted graph. Let $e = uv$ be any edge of $G$. Then $e$ is a weighted monophonic path between $u$ and $v$.*

***Proof*** Since the graph is simple, it is free from parallel edges and self loops. So no edge cannot have a chord, and hence every edge will be a monophonic path between its end vertices. This proves the proposition.

We can show that the intersection of two weighted monophonic convex sets is again weighted monophonic convex, as seen in Theorem 5.9.12.

**Theorem 5.9.12** *Let $G = (V, E, w)$ be a simple connected weighted graph. Then the intersection of two weighted monophonic convex sets of $G$ is again weighted monophonic convex.*

***Proof*** Let $G = (V, E, w)$ be a simple connected weighted graph. Let $S$ and $T$ be two weighted monophonic convex sets of $G$. We have to prove that $S \cap T$ is weighted monophonic convex. Let $u$ and $v$ be any two vertices in $S \cap T$. This means $u$ and $v$ are two vertices in both $S$ and $T$, which are weighted monophonic convex. Therefore all vertices in all weighted $u - v$ monophonic paths are in both $S$ and $T$, and hence in $S \cap T$. This shows that $S \cap T$ is weighted monophonic convex.

We can not expect that the union of two weighted monophonic convex sets is again weighted monophonic convex. In the weighted graph shown in Fig. 5.21, $\{a, b\}$ and $\{c, d\}$ are weighted monophonic convex, but their union $\{a, b, c, d\}$ is not weighted monophonic convex. Since any monophonic path between $a$ and $c$ is passing through the vertex $e$.

In the next section, we present the notion of weighted monophonic blocks and some of their properties. It is proved that every weighted geodetic block is a weighted monophonic block.

## 5.10 Monophonic Blocks

**Definition 5.10.1** Let $G$ and $H$ be a connected weighted graph and a connected weighted subgraph of $G$ respectively. Then $H$ is called a **weighted monophonic block of** $G$, if for every $u, v \in V(H)$, there exists exactly one weighted $u - v$ monophonic path in $G$ [2, 3].

**Remark 5.10.2** Let $G = (V, E, w)$ be a connected weighted graph. Any edge $e = uv$ of $G$ is a weighted monophonic path between its end vertices $u$ and $v$, by Proposition 5.9.11. Definition 5.10.1 along with the mentioned proposition, we can see that, each edge of a weighted monophonic block is a unique weighted monophonic path between its end vertices.

In the following example, we illustrate weighted monophonic blocks.

**Example 5.10.3** Consider the weighted graph in Fig. 5.22.

In the weighted graph $G_2$, there exists exactly a weighted monophonic path between any two vertices. Hence, it is a weighted monophonic block. On the other hand, $G_1$ is not a weighted monophonic block. For, $a - b$ and $a - c - b$ are two weighted monophonic paths between the vertices $a$ and $b$.

In the next theorem, we prove that all weighted geodetic blocks are weighted monophonic blocks. But the converse of the result is not true in general.

**Theorem 5.10.4** *Let $G = (V, E, w)$ be a connected weighted graph. If $H$ is a weighted geodetic block of $G$, then it is a weighted monophonic block.*

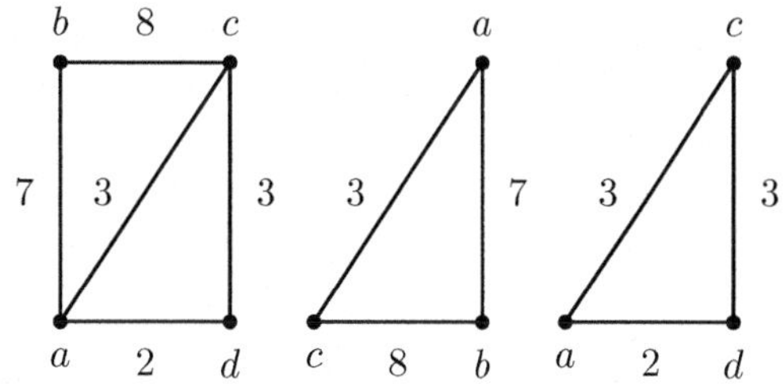


**Fig. 5.22** Weighted monophonic blocks

***Proof*** Let $G = (V, E, w)$ be a connected weighted graph. Suppose that $H$ is a weighted geodetic block of $G$. Then $H$ is complete and each edge $e = uv$ of $H$ is a unique weighted $u - v$ geodesic in $G$. We have to prove that $H$ is a weighted monophonic block of $G$. It is enough, if we prove, for any pair of vertices $u, v$ of $H$, there exists a unique weighted monophonic path. Now each edge of a weighted graph is a weighted monophonic path between its end vertices, and since $H$ is complete, we complete the proof by showing that each edge of $H$ is a unique weighted monophonic path between its end vertices. Let $f = xy$ be any edge of $H$. Then $f$ is a weighted monophonic $x - y$ path. We have to prove the uniqueness of $f$. If possible, let there be another weighted monophonic $x - y$ path, say, $P$ in $G$. Since $P$ is a weighted $x - y$ monophonic path, it has no chords. But since $f$ is the only weighted $x - y$ geodesic in $G$, it is a chord of $P$. This is a contradiction. Hence $f$ is the only weighted $x - y$ monophonic path in $G$. Therefore $H$ is a weighted monophonic block of $G$.

Due to the lack of completeness of the subgraph, all weighted monophonic blocks are not weighted geodetic blocks of $G$. Thus the converse of Theorem 5.10.4 is not true. In the following theorem, we show that, every complete weighted monophonic block is a weighted geodetic block.

**Theorem 5.10.5** *Let $G = (V, E, w)$ be a connected weighted graph and let $H$ be a complete weighted subgraph of $G$. If $H$ is a weighted monophonic block of $G$, then it is a weighted geodetic block of $G$.*

***Proof*** Let $G = (V, E, w)$ be a connected weighted subgraph of $G$. Let $H$ be a complete weighted subgraph of $G$. Suppose that $H$ is a weighted monophonic block of $G$. Then , between any pair of vertices $u, v$ in $H$, there exists a unique weighted monophonic path in $G$. Now each edge is a weighted monophonic path between its end vertices (Proposition 5.9.11). Since $H$ is complete, each pair of its end vertices are adjacent. Thus each edge is a unique weighted monophonic path path between its end vertices in $G$.

We have to prove that, $H$ is a weighted geodetic block of $G$. It is enough, if we prove each edge of $H$ is a unique weighted geodesic between its end vertices.

Let $e = uv$ be any edge of $H$. First we prove that, $e$ is weighted $u - v$ geodesic, and then its uniqueness. If possible, let $e$ be not a weighted $u - v$ geodesic in $G$. Then there exists another weighted $u - v$ geodesic, say, $P$. Then $P$ is a weighted monophonic path also (Proposition 5.9.9). Thus both $P$ and $e$ are weighted $u - v$ monophonic paths in $G$. It is a contradiction to the fact that, $e$ is a unique weighted $u - v$ monophonic path in $G$. Hence edge $e = uv$ is a weighted $u - v$ geodesic in $G$. Now it remains to prove the uniqueness of $e$. Suppose the contrary. Let $P$ be any other weighted $u - v$ geodesic in $G$. Since, $P$ and $e$ are both weighted $u - v$ geodesics, the weighted distance between $u$ and $v$ can be calculated along both of them. This means, $S(P) * l(P) = S(e) * l(e) = w(e)$. This implies that $e$ is not a chord of $P$. Hence $P$ is another weighted monophonic $u - v$ path in $G$, a contradiction to the assumption that, $H$ is a weighted monophonic block of $G$. Thus the proof is complete.

Here we present a characterization for weighted monophonic blocks.

**Theorem 5.10.6** *Let $H$ be a subgraph of a connected weighted graph $G$. Suppose that, $H$ is either a weighted tree or a complete graph. Then $H$ is a weighted monophonic block of $G$ if and only if every subsets $S$ of $V(H)$, where $S$ is a two element set consisting of the end vertices of some edge in $H$, are weighted monophonic convex sets of $H$.*

***Proof*** Let $G$ be a connected weighted graph and let $H$ be a weighted monophonic block of $G$. Let $ab$ be any edge of $H$, and take $S = \{a, b\}$. We need to prove that $S$ is a weighted monophonic convex set of $G$. Any edge is a weighted monophonic path between its end vertices (Proposition 5.9.11), and since $H$ is a weighted monophonic block, there exists no other weighted monophonic paths between the vertices $a$ and $b$. Thus $S$ is weighted monophonic convex.

Conversely assume that each $S = \{a, b\}$ where $a$ and $b$ are the end vertices of some edge of $H$ is weighted monophonic convex. We need to prove that $H$ is a weighted monophonic block of $G$. Let $u, v$ be any two vertices of $H$. We must find a unique weighted monophonic $u - v$ path.

**Case 1.** $H$ is complete.

In this case each pair of its vertices are adjacent, in particular $u$ and $v$. By our assumption, $\{u, v\}$ is weighted monophonic convex. This means the edge $uv$ is the only weighted monophonic path between $u$ and $v$. Thus $H$ is a weighted monophonic block of $G$.

**Case 2.** $H$ is a weighted tree.

If $u$ and $v$ are adjacent, then the direct edge $uv$ is the only connection between $u$ and $v$. So edge $uv$ is the unique weighted monophonic path. If $u$ and $v$ are not adjacent, then there exist adjacent vertices $v_1, v_2, \ldots, v_n$ such that $v_1$ is adjacent to $u$ and $v_n$ is adjacent to $v$. Since $u$ and $v$ are not adjacent, and $H$ is a weighted tree, the path $u - v_1 - v_2 - \cdots - v_n - v$ will be the unique $u - v$ path and has no chords. Hence it will be the unique weighted $u - v$ monophonic path. Thus in both cases $H$ is a weighted monophonic block of $G$.

By Theorem 3.5.6, it is obvious that a weighted tree satisfies all the conditions of last theorem. Now we have the following corollary.

**Corollary 5.10.7** *All weighted trees are weighted monophonic blocks.*

*Theorem 5.10.8 is valid for the class of weighted trees only.*

**Theorem 5.10.8** *If $G = (V, E, w)$ is a weighted tree, then there exists a sequence of sets $V = V_n \supset V_{n-1} \supset \cdots \supset V_1$, where for each $i$, $V_i$ is weighted monophonic convex and $| V_i |= i$.*

***Proof*** Let $G$ be a weighted tree with $n$ vertices. Then $V(G) = V_n$ is a weighted monophonic convex set, as there exists only one path between any pair of vertices, and hence that path is weighted monophonic. Let $v_1$ be a pendant vertex of $G$ and $V_{n-1} = V \setminus \{v_1\}$. Then clearly $V_{n-1}$ is weighted monophonic convex in $G$ as $V_n$. Let

$v_2$ be any pendant vertex of$G \setminus v_1$ and $V_{n-2} = V_n \setminus \{v_1, v_2\}$. Then$V_{n-2}$ is weighted monophonic convex in $G$, and so on. The sets $V_i$ thus formed, have the stated nested property.

Next section presents the idea of weighted monophonic boundary and interior vertices.

## 5.11 Monophonic Boundary and Interior

Following are the definitions of boundary and interior vertices in a weighted graph.

**Definition 5.11.1** Let $G$ be a weighted graph and $S$ be a weighted monophonic convex set of $G$. A vertex $u \in S$ is called a **weighted monophonic boundary vertex** of $S$ if and only if $S \setminus \{u\}$ is a weighted monophonic convex set of $G$.

**Definition 5.11.2** A vertex $v \in S$ is a **weighted monophonic interior vertex** of $S$ if and only if $S \setminus \{v\}$ is not a weighted monophonic convex set of $G$.

Next theorem presents a characterization for the weighted monophonic boundary and interior vertices in a weighted graph.

**Theorem 5.11.3** *Let $G$ be a connected weighted graph, and $S$ be a weighted monophonic convex set of $G$. Then*
*(a) A vertex $u \in S$ is a weighted monophonic boundary vertex of $S$ if and only if $u$ does not lie on any weighted $v - w$ monophonic paths for all $v, w \in S \setminus \{u\}$.*
*(b) A vertex $u \in S$ is a weighted monophonic interior vertex of $S$ if and only if there exists $v, w \in S \setminus \{u\}$ such that $u$ lies on a weighted monophonic $v - w$ path.*

***Proof*** $(a)$ Let $u$ be a weighted monophonic boundary vertex of the weighted monophonic convex set $S$. Then $S \setminus \{u\}$ is weighted monophonic convex. Let $v, w \in S \setminus \{u\}$. Then all the vertices in all weighted $v - w$ monophonic paths are in $S \setminus \{u\}$ itself. Hence $u$ does not lie in on any weighted $v - w$ monophonic paths.

Conversely $u \in S$ does not lie on any weighted $v - w$ monophonic path for all $v, w \in S \setminus \{u\}$. Then $S \setminus \{u\}$ is a weighted monophonic convex set of $G$. Hence $u$ is weighted monophonic boundary point of $S$.

$(b)$ Let $u$ be a weighted monophonic interior vertex of the weighted monophonic convex set $S$. Then $S \setminus \{u\}$ is not weighted monophonic convex. That means $S$ is weighted monophonic convex and $S \setminus \{u\}$ is not. Thus there exists a weighted $v - w$ monophonic path that is passing through the vertex $u$.

Conversely, assume that $u$ lies on some weighted $v - w$ monophonic path for $v, w \in S \setminus \{u\}$. Then $S \setminus \{u\}$ is not weighted monophonic convex, and hence $u$ is a weighted monophonic interior vertex of $S$.

**Theorem 5.11.4** *Let $G = (V, E, w)$ be a connected weighted graph and $S$ be a weighted monophonic convex subset of $V$. Let $u \in V \setminus S$. If $u$ is pendant and adjacent with exactly one vertex of $S$, then $S \cup \{u\}$ is weighted monophonic convex.*

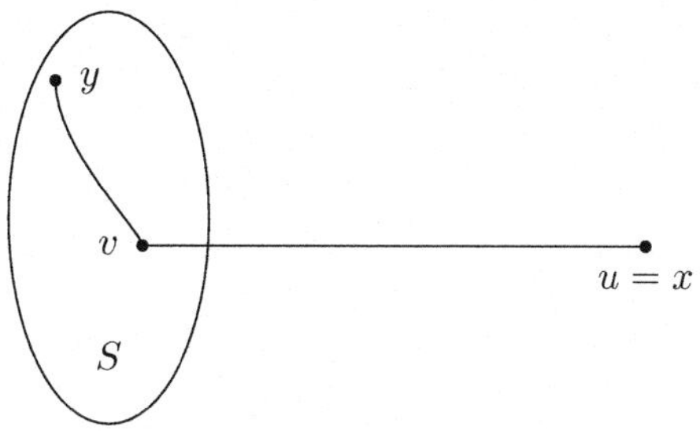


**Fig. 5.23** Case II. Illustration

***Proof*** Let $G = (V, E, w)$ be a connected weighted graph, and let $S$ be a weighted monophonic convex set of $V$. Suppose that $u \in V \setminus S$ be a pendant vertex of $G$, and adjacent with exactly one vertex, say, $v$ of $S$. We have to prove that $S \cup \{u\}$ is weighted monophonic convex.

Let $S \cup \{u\} = T$. We prove that $T$ is weighted monophonic convex. Let $x, y$ be any two vertices of $T$. We complete the proof in two cases.

**Case 1.** $x \neq u \neq y$.

In this case, both $x$ and $y$ are in $S$, which is weighted monophonic convex. So $T$ is weighted monophonic convex.

**Case 2.** *Either one of $x$ or $y$ is $u$.*

Refer Fig. 5.23. Let $u = x$. The direct edge $uv$ the unique path between $u$ and $v$. So any $u - y$ path contains the edge $uv$. Since $v, y$ are two vertices in $S$, which is weighted monophonic convex, all vertices in all weighted monophonic $v - y$ paths are in $S$ itself. Now the union of edge $uv$ with all weighted monophonic $v - y$ paths are again weighted monophonic. So all vertices in all weighted $x - y$ monophonic paths are in $T$. This proves $T$ weighted monophonic convex (Fig. 5.24).

**Theorem 5.11.5** *Let $G = (V, E, w)$ be a connected weighted graph and let $S$ be a weighted monophonic convex set of $G$ and $u \in V(G) \setminus S$. Then $S \cup \{u\}$ is weighted monophonic convex if and only if $u$ is adjacent to some vertex $v \in S$ and $u, v$ do not lie on a cycle.*

***Proof*** Suppose that $S \cup \{u\}$ is weighted monophonic convex. Then there exists a vertex $v \in S$, which is adjacent to $u$. Thus if $w \in S$, then all vertices in all weighted

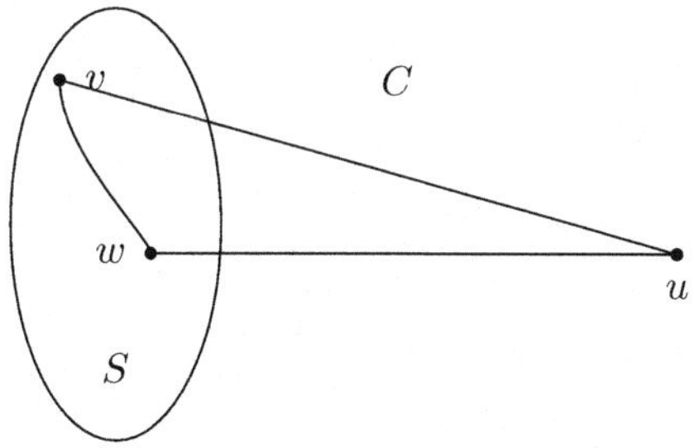


**Fig. 5.24** Cycle $C$ in Theorem 5.11.5

$u - w$ monophonic paths (except $u$) belong to $S$. Otherwise $S \cup \{u\}$ will not be weighted monophonic convex.

If $u, v$ both lie on a cycle, say, $C$, then it follows that $S$ contains a vertex $w$ of $C$ which is different from $u$ and $v$, and since $v, w \in S$, and $S$ is weighted monophonic convex and hence $u \in S$. This is a contradiction and our assumption is wrong. Hence $u, v$ do not lie on a cycle.

Conversely assume that the two vertices $u$ and $v$ do not lie on a cycle, which implies that $u$ and $w$ do not lie on a cycle for all $w \in S$. As a consequence, for all $w \in S$, all vertices (except $u$) on any weighted $u - w$ monophonic path belong to $S$. It follows that $S \cup \{u\}$ is weighted monophonic convex.

## References

1. Chartrand, G., Saba, F., Zou, H.B.: Edge rotations and distance between graphs. Casopis pro pestovani matematiky **110**(1), 87–91 (1985)
2. Chartrand, G., Erwin, D., Johns, G.L., Zhang, P.: On boundary vertices in graphs. J. Combin. Math. Combin. Comput. **48**, 39–53 (2004)
3. Chung, F.R.K., Graham, F.C.: Spectral Graph Theory, vol. 92. American Mathematical Society (1997)

# Chapter 6
# Degree Sequences and Saturation

Sequences can be used as representatives of weighted graphs. In this chapter, three kinds of sequences are introduced. Their properties are studied in different weighted graph structures. Due to the scope and wide applications, detailed study and characterizations are carried out in partial trees and partial blocks. Mathematical manipulations at the studies on weighted graphs can be simplified with the help of these sequences. This chapter is based on the work by Jill et al. [1].

## 6.1 Sequences in a Weighted Graph

Based on the $\alpha$ and $\beta$ classification of strong edges of weighted graphs, we introduce three types of sequences. This section considers simple undirected weighted graphs only.

**Definition 6.1.1** Let $G = (V, E, w)$ be a connected weighted graph with $V = \{v_1, v_2, v_3, \ldots, v_p\}$ in some order. Then a finite sequence $\alpha_s(G) = (n_1, n_2, \ldots, n_p)$ is called the $\alpha$-**sequence** of $G$ if $n_i$ represents the number of $\alpha$-strong edges incident at $v_i$. $n_i$= 0, if no $\alpha$-strong edges is incident at $v_i$. Similarly, a finite sequence $\beta_s(G) = (n_1, n_2, n_3, \ldots, n_p)$ is called the $\beta$ -**sequence** of $G$ if $n_i$ represents the number of $\beta$-strong edges incident at $v_i$. $n_i = 0$, if no $\beta$-strong edge is incident at $v_i$.

If there is no confusion persist, we use the notation $\alpha_s$ and $\beta_s$ for $\alpha_s(G)$ and $\beta_s(G)$ respectively.

**Definition 6.1.2** Let $G = (V, E, w)$ be a connected weighted graph with $V = \{v_1, v_2, \ldots, v_p\}$. Then a finite sequence $S_s = (n_1, n_2, n_3, \ldots, n_p)$ is called the **strong sequence** of $G$ if $n_i$ represents the number of $\alpha$ or $\beta$-strong edges incident at $v_i$. $n_i = 0$, if no $\alpha$ or $\beta$-strong edge is incident at $v_i$.



S. Mathew et al., *Weighted and Fuzzy Graph Theory*, Studies in Fuzziness and Soft Computing 429, https://doi.org/10.1007/978-3-031-39756-1_6

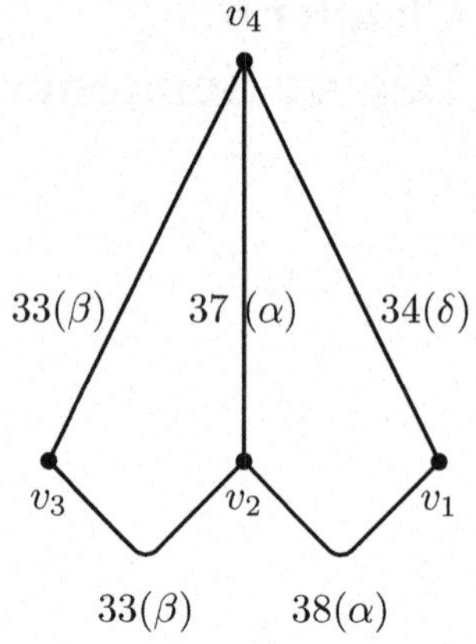


**Fig. 6.1** Sequences -Illustration 1

For the ease of writing, we use the notation $S_s$ instead of $S_s(G)$. We find these sequences in the following illustration.

**Definition 6.1.3** Consider the graph in Fig. 6.1. Here $V = \{v_1, v_2, v_3, v_4\}$. The $\alpha$-sequence and $\beta$-sequence of $G$ are $\alpha_s = (1, 2, 0, 1)$ and $\beta_s = (0, 1, 2, 1)$ respectively. Strong sequence is given by $s_s = (1, 3, 2, 2)$. Consider another example in Fig. 6.2.

**Example 6.1.4** In Fig. 6.2, let $V = \{v_1, v_2, v_3, v_4, v_5, v_6, v_7, v_8\}$. Now $\alpha_s = (2, 2, 2, 2, 1, 1, 1, 1)$, $\beta_s = (1, 1, 1, 1, 0, 0, 0, 0)$ and $s_s = (3, 3, 3, 3, 1, 1, 1, 1)$.

## 6.2 Characterization of Partial Blocks

In this section, we present characterizations of partial blocks in terms of strong sequences. Some necessary conditions for a partial block are also given.

**Definition 6.2.1** A sequence of integers is said to be a **binary sequence** if it consists of entries 0 and 1 only.

**Theorem 6.2.2** *If a connected weighted graph $G = (V, E, w)$ is a partial block, then $\alpha_s(G)$ is a binary sequence.*

***Proof*** Suppose that $G = (V, E, w)$ is a partial block. We have to prove that $\alpha_s(G)$ is binary. That is, we need to prove that $\alpha_s(G)$ contains only 0's and 1's. If possible, suppose the contrary. If possible, Suppose that there exists an entry which is at least 2 in $\alpha_s(G)$. Let $n_i = 2$. This means, there are 2 different $\alpha$-strong edges incident on the vertex $v_i$. Now an edge $e$ in a weighted graph $G$ is a partial bridge if and only if it is an $\alpha$-strong edge [2]. Also if a vertex is common to more than one partial bridges, then it is a partial cut vertex [3]. Therefore we see that $v_i$ is a partial cut

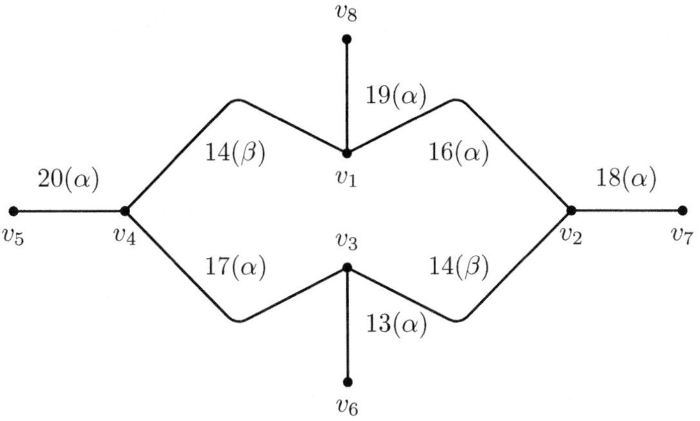


**Fig. 6.2** Sequences—Illustration 2

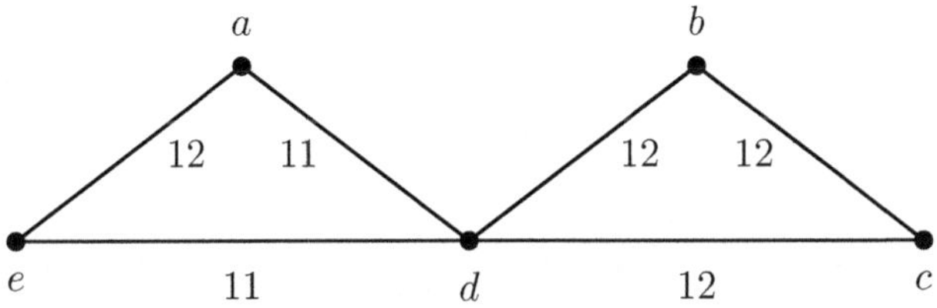


**Fig. 6.3** A weighted graph with a cut vertex

vertex of $G$, which is a contradiction to our assumption that the weighted graph $G$ is free from partial cut vertices, as it is a partial block. So our assumption is wrong. Hence $n_i < 2$. This means $n_i = 0$ or 1. Thus $\alpha_s(G)$ is binary. ∎

**Remark 6.2.3** The above condition is not sufficient. It is clear from the following example.

**Example 6.2.4** Consider the weighted graph in Fig. 6.3, where $V = \{a, b, c, d, e\}$.

In this example we get $\alpha_s(G)$ as a binary sequence $(1, 0, 0, 0, 1)$. The vertex $d$ is a cut vertex and hence a partial cut vertex. Then the graph is not a partial block.

In the next characterization, we restrict the underlying graph $G^*$ of $G$ to be a block ($G^*$ has no cut vertices).

**Theorem 6.2.5** *Let $G = (V, E, w)$ be a connected weighted graph such that $G$ has no $\delta$-edges, all the $\beta$-strong edges posses equal weights and the underlying unweighted graph $G^*$ is a block, then $G$ is a partial block if and only if $\alpha_s(G)$ is binary.*

**Fig. 6.4** Paths in Theorem 6.2.5

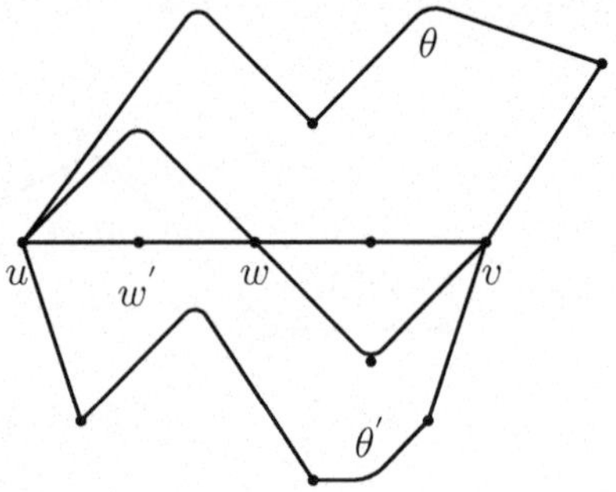


***Proof*** Let $G = (V, E, w)$ be a connected weighted graph such that $G$ has no $\delta$-edges, all the $\beta$-strong edges posses equal weights and the underlying unweighted graph $G^*$ is a block. If $G$ is a partial block, then by Theorem 6.2.2, $\alpha_s(G)$ is binary.

Conversely suppose that, $\alpha_s(G)$ is a binary sequence. We have to prove that $G$ is a partial block. That is we have to prove that $G$ has no partial cut vertices. If possible, let $w$ be a partial cut vertex of $G$. Then there exist two vertices $u$ and $v$ in $G$ such that $u \neq w \neq v$ and $CONN_{G-w}(u, v) < CONN_G(u, v)$. Since $G^*$ is a block, it has no cut vertices, and hence $w$ is not a cut vertex of $G$. Therefore we can consider many possible $u - v$ paths which are not passing through the vertex $w$. Now from the above inequality, it is clear that weights of all edges in all $u - w - v$ paths are strictly greater than the weights of all edges in all possible $u - v$ paths which are not passing through the vertex $w$.

Also given that, $G$ has no $\delta$-edges and all the $\beta$-strong edges are of equal weight. This means that, all edges in the $u - w - v$ paths are $\alpha$-strong. For, if let $w'w$ be a $\beta$-strong edge such that weight of $w'w$ is $\xi$. Then the strength of all $u - w - v$ paths as well as that of all $u - w$ paths are $\xi$ ( See Fig. 6.4). Clearly this is a contradiction to the inequality, $CONN_{G-w}(u, v) < CONN_G(u, v)$. So all the edges in all the $u - w - v$ paths are $\alpha$-strong. Thus all the internal vertices of all the $u - w - v$ paths are incident with at least 2 $\alpha$-strong edges, which shows that $\alpha_s(G)$ is not binary. It is a contradiction. So our assumption is wrong. Hence $G$ is a partial block. This completes the proof of the theorem. ∎

Next theorem characterizes partial blocks on the basis of $\alpha$ sequences and strong sequences.

**Theorem 6.2.6** *Let $G = (V, E, w)$ be a connected weighted graph. Then $G$ is a partial block if and only if the following 2 conditions are satisfied.*

1. $\alpha_s(G)$ is a binary sequence.
2. For any given pair of vertices $u$ and $v$, there exists a cycle $C$, containing $u$ and $v$ such that $S_s(C)$ contains only entries which are at least 2.

***Proof*** Let $G = (V, E, w)$ be a connected weighted graph. First suppose hat $G$ is a partial block. We have to prove conditions 1 and 2. The proof of 1 is the same as that of Theorem 6.2.5. Now we prove condition 2 as follows. Let $u$ and $v$ be any two vertices of $G$. We have to prove that $u$ and $v$ lie on a common cycle $C$ such that $S_s(C)$ contains entries which are at least 2 only. That is we need to prove that $u$ and $v$ lie on a common strong cycle $C$. If possible, suppose the contrary. Let there be no common strong cycle containing both $u$ and $v$. There are two cases arise.

**Case 1.** *$e = uv$ is a strong edge.*

Since $u$ and $v$ are not on any strong cycle and edge $uv$ is strong, the edge $uv$ lies in every $MST$ of $G$. Now every edge of an $MST$ is a partial bridge [3]. Hence $uv$ is a partial bridge.

If $u$ is an end vertex of all maximum spanning trees, then clearly $v$ is a common vertex of at least 2 partial bridges, and hence it will be a partial cut vertex of $G$, which is a contradiction to our assumption that $G$ is free from partial cut vertices as it is a partial block. On the other hand if $v$ is an end vertex of all $MSTs$, then $u$ will be a partial cut vertex of $G$, a contradiction to our assumption.

Now suppose that $u$ is an end vertex of $MST$, $T_1$ and $v$ is an end vertex of $MST$, $T_2$. Since $T_1$ is a spanning tree, $v$ will be an internal vertex of $T_1$. Let $w$ be a strong neighbor of $u$ in $T_2$. Clearly there is a strong path $P$ in $T_1$ from $u$ to $w$ through $v$. This path $P$ together with the strong edge $uw$ forms a strong cycle $C$ in $G$, which is also a contradiction.

**Case 2.** *$e = uv$ is a $\delta$-edge.*

If $e = uv$ is a $\delta$-edge, then $CONN_{G-e}(u, v) > w(e)$. Let $e$ be lying on a cycle $C$. Then there exists a strong path $C - e$ between $u$ and $v$ (See Fig. 6.5). If $C - e$ is not a strong path, then we can extend $C - e$ to a strong path by adding more number of edges. If we are not able to find such a strong path, it will be a contradiction to the definition of $\delta$-edges. Also this strong path $C - e$ is unique.

Therefore it belongs to all $MST's$ and all internal vertices of $P$ are partial cut vertices of $G$, which is a contradiction.

Therefore in all cases, our assumption is wrong. Hence there exists a common strong cycle $C$, containing any given pair of vertices $u$ and $v$. Since $C$ is strong, $S_s(C)$ contains entries which are at least 2 only.

Conversely suppose the conditions 1 and 2 hold. We have to prove that $G$ is a partial block. That is we need to prove that $G$ has no partial cut vertices. By condition 1, it is clear that at most one $\alpha$-strong edge is incident at every vertex. By condition 2, any given pair of vertices $u$ and $v$ lie on a common strong cycle. So if we delete a vertex from $G$, then the strength of connectedness between any pair of vertices remains the same. Thus no vertex of $G$ can be a partial cut vertex. So $G$ is a partial block. ■

## 6.3 Characterization of Partial Trees

In this section, we present some necessary conditions for a weighted graph to hold the structure of a partial tree.

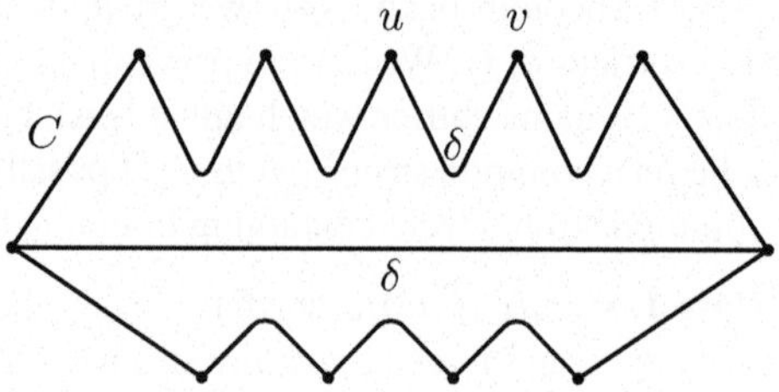


**Fig. 6.5** Paths in Theorem 6.1.6

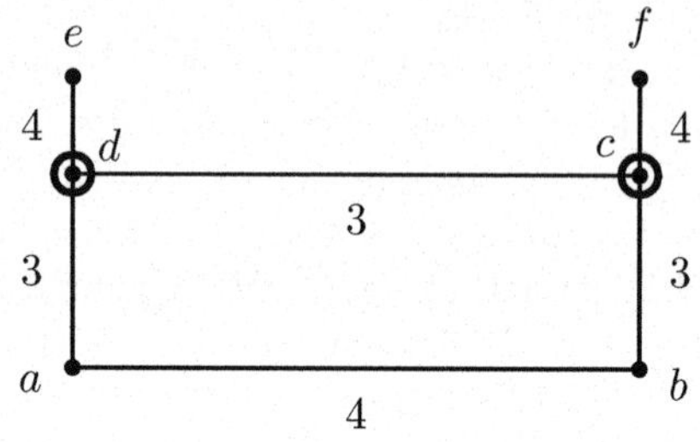


**Fig. 6.6** A weighted graph with 2 cut vertices

**Theorem 6.3.1** *If $G = (V, E, w)$ is a partial tree and $\mid V \mid= p$, then $\alpha_s(G) \in (\mathbf{Z}^+)^p$.*

***Proof*** By the definition of $\alpha_s(G)$, it is clear that all of its elements are greater than or equal to zero. We want to prove that all the elements in $\alpha_s(G)$ is at least unity. Suppose the contrary. Let the $i^{th}$ element in $\alpha_s(G)$, say, $n_i$ be zero. Since $n_i = 0$, the corresponding vertex $v_i$ will not be incident with any $\alpha$-strong edge. This will result in the disconnection of the maximum spanning tree $F$ of $G$. which is a contradiction to the definition of $F$. So our assumption is wrong and hence all the elements of $\alpha_s(G)$ are at least unity. This completes the proof of the theorem. ■

The condition in the above theorem is not sufficient as seen from the following example.

**Example 6.3.2** Consider the following weighted graph in Fig. 6.6. Here, $\alpha_s(G) = (1, 1, 1, 1, 1, 1)$. Note that vertices $c$ and $d$ are cut vertices of the graph. Hence the graph is not a partial tree.

Next result helps us to obtain the number of partial cut vertices of a partial tree from the corresponding $\alpha$ sequence.

**Theorem 6.3.3** *Let $G = (V, E, w)$ be a connected weighted graph such that $\mid V \mid= p$. Let $t$ be a positive integer such that $t \leq p$. If $\alpha_s(G)$ contains $t$ elements which are at least 2, then $G$ has exactly $t$ partial cut vertices.*

***Proof*** Let $G = (V, E, w)$ be a partial tree. Let $F$ be the spanning tree of $G$ with the property given in the definition of partial trees. Then the internal vertices of $F$ are the partial cut vertices of $G$ [3]. Also we know that, if a vertex is common to more than one $\alpha$-strong edge, then it is a partial cut vertex [3]. So the vertex of $G$ which corresponds to an entry in $\alpha_s(G)$ which is at least 2 must be a partial cut vertex. This completes the proof of the theorem. ■

If the condition in the above theorem was sufficient, it would be easier to identify the partial cut vertices of $G$. But, Example 6.3.4 shows that the condition in Theorem 6.3.3 is not sufficient.

**Example 6.3.4** Consider the weighted graph in Fig. 6.3 given in Example 6.2.4. Here the vertex $d$ is a cut vertex and hence is a partial cut vertex. But 0 is the entry in the $\alpha$ sequence corresponding to vertex $d$.

The condition will be sufficient if we restrict the structure of the underlying unweighted graph $G^*$ as a block satisfying two conditions on its edges.

**Theorem 6.3.5** *Let $G = (V, E, w)$ be a connected weighted graph such that $\mid V \mid = p$ and the underlying unweighted graph $G^*$ is a block such that $G$ is free from $\delta$-edges and all the $\beta$-strong edges posses equal weight, then the partial cut vertices of $G$ are exactly those vertices whose entry in $\alpha_s(G)$ is at least 2.*

**Definition 6.3.6** A **zero sequence** is a real sequence with all its entries are zero. It is denoted by $(0)$.

Theorem 6.3.7 is a characterization of partial trees by the nature of corresponding $\beta$ sequence.

**Theorem 6.3.7** *A connected weighted graph $G$ is a partial tree if and only if $\beta_s(G) = (0)$.*

***Proof*** Let $G = (V, E, w)$ be a connected weighted graph. Suppose that $G$ is a partial tree. If $G$ is a weighted tree, then all the edges of $G$ are partial bridges. Now an edge $e = uv$ in $G$ is a partial bridge if and only if it is $\alpha$-strong [2]. Thus all the edges of $G$ are $\alpha$-strong. Therefore $G$ has no $\beta$-strong edges and hence $\beta_s(G) = (0)$.

If $G$ is not a weighted tree, then $G$ has a weighted cycle, say, $C$. Since $G$ is a partial tree, there exists an edge $e = uv$ such that $w(e) < CONN_{G-e}(u, v)$, where $G - e$ is the subgraph of $G$ obtained by deleting the edge $e$ from $G$ [2]. This means $e$ is a $\delta$-edge. If $G - e$ is a weighted spanning tree of $G$, all the edges in $G - e$ are $\alpha$-strong. Hence $\beta_s(G) = (0)$. If $G - e$ is not a weighted spanning tree of $G$, then continue the above procedure of deleting $\delta$-edges from $G - e$ until we get weighted spanning tree.

Conversely suppose that $\beta_s(G) = (0)$. We have to prove that $G$ is a partial tree. If $G$ has no weighted cycles, then $G$ is a weighted tree and hence a partial tree. Suppose that $G$ has a cycle, say, $C$. Then $C$ contains only $\alpha$-strong and $\delta$-edges. Also note that all the edges of $C$ cannot be $\alpha$-strong, since otherwise it will contradict the definition of $\alpha$-strong edges. Thus there exists at least one $\delta$-edge in $C$. if we delete $e$ from $C$, we get a maximum spanning tree of $G$. If not, remove one $\delta$-edge from the existing weighted cycles in $G$. Continue this procedure until we get a weighted spanning tree of $G$. Hence $G$ is a partial tree. ■

Consider the case when $G$ is a partial tree. Let $F$ be the spanning tree of $G$. Then the edges of $F$ are the partial bridges of $G$. In fact, edges of $F$ are the $\alpha$-strong edges of $G$. $\beta_s(F)$ contains only zeros since $F$ has no $\beta$-strong edges and thus $\beta_s(G) = \beta_s(F)$.

Following theorem presents a necessary and sufficient condition for a weighted graph $G$ to be a partial tree.

**Theorem 6.3.8** *A connected weighted graph $G = (V, E, w)$ is a partial tree if and only if $\alpha_s(G) = \alpha_s(F)$, where $F$ is the corresponding maximum spanning tree of $G$.*

***Proof*** Suppose that $G = (V, E, w)$ is a partial tree. If $G$ is itself a weighted tree, then $F$ and $G$ are isomorphic and hence $\alpha_s(G) = \alpha_s(F)$. If $G$ is not a weighted tree, then $G$ contains a cycle $C$, since $G$ is a partial tree, there exists an edge $e = uv$ in $C$ such that $w(e) < CONN_{G-e}(u, v)$, where $G - e$ is the weighted subgraph obtained by deleting the edge $e$ from $G$. If $G - e$ is a weighted tree, then $G - e$ and $F$ are isomorphic and since $e$ is a $\delta$-edge, we get $\alpha_s(G) = \alpha_s(F)$. If not, continue the above procedure of deleting $\delta$-edges from cycles in $G - e$ until we get a maximum spanning tree $F$ of $G$ such that $\alpha_s(G) = \alpha_s(F)$.

Conversely assume that $\alpha_s(G) = \alpha_s(F)$, where $F$ is the corresponding maximum spanning tree of $G$. We want to prove $G$ is a partial tree. If possible, let $G$ be not a partial tree. Then there exists at least one $\beta$-strong edge in $G$. Let $uv$ be a $\beta$-strong edge in $G$. Then there exists at least one another $u - v$ path, say, $P$ in $G$ such that $w(xy) \geq w(uv)$ for every edge $xy$ in $P$. Now the union of $P$ and edge $uv$ is a cycle in $G$. Suppose that the number of $\alpha$-strong edges incident at $u$ in $G$ is $t$. Now to get $F$, delete the edge $uv$ from $G$, which has the least weight in $C$. Then the number of $\alpha$-strong edges incident at $u$ in $F$ is $t + 1$. Which is a contradiction to our assumption that $\alpha_s(G) = \alpha_s(F)$. Hence our assumption is wrong. Thus $G$ is a partial tree. ■

Let $G$ be a connected weighted graph which is a weighted tree but not a partial tree. Then, $G$ has no cycles and every edge is an $\alpha$-strong edge. Thus the sum of entries of $\alpha_s(G)$ is equal to twice the number of edges of $G$.

Next theorem is also a characterization of partial trees. This characterization is guaranteed by the uniqueness of the maximum spanning tree $F$ of $G$.

**Theorem 6.3.9** *A connected weighted graph $G = (V, E, w)$ is a partial tree if and only if $\alpha_s(F)$ is same for all maximum spanning trees of $G$.*

***Proof*** We know that a connected weighted graph is a partial tree if and only if it has a unique maximum spanning tree [2]. Moreover the spanning tree $F$ in the definition of partial trees, is a maximum spanning tree. So the proof is complete. ■

In the next result, we characterize partial trees by using $\alpha$-sequence and the number of disjoint cycles in it.

**Theorem 6.3.10** *Let $G = (V, E, w)$ be a connected weighted graph with exactly $k$-edge disjoint cycles. Then $G$ is a partial tree if and only if $\sum_{n_i \in \alpha_s(G)} n_i = 2(e - k)$, where $e$ is the total number of edges of $G$.*

***Proof*** Let $G = (V, E, w)$ be a connected weighted graph with exactly $k$-edge disjoint cycles. Suppose that $G$ is a partial tree. Then it has no $\beta$-strong edges. This means the only edges present in $G$ are $\alpha$-strong and $\delta$. Now given that $G$ has $k$-edge disjoint cycles. Consider an arbitrary cycle, say, $C$. Let the minimum of the weights of all edges in $C$ be $\theta$ and let it be assigned to the edge $uv$. Now all the edges of $C$ have weights strictly greater than $\theta$. For if, let there be another edge $ab$ in $C$ with weight $\theta$. Then $CONN_{G-(u,v)}(u, v) = \theta = w(uv)$, which implies that $uv$ is a $\beta$-strong edge. Hence our assumption is wrong. So the minimum weight in a cycle is assigned to exactly one edge in $C$. So the edge $uv$ is the only $\delta$-edge in $C$ and all other edges are $\alpha$-strong. This situation is same in all the $k$ cycles in $G$, which are disjoint. Thus the total number of $\delta$-edges of $G = k$, and hence total number of $\alpha$-strong edges of $G = e - k$. Hence $\sum_{n_i \in \alpha_s(G)} n_i = 2(e - k)$.

Conversely assume that $\sum_{n_i \in \alpha_s(G)} n_i = 2(e - k)$ where $e$ is the total number of edges of $G$. We have to prove that $G$ is a partial tree. It is enough, if we prove $G$ has no $\beta$-strong edges. Suppose the contrary. Let there be a $\beta$-strong edge, say, $e$ exists in $G$. Then clearly $e$ lies on a cycle, say, $C$. Then there will be at least one edge, say $e'$ in $C$, which is different from $e$ and with weight $w(e)$. That means $e'$ is also a $\beta$-strong edge. Thus there will be at most $e - k - 1$ number of $\alpha$-strong edges in $G$. It is a contradiction to our assumption $\sum_{n_i \in \alpha_s(G)} n_i = 2(e - k)$. So $G$ is a partial tree. ■

If $G = (V, E, w)$ is a simple connected weighted graph such that the number of edges is at most the number of vertices in $V$, then all cycles of $G$ will be edge disjoint. Thus we have the following results.

**Corollary 6.3.11** *Let $G = (V, E, w)$ be a simple connected weighted graph such that $|E| \leq |V|$ and has exactly k-cycles. Then G is a partial tree if and only if $\sum_{n_i \in \alpha_s(G)} n_i = 2(e - k)$, where e is the total number of edges of G.*

**Corollary 6.3.12** *A connected weighted graph $G = (V, E, w)$ is a weighted tree, not a partial tree if and only if $\sum_{n_i \in \alpha_s(G)} n_i = 2e$ where $e = |E|$.*

## 6.4 Vertex and Edge Saturation Counts

Now we discuss the concepts of vertex and edge saturation counts in weighted graphs and their properties.

**Definition 6.4.1** Let $G = (V, E, w)$ be a connected weighted graph without loops and multiple edges. Then the **strong vertex count** of $G$ is defined and denoted by $S_V(G) = \frac{number\ of\ strong\ edges\ of\ G}{number\ of\ verties\ of\ G} = \frac{number\ of\ \alpha\ or\ \beta-strong\ edges}{|V|}$, and the **strong edge count** of $G$ is defined and denoted by

$$S_E(G) = \frac{number\ of\ strong\ edges\ of\ G}{number\ of\ edges\ of\ G} = \frac{number\ of\ \alpha\ or\ \beta-\ strong\ edges}{|E|}$$

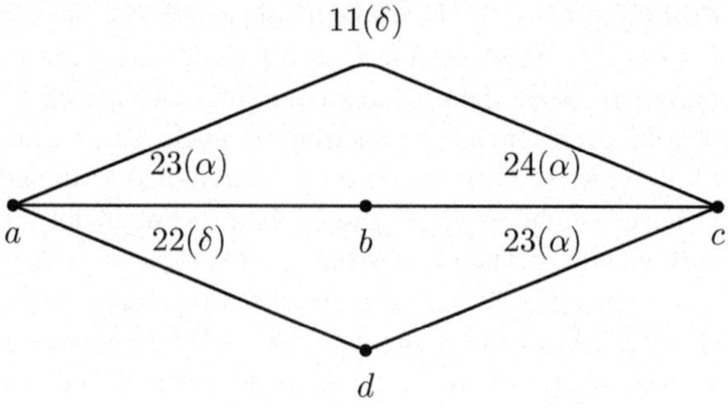


**Fig. 6.7** Strong saturation count—illustration

**Table 6.1** Vertex and edge counts in Example 6.4.3

| $\alpha_V(G)$ | $\beta_V(G)$ | $S_V(G)$ | $\alpha_E(G)$ | $\beta_E(G)$ | $S_E(G)$ |
|---|---|---|---|---|---|
| $\frac{3}{4}$ | $\frac{0}{4}$ | $\frac{3}{4}$ | $\frac{3}{5}$ | $\frac{0}{5}$ | $\frac{3}{5}$ |

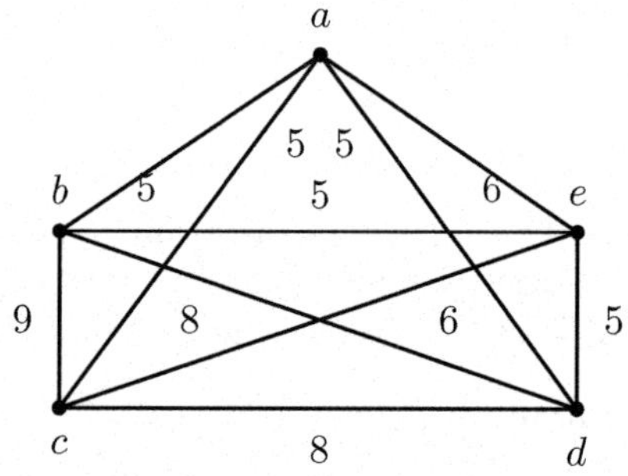


**Fig. 6.8** A complete weighted graph with non zero $\beta$ saturation count

**Remark 6.4.2** If the numerator in the Definition 6.4.1 is replaced with the number of $\alpha$-strong edges of $G$, then we get $\alpha$-**vertex count** $\alpha_V(G)$ and $\alpha$-**edge count** $\alpha_E(G)$ accordingly. Similar replacement with the number of $\beta$-strong edges leads to $\beta$-**vertex count** $\beta_V(G)$ and $\beta$-**edge count** $\beta_E(G)$ respectively.

**Example 6.4.3** Consider the graph in Fig. 6.7. Number of $\alpha$-strong edges and $\beta$-strong edges are 3 and 0 respectively. Corresponding vertex counts and edge counts are shown in Table 6.1.

**Example 6.4.4** Consider another example given in Fig. 6.8.

Corresponding saturation counts are calculated and shown in Table 6.2.

Since all edges of a weighted tree are $\alpha$-strong, $\alpha_V(G)= \frac{n-1}{n}$ and $\alpha_E(G)=\frac{n-1}{n-1}=1$. Note that the number of $\alpha$-strong edges never exceeds the number of vertices in any other basic weighted graph structure. If we assign equal weights to all the edges, each possible edge of a complete weighted graph will become $\beta$-strong so that $\beta_V(G)= \frac{nC_2}{n}$ and $\beta_E(G)= \frac{nC_2}{nC_2} =1$.

**Table 6.2** Vertex and edge counts in Example 6.4.4

| $\alpha_V(G)$ | $\beta_V(G)$ | $S_V(G)$ | $\alpha_E(G)$ | $\beta_E(G)$ | $S_E(G)$ |
|---|---|---|---|---|---|
| $\frac{4}{5}$ | $\frac{1}{5}$ | 1 | $\frac{4}{10}$ | $\frac{1}{10}$ | $\frac{1}{2}$ |

We have the following propositions based on the observations discussed above.

**Proposition 6.4.5** *Let $G = (V, E, w)$ be a connected weighted graph on n vertices. Then we have the following inequalities.*

(1) $0 \le \alpha_V(G) \le \frac{n-1}{n}$ (2) $0 \le \beta_V(G) \le \frac{nC_2}{n}$ (3) $0 \le S_V(G) \le \frac{nC_2}{n}$

(4) $0 \le \alpha_E(G) \le 1$ (5) $0 \le \beta_E(G) \le 1$ (6) $0 \le S_E(G) \le 1$.

Note that the strong saturation count is the sum of $\alpha$ and $\beta$ saturation counts in a weighted graph; the proof of third and sixth inequalities are immediate.

In the following propositions, we make a comparison between the edge count and vertex count in a connected weighted graph.

**Proposition 6.4.6** *Let $G = (V, E, w)$ be a connected weighted graph on n vertices. Then we have the following inequalities.*

1. $0 \le \beta_E(G) \le \beta_V(G)$.
2. $0 \le S_E(G) \le S_V(G)$.

Further more, if $G$ is a weighted graph which is not a weighted tree, the following relation holds;

3. $0 \le \alpha_E(G) \le \alpha_V(G)$.

**Proposition 6.4.7** *Let $G = (V, E, w)$ be a weighted tree, then $0 \le \alpha_V(G) < \alpha_E(G)$.*

Results on saturation counts of specific structures like weighted trees and partial trees are accomplished here. Some useful characterizations are also derived.

Theorem 6.4.8 is a characterization for the weighted tree structure in terms of $\alpha$-vertex count and $\alpha$-edge count.

**Theorem 6.4.8** *Let $G = (V, E, w)$ be a connected weighted graph on n vertices, where $n \ge 2$. Then the following statements are equivalent.*

1. $G$ is a weighted tree.
2. $\alpha_V(G) = \frac{n-1}{n}$ and $\alpha_E(G) = 1$.
3. $n\, \alpha_V(G) = (n-1)\alpha_E(G)$.

***Proof*** (1) $\Rightarrow$ (2)

Suppose that $G$ is a weighted tree on $n$ vertices, where $n \ge 2$. Then $G$ is connected, acyclic and has exactly $(n-1)$ edges. Note that any edge $e = uv$ in $G$ is the unique path between $u$ and $v$ [2]. Hence $e$ is $\alpha$-strong [3]. Thus all the $(n-1)$ edges of $G$ are $\alpha$-strong. Hence $\alpha_V(G) = \frac{n-1}{n}$ and $\alpha_E(G) = 1$.

(2) $\Rightarrow$ (3)

Suppose that $\alpha_V(G) = \frac{n-1}{n}$ and $\alpha_E(G) = 1$. Then $n\ \alpha_V(G) = (n-1)$. That is, $n\ \alpha_V(G) = (n-1) * 1$. That is $n\ \alpha_V(G) = (n-1)\alpha_E(G)$ (Because $\alpha_E(G) = 1$).

(3) $\Rightarrow$ (1)

Suppose that $n\ \alpha_V(G) = (n-1)\ \alpha_E(G)$. That is $\alpha_V(G) = \frac{n-1}{n}\ \alpha_E(G)$.

We have to prove that $G$ is a weighted tree. The proof will complete, if we show that $G$ has $(n-1)$ edges.

We have $\frac{\alpha_V(G)}{\alpha_E(G)} = \frac{(n-1)}{n}$. This implies,

$$\frac{number\ of\ \alpha-\ strong\ edges}{\mid V \mid} / \frac{number\ of\ \alpha-\ strong\ edges}{\mid E \mid} = \frac{(n-1)}{n}.$$

Hence, $\frac{|E|}{|V|} = \frac{n-1}{n}$. Thus, Number of edges = $\mid E \mid = (n-1)$. ■

Following theorem is a characterization for partial trees.

**Theorem 6.4.9** *Let $G = (V, E, w)$ be a connected weighted graph. Then $G$ is a partial tree if and only if $\alpha_V(G) = S_V(G)$ and $\alpha_E(G) = S_E(G)$.*

***Proof*** Let $G = (V, E, w)$ be a connected weighted graph. Suppose that $G$ is a partial tree. If $G$ is a weighted tree, then all the edges of $G$ are partial bridges [3]. Now an edge $e = uv$ in $G$ is a partial bridge if and only if it is $\alpha$-strong [3]. Thus all the edges of $G$ are $\alpha$-strong. Therefore $G$ has no $\beta$-strong edges and hence $\beta_V(G) = \frac{0}{|V|} = 0$ and $\beta_E(G) = \frac{0}{|E|} = 0$. Consequently $\alpha_V(G) = S_V(G)$ and $\alpha_E(G) = S_E(G)$.

If $G$ is not a weighted tree, then $G$ has a weighted cycle, say, $C$. Since $G$ is a partial tree, there exists an edge $e = uv$ such that $w(e) < CONN_{G-e}(u, v)$, where $G - e$ is the subgraph of $G$ obtained by deleting the edge $e$ from $G$ [3]. This means $e$ is a $\delta$-edge. If $G - e$ is a weighted spanning tree of $G$, all the edges in $G - e$ are $\alpha$-strong. Hence $\beta_V(G) = \frac{0}{|V|} = 0$ and $\beta_E(G) = \frac{0}{|E|} = 0$. Consequently $\alpha_V(G) = S_V(G)$ and $\alpha_E(G) = S_E(G)$. If $G - e$ is not a weighted spanning tree of $G$, then continue the above procedure of deleting $\delta$-edges from $G - e$ until we get weighted spanning tree.

Conversely suppose that $\alpha_V(G) = S_V(G)$ and $\alpha_E(G) = S_E(G)$. We have to prove that $G$ is a partial tree. If $G$ has no weighted cycles, then $G$ is a weighted tree and hence is a partial tree. Suppose that $G$ has a cycle, say, $C$. Then $C$ contains only $\alpha$-strong and $\delta$-edges. Also note that all the edges of $C$ cannot be $\alpha$-strong, since otherwise it will contradict the definition of $\alpha$-strong edges. Thus there exists at least one $\delta$-edge in $C$. If we delete $e$ from $C$, we get a maximum spanning tree of $G$. If not, remove one $\delta$-edge from the existing weighted cycles in $G$. Continue this procedure until we get a maximum spanning tree of $G$. Hence $G$ is a partial tree. ■

Theorem 6.4.10 is also a characterization of partial trees which connects the $\alpha$-vertex counts of a partial tree and the corresponding maximum spanning tree.

**Theorem 6.4.10** *A connected weighted graph $G = (V, E, w)$ is a partial tree if and only if $\alpha_V(G) = \alpha_V(F)$, where $F$ is the corresponding maximum spanning tree of $G$.*

***Proof*** Suppose that $G = (V, E, w)$ is a partial tree. If $G$ itself is a weighted tree, then $F$ and $G$ are isomorphic and $\alpha_V(G) = \frac{number\ of\ \alpha-strong\ edges\ in\ G}{|V|} = \frac{number\ of\ \alpha-strong\ edges\ in\ F}{|V|} = \alpha_V(F)$. If $G$ is not a weighted tree, then $G$ contains a cycle say $C$. Since $G$ is a partial tree, there exists an edge $e = uv$ in $C$ such that $w(e) < CONN_{G-e}(u, v)$, where $G - e$ is the weighted subgraph obtained by deleting the edge $e$ from $G$. If $G - e$ is a weighted tree, then $G - e$ and $F$ are isomorphic and since $e$ is a $\delta$-edge, we get $\alpha_V(G) = \alpha_V(F)$. If not, continue the above procedure of deleting $\delta$-edges from cycles in $G - e$ until we get a maximum spanning tree $F$ of $G$ such that $\alpha_V(G) = \alpha_V(F)$.

Conversely assume that $\alpha_V(G) = \alpha_V(F)$, where $F$ is the corresponding maximum spanning tree of $G$. We want to prove $G$ is a partial tree. If possible, let $G$ be not a partial tree. Then there exists at least one $\beta$-strong edge in $G$. Let $uv$ be a $\beta$-strong edge in $G$. Then there exists at least one another $u - v$ path, say, $P$ in $G$ such that $w(xy) \geq w(uv)$ for every edge $xy$ in $P$. Now the union of $P$ and edge $uv$ is a cycle in $G$. Suppose that the number of $\alpha$-strong edges incident at $u$ in $G$ is $t$. Now to get $F$, delete the edge $uv$ from $G$, which has the least weight in $C$. Then the number of $\alpha$-strong edges incident at $u$ in $F$ is $t + 1$. Let $t'$ be the remaining number of $\alpha$-strong edges of $G$. Hence $\alpha_V(G) = \frac{t+t'}{|V|}$ and $\alpha_V(F) = \frac{t+1+t'}{|V|}$. Which is a contradiction to our assumption that $\alpha_V(G) = \alpha_V(F)$. Hence our assumption is wrong. Thus $G$ is a partial tree. ■

We have another characterization for partial trees.

**Theorem 6.4.11** *Let $G = (V, E, w)$ be a connected weighted graph with $n$ vertices $(n > 2)$ and $q$ edges. Then the following statements are equivalent.*

*(1)* $G$ is a partial tree.
*(2)* $n\,\alpha_V(G) = q\,\alpha_E(G)$.

***Proof*** (1) ⇒ (2)

Suppose $G$ is a partial tree. Then $\alpha_V(G) = \frac{number\ of\ \alpha-strong\ edges}{|V|} = \frac{n-1}{n}$ and $\alpha_E(G) = \frac{number\ of\ \alpha-strong\ edges}{|E|} = \frac{n-1}{q}$.

(2) ⇒ (1)

Suppose that $n\,\alpha_V(G) = q\,\alpha_E(G)$. We have to prove that $G$ is a partial tree. If possible suppose the contrary. Let $G$ be not a partial tree. We know that a weighted graph $G$ is a partial tree if and only if it has no $\beta$-strong edges [2, 3]. Now since $G$ is not a partial tree, there exists at least one $\beta$-strong edge in $G$. Let $uv$ be a $\beta$-strong edge in $G$. Then there exists at least one another $u - v$ path, say, $P$ in $G$ such that $w(xy) \geq w(uv)$ for every edge $xy$ in $P$. Now the union of $P$ and edge $uv$ is a cycle in $G$. Suppose that the number of $\alpha$-strong edges incident at $u$ in $G$ is $t$. Now to get $F$ (the MST in the definition of partial trees), delete the edge $uv$ from $G$, which has the least weight in $C$. Then the number of $\alpha$-strong edges incident at $u$ in $F$ is $t + 1$. Let $t'$ be the remaining number of $\alpha$-strong edges of $G$. Hence $\alpha_V(G) = \frac{t+t'}{|V|}$ and $\alpha_V(F) = \frac{t+1+t'}{|V|}$. Also $\alpha_E(G) = \frac{t+t'}{q}$ and $\alpha_E(F) = \frac{t+t'+1}{n-1}$.

Now the assumption, $n\ \alpha_V(G) = q\ \alpha_E(G)$ becomes $n(\alpha_E(F) - \frac{1}{n}) = (n-1)$ $(\alpha_E(F) - 1)$, where $\alpha_V(F) = \frac{(n-1)}{n}$ and $\alpha_E(F) = 1$. Therefore from the assumption, we get $n = 2$, a contradiction to the given condition. So our assumption is wrong. Hence $G$ is a partial tree. ■

## 6.5 Saturation

Saturation plays a significant role in the connectivity and centrality of graphs. This section introduces certain types of saturation. A discussion on the peculiarity of counts of such saturated structures is included. Characterization of different weighted graph structures based on these counts are also accomplished. Certain bounds for these saturation counts are obtained. A discussion on the saturated partial blocks and weighted cycles is also made.

**Definition 6.5.1** Let $G = (V, E, w)$ be a connected weighted graph. Then $G$ is called $\alpha$-**saturated** , if at least one $\alpha$-strong edge is incident at every vertex. $G$ is called $\beta$-**saturated** , if at least one $\beta$-strong edge is incident on every vertex.

**Remark 6.5.2** In a weighted graph, each vertex is incident with at least a single strong edge. This is because of the existence of strong path between any pair of vertices [3]. We can have the following definition based on this discussion.

**Definition 6.5.3** Let $G = (V, E, w)$ be a weighted graph. Then $G$ is called **saturated**, if it is both $\alpha$-saturated and $\beta$-saturated. That is at least one $\alpha$-strong edge and one $\beta$-strong edge is incident at every vertex of $G$. Also a graph which is not saturated is called **unsaturated**.

**Example 6.5.4** Consider the weighted graph in Fig. 6.9. Here, each vertex is both $\alpha$-saturated and $\beta$-saturated. Hence this weighted graph is saturated. While in the

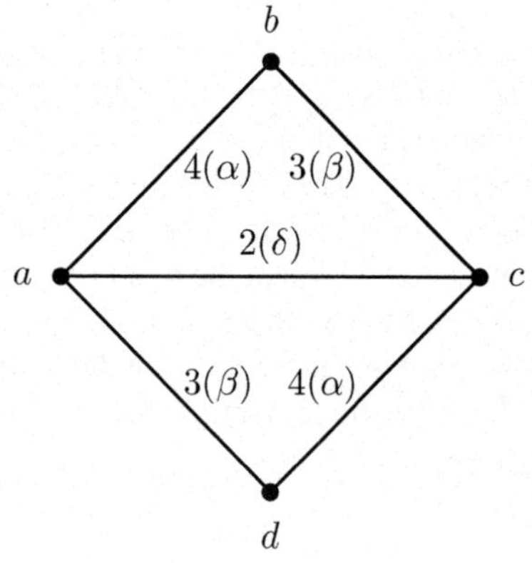


**Fig. 6.9** A saturated weighted graph—Illustration

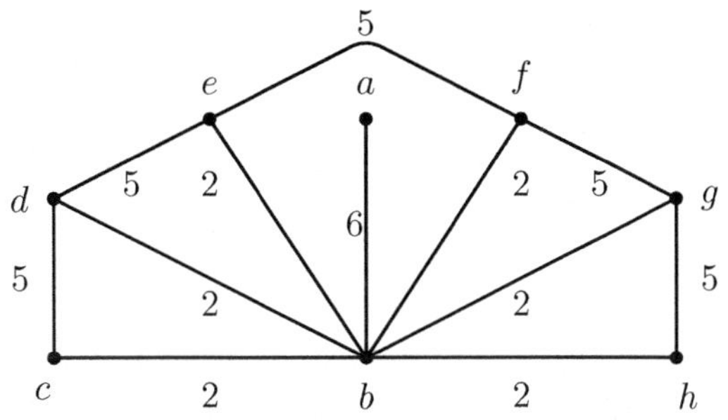


**Fig. 6.10** An unsaturated weighted graph—illustration

following example, all vertices other than $'a'$ are incident with both $\alpha$ and $\beta$-strong edges. Thus the weighted graph in Figure. 6.10 is $\alpha$-saturated, but not saturated.

**Example 6.5.5** Example for an unsaturated weighted graph is given in Fig. 6.10. Let $G$ be a weighted graph on $p$ vertices. Consider a finite sequence $\alpha_s(G) = (m_1, m_2, m_3, \ldots, m_p)$. If $m_i$ is the number of $\alpha$-strong edges incident at $v_i$ and is zero if there is no $\alpha$-strong edge incident at $v_i$. Recall that this kind of sequence is called $\alpha$-sequence of $G$. Similar definitions for $\beta$ and strong sequences hold. Note that $\sum_{m_i \in \alpha_S(G)} m_i + \sum_{m_i \in \beta_S(G)} m_i = \sum_{m_i \in S_S(G)} m_i$.

Next theorem reveals the relationship between the nature of sequences of weighted graphs and their saturation.

**Theorem 6.5.6** *Let $G = (V, E, w)$ be a weighted graph with p vertices. Then G is,*

1. *$\alpha$-saturated if and only if the least entry in $\alpha_S(G)$ is unity. Or* $\sum_{n_i \in \alpha_S(G)} n_i \geq p$.
2. $\beta$-saturated if and only if the least entry in $\beta_S(G)$ is unity. Or $\sum_{n_i \in \beta_S(G)} n_i \geq p$.
3. Saturated, only if the least entry in $S_S(G)$ is two. Or if $\sum_{n_i \in S_S(G)} n_i \geq 2p$.

***Proof*** Let $G = (V, E, w)$ be a weighted graph with order $p$. Let $\alpha_S(G)$, $\beta_S(G)$ and $S_S(G)$ be the $\alpha$, $\beta$ and strong sequences of $G$ respectively. Suppose that $G$ is $\alpha$-saturated. Then all the vertices are incident with at least one $\alpha$-strong edge. This means unity is the least entry in $\alpha_S(G)$. Thus $\sum_{n_i \in \alpha_S(G)} n_i \geq \underbrace{1 + 1 + 1 + \cdots + 1}_{p\ times}$ $= p$.

Conversely assume that, the least entry in $\alpha_S(G)$ is unity. This means all the $p$ vertices of $G$ are incident with at least one $\alpha$-strong edge, and hence $G$ is $\alpha$-saturated.

In a similar manner, we can prove the result for $\beta$-saturated weighted graphs.

Now we have to prove the result for saturated weighted graphs. Suppose that $G$ is saturated. That is $G$ is both $\alpha$-saturated and $\beta$-saturated. That is each vertex of $G$ is incident with at least one $\alpha$-strong edge and with at least one $\beta$-strong edge. Thus the least entry in $S_S(G)$ is at least two. ■

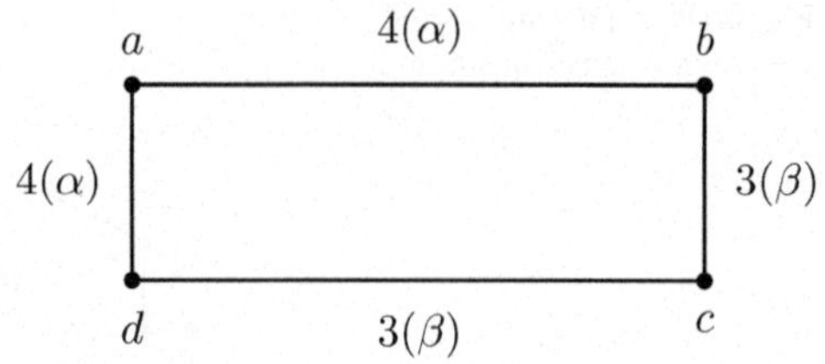


**Fig. 6.11** An unsaturated graph

Note that, the condition on saturated graphs in Theorem 6.5.6 is necessary. It is possible to have a weighted graph in which the least entry in $S_S(G)$ is at least two, but the graph may not be saturated. This is clear from Example 6.5.7.

**Example 6.5.7** For the weighted graph in Fig. 6.11, the number of $\alpha$-strong edges incident with vertex $'a'$ is 2 and no $\beta$-strong edges are incident with the same vertex. Corresponding strong sequence of $G$ is $S_S(G) = (2, 2, 2, 2)$, but it is not saturated.

Lower bounds of various saturation counts in weighted graphs are described in Theorem 6.5.8.

**Theorem 6.5.8** Let $G = (V, E, w)$ be a weighted graph with $n$ vertices. Then

*1.* If $G$ is $\alpha$-saturated then, $\alpha_V(G) \geq 0.5$.
*2.* If $G$ is $\beta$-saturated then, $\beta_V(G) \geq 0.5$.
*3.* If $G$ is saturated then, $S_V(G) \geq 1$.

***Proof*** Suppose that $G$ is $\alpha$-saturated. Then each vertex is incident with at least one $\alpha$-strong edge. This means $G$ contains at least $\frac{n}{2}$ $\alpha$ strong edges. Hence $\alpha_V(G) \geq \frac{n/2}{n} = 0.5$. In the same manner, we can prove the result for $\beta$-saturated graphs.

Now we have to prove the result for saturated graphs. Suppose that $G$ is a saturated weighted graph. Then $G$ is both $\alpha$-saturated and $\beta$-saturated. That is each vertex of $G$ is incident with at least one $\alpha$-strong edge and with at least one $\beta$-strong edge. Thus the number of strong edges of $G$ is equal to the sum of number of $\alpha$-strong edges of $G$ and the number of $\beta$ strong edges of $G$ which is greater than or equal to $\frac{n}{2} + \frac{n}{2} = n$. Hence $S_V(G) \geq \frac{n}{n} = 1$. ■

A weighted cycle has the structure of a cycle and such that its edges are assigned with some weights. A characterization of saturated weighted cycles is presented in the following theorem .

**Theorem 6.5.9** Let $C_n$ be a weighted cycle, which is not a partial tree. Then it is saturated if and only if the following two conditions are satisfied.

*1.* $n = 2k$, where $k$ is an integer.
2. $\alpha$-strong and $\beta$-strong edges appears alternatively on $C_n$.

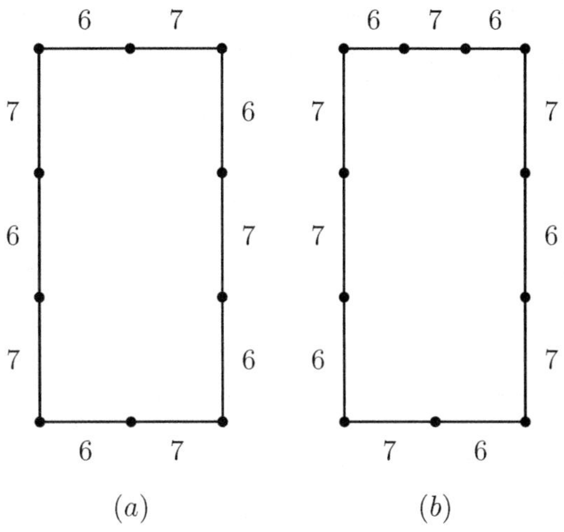


**Fig. 6.12** (*a*) A cycle on 10 vertices (*b*) A cycle on 11 vertices

***Proof*** Let $C_n$ be a weighted cycle and not a partial tree. Then it has no $\delta$-edges. That is all the edges appear in $C_n$ are either $\alpha$-strong or $\beta$ strong. Suppose that $C_n$ is saturated. Then it is both $\alpha$-saturated and $\beta$-saturated. That is, each of its vertices are incident with at least one $\alpha$-strong edge and with at least one $\beta$-strong edge. Which implies that the number of $\alpha$-strong edges $=$ number of $\beta$-strong edges $= k$, where $k$ is a positive integer and $k + k = n$. Thus $n = 2k$. Also each vertex is incident with both $\alpha$-strong and $\beta$-strong edges happens only when they appear alternatively on $C_n$.

Now conversely suppose that $C_n$ is an even cycle and $\alpha$-strong and $\beta$-strong edges appear alternatively in $C_n$. This means all the vertices are incident with exactly one $\alpha$-strong and exactly with one $\beta$-strong edges. Thus $C_n$ is both $\alpha$-saturated and $\beta$-saturated. Hence $C_n$ is saturated. ■

In the following example, there are two cycles in which those edges with weight 7 are $\alpha$-strong and that with weights 6 are $\beta$-strong.

**Example 6.5.10** In Fig. 6.12a, all the vertices are incident with both $\alpha$-strong and $\beta$-strong edges. Thus the weighted cycle is a saturated cycle. Also, note that the number of vertices is even in Fig. 6.12a. While in Fig. 6.12b, there are odd number of vertices. Moreover, it has a vertex which is incident with two $\alpha$-strong edges but no $\beta$-strong edge. Thus the weighted cycle is $\alpha$-saturated, but not $\beta$-saturated and hence it is not saturated.

Following theorem presents another characterization for weighted saturated cycles.

**Theorem 6.5.11** Let $G = (V, E, w)$ be weighted cycle. Then the following are equivalent.

*1.* $G$ is either saturated or $\beta$-saturated.
*2.* $G$ is a partial block.
*3.* $G$ is a strongest strong cycle (SSC).
*4.* $G$ is a locamin cycle.

***Proof*** In [4], the equivalence of conditions 2, 3 and 4 are proved. Now we stop the proof by proving condition 1 is implies and implied by condition 2.

$1 \Rightarrow 2$.

Suppose that the weighted cycle $G$ saturated. Then it is both $\alpha$-saturated and $\beta$-saturated. That is each vertex of $G$ is incident with at least one $\alpha$-strong edge and with at least one $\beta$-strong edge. Since, $G$ is a cycle, Each vertex is incident with exactly two edges. Thus, exactly one $\alpha$-strong edge and exactly one $\beta$-strong edge is incident on every vertex. Hence removal of any vertex from $G$ will not reduce the strength of connectedness between any other vertices. This implies that no vertex of $G$ is a partial cut vertex, and hence, $G$ is a partial block.

Also suppose that the weighted cycle $G$ is $\beta$-saturated. That is each vertex of $G$ is incident with at least one $\beta$-strong edge. We have to prove that $G$ is a partial block. We claim that $G$ has no partial cut vertices. Suppose the contrary. Let $x$ be a partial cut vertex of $G$. Then there exists two vertices $u$ and $v$ such that $u \neq x \neq v$ such that $u - x - v$ path has more strength than $x - v$ path. This means all the edges in the $u - x - v$ path have weights more than the strength of the $x - y$ path. This means, all the edges in the $u - x - y$ path are $\alpha$-strong. Thus, the two edges, which are incident on $x$ will be $\alpha$-strong. It is a contradiction to our assumption that, $G$ is $\beta$-saturated. Hence our assumption is wrong. Hence our claim is true. Thus $G$ has no partial cut vertices and hence $G$ is a partial block.

$2 \Rightarrow 1$.

Suppose that the weighted cycle $G$ is a partial block. We claim that $G$ has no $\delta$-edges. Suppose the contrary. Let $e = uv$ be a $\delta$-edge in $G$. Then all the other edges in $G$ will be $\alpha$-strong and thereby $G$ will have exactly $n - 2$ partial cut vertices, which is a contradiction to our assumption that, $G$ is a partial block. So our assumption is wrong and the claim is true. Thus only $\alpha$-strong and $\beta$-strong edges can be appeared in $G$.

If $G$ contains both $\alpha$-strong and $\beta$-strong edges, then they must be appeared alternatively. Otherwise the partial block structure will be lost. If the number of $\alpha$-strong edges = number of $\beta$-strong edges $= \frac{n}{2}$, then $G$ is both $\alpha$-saturated and $\beta$-saturated, and hence, saturated. If number of $\alpha$-strong edges $<$ number of $\beta$-strong edges, then $G$ will be $\beta$-saturated only. The case where, number of $\alpha$-strong edges $>$ number of $\beta$-strong edges, will not happen, as it looses the partial block structure. If all the edges in $G$ are $\beta$-strong, then it is $\beta$-saturated.

Thus in all the cases, $G$ is either saturated or $\beta$-saturated. ■

**Remark 6.5.12** If $G = (V, E, w)$ is a saturated cycle or a $\beta$-saturated cycle, then $\alpha_V(G) = 0.5$ and $\beta_V(G) = 0.5$ or $.5 \leq \beta_V(G) \leq 1$. Following result is based on these values.

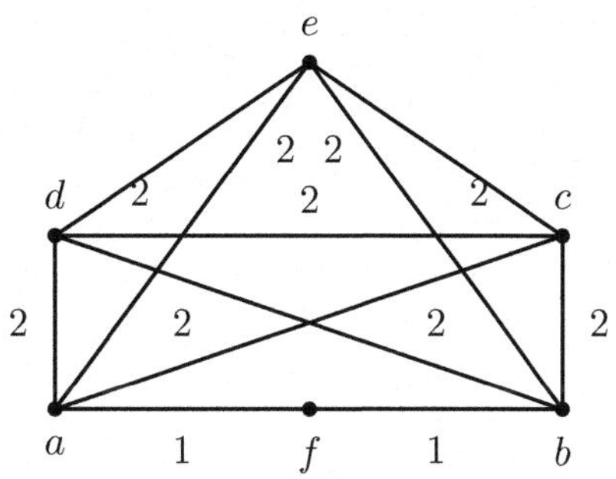


**Fig. 6.13** An unsaturated partial block-illustration

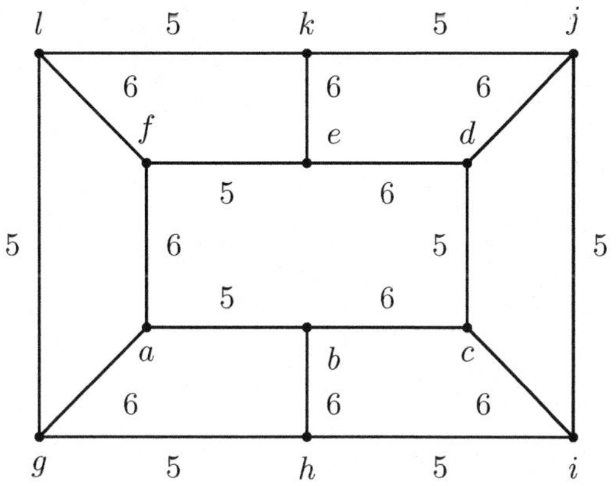


**Fig. 6.14** A saturated graph which is not a partial block

**Theorem 6.5.13** Let $G = (V, E, w)$ be weighted cycle. Then the following are equivalent.

1. $\alpha_V(G) = 0.5$ and $\beta_V(G) = 0.5$ or $.5 \leq \beta_V(G) \leq 1$.
2. $G$ is either saturated or $\beta$-saturated.
3. $G$ is a partial block.
4. $G$ is a strongest strong cycle (SSC).
5. $G$ is a locamin cycle.

It is not always true that all partial blocks are saturated. Also all saturated graphs are not partial blocks. For, consider Example 6.5.14.

**Example 6.5.14** For the weighted graph in Fig. 6.13, vertex $f$ is incident with $\delta$ edges of unit weight. Those edges with weight two are $\beta$-strong. Thus the graph is unsaturated. Moreover it is a partial block.

Consider the weighted graph shown in Fig. 6.14, edges with weights 5 and 6 are $\beta$-strong and $\alpha$-strong respectively. This graph is both $\alpha$-saturated and $\beta$-saturated, and hence saturated. Note that it is not a partial block.

It is notable that $\alpha$-vertex count of a partial block has an upper bound as proved in the following theorem.

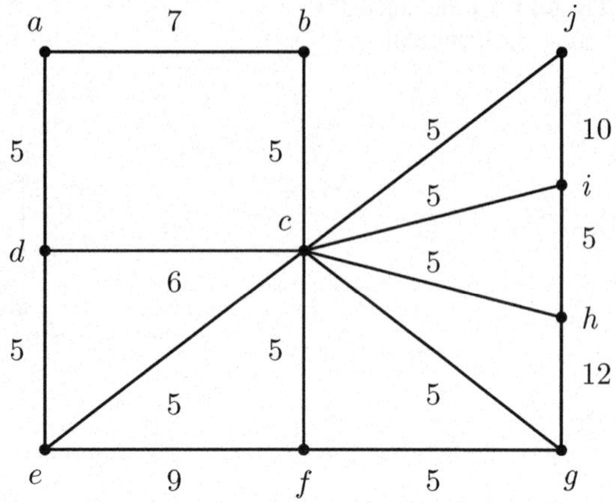


**Fig. 6.15** A partial block

**Theorem 6.5.15** Let $G = (V, E, w)$ be a partial block. Then $\alpha_V(G) \le \frac{1}{2}$.

*Proof* Let $G = (V, E, w)$ be a partial block on $n$ vertices. We know that if a vertex $w$ is common vertex for more than one $\alpha$-strong edges, then $w$ is a partial cut vertex, and hence $G$ cannot be a partial block. Now $\alpha_V(G) = \frac{\text{number of } \alpha-\text{strong edges in } G}{|V|} \le \frac{n/2}{n} = \frac{1}{2}$. ■

As a consequence of Theorem 6.5.15, we can have the following two corollaries.

**Corollary 6.5.16** A partial block is $\alpha$-saturated if and only if $\alpha_V(G) = 0.5$

*Proof* Let $G = (V, E, w)$ be a partial block. Let $G$ be $\alpha$-saturated. Then each vertex of $G$ is incident with at least one $\alpha$-strong edge. Also, as partial blocks are free from partial cut vertices, exactly one $\alpha$-strong edge is incident with every vertex. Therefore $\alpha_V(G) = \frac{n/2}{n} = 0.5$ ■

**Corollary 6.5.17** There exists no $\alpha$-saturated partial blocks of odd order.

*Proof* Let $G = (V, E, w)$ be an $\alpha$-saturated partial block with odd number of vertices. Then $\alpha_V(G) \neq 0.5$. Then by the Corollary 6.5.16, we see that $G$ is not a partial block. This completes the proof. ■

Theorems 6.5.11 and 6.5.13 are characterization of partial blocks whose underlying structure is a cycle. Following theorem says that a weighted graph satisfying certain conditions with specific saturation counts form a partial block.

**Theorem 6.5.18** Let $G = (V, E, w)$ be a saturated, connected, weighted graph on $n$ vertices such that

1. $\alpha_V(G) = 0.5$
2. $S_E(G) = 1$
3. All the $\beta$-strong edges are of equal weights.

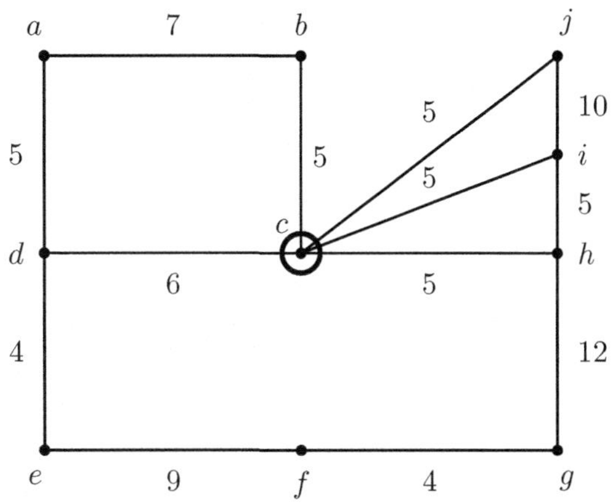


**Fig. 6.16** A non p-block ($\beta$-edges with different weights)

4. $G^*$ is a block
   then $G$ is a partial block.

***Proof*** Suppose that $G$ is a saturated, connected, weighted graph on $n$ vertices with the given four conditions. Since $G$ is saturated, it is both $\alpha$-saturated and $\beta$-saturated. Now from the first condition, $\alpha_V(G) = 0.5$, we have, number of $\alpha$-strong edges is equal to $n/2$. Thus each vertex of $G$ is incident with exactly one $\alpha$-strong edge. Now the second condition $S_E(G) = 1$, means that all the edges of $G$ are strong. This means $G$ has no $\delta$-edges. Also from condition 3, we get, all the $\beta$-strong edges have same weight and condition 4 guarantees that $G$ has no cut vertices.

We have to prove that, $G$ is a partial block. That is, we need to prove $G$ has no partial cut vertices. If possible, suppose the contrary. Let $u$ be a partial cut vertex of $G$. Then there exists two vertices $x$ and $y$ such that $x \neq u \neq y$ and $CONN_G(x, y) > CONN_{G-u}(x, y)$. This means all $x - u - y$ paths have strengths greater than all the $x - y$ paths which are not passing through $u$. But each vertex of $G$ is incident with at least one $\beta$-strong edge, and all the $\beta$-strong edges have equal weight. Now the above inequality will be true only when, all the $\beta$-strong edges in all $x - u - y$ paths have higher weights than the other $\beta$-strong edges. This is a contradiction to the condition 4. Hence our assumption is wrong. So $G$ is a partial block. ■

Theorem 6.5.18 is illustrated in the following example (Fig. 6.15). All the edges whose weights are not equal to 5 are $\alpha$-strong and that of weight 5 are $\beta$-strong.

**Example 6.5.19** Consider the weighted graphs in Figs. 6.15, 6.16 and 6.17.

we can not relax the third condition of Theorem 6.5.18. As an example, consider Fig. 6.16. In which all the edges with weights 4 and 5 are $\beta$-strong. But this weighted graph is not a partial block since the encircled vertex is a partial cut vertex.

Similarly we can not relax the second condition. For, consider Fig. 6.17. Here also the encircled vertex is a partial cut vertex. Thus the graph is not a partial block.

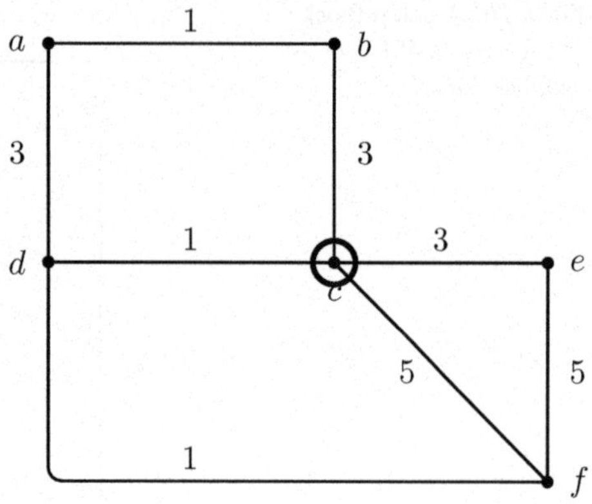


**Fig. 6.17** A non p-block with $\delta$-edges

Now we compute the vertex and edge saturation counts for a precisely weighted graph.

Recall that a PWG has at most one $\alpha$-strong edge and it will not contain any $\delta$ edges. Also, all PWG's are partial blocks [3]. Based on all these facts, we have the following theorem.

**Theorem 6.5.20** Let G be a PWG on $n$ vertices. Then we have the following inequalities.

*1.* $0 \leq \alpha_V(G) \leq \frac{1}{n}$.
2. $\frac{n^2-n-2}{2n} \leq \beta_V(G) \leq \frac{n-1}{2}$.

***Proof*** Let $G$ be a PWG on $n$ vertices. Then $G$ has at most one $\alpha$-strong edge. Hence $0 \leq \alpha_V(G) \leq \frac{1}{n}$ is clear. Since $G$ has no $\delta$-edges, the minimum number of $\beta$-strong edges in $G$ is $nC_2 - 1 = \frac{n(n-1)}{2} - 1$. Thus $\beta_V(G) \geq \frac{\frac{n(n-1)}{2}-1}{n} = \frac{n^2-n-2}{2n}$. Hence $\frac{n^2-n-2}{2n} \leq \beta_V(G) \leq \frac{n-1}{2}$ is very clear. ∎

From the definition of PWG itself, it is clear that all PWG's are unsaturated.

## References

1. Mathew, J.K., Mathew, S.: Sequences in a weighted graph and characterization of partial trees. Int. J. Sci. Eng. Res. **5**(9), 20–22 (2014)
2. Mathew, S., Sunitha, M.S.: Cycle connectivity in weighted graphs. Proyecciones J. Math. **30**(1), 1–17 (2001)
3. Mathew, S.: On totally weighted interconnection networks. J. Interconnect. Netw. **14**, 1350004 (1–16) (2013)
4. Mathew, S., Sunitha, M.S.: Partial trees in weighted graphs -1. Proyecciones J. Math **30**(2), 163–174 (2011)

# Chapter 7
# Intervals and Gates

## 7.1 Distances in Weighted Graphs

Consider a connected weighted graph $G = (V, E, w)$. Let $u \in V$ and $d$ be any metric. Let us use $e_d(u)$ to denote the eccentricity of $u$ which is defined by $e_d(u) = max_v d(u, v)$. If $v \in V$ is such that $e_d(v) = \wedge\{e_d(u) \mid u \in V\} = r$, then $v$ is called the central vertex of $G$. Here $\wedge$ denotes the minimum. Also, $r$ is the radius of $G$ with respect to the metric $d$. The center of $G$ denoted by $< C_d(G) >$ is the subgraph of $G$ induced by the central vertices of $G$ with respect to $d$. An eccentric vertex of $u$ with respect to $d$ is that vertex $v$ such that $e(u) = d(u, v)$.

**Definition 7.1.1** Let $G = (V, E, w)$ be a weighted graph. The $w$**-distance** between two distinct vertices $u$ and $v$ in $G$, denoted by $d_w(u, v)$, is defined as the smallest $w$**-length** of any $u - v$ path, where $w$-length of a path $P = u_0, u_1, u_2, \ldots, u_n$ is defined as $l_w(P) = \sum_{i=1}^{n} \frac{1}{w(u_{i-1}, u_i)}$. Also $d_w(u, u) = 0$ for every vertex $u$ in $G$.

Theorem 7.1.2 proves that $d_w$ is a metric.

**Theorem 7.1.2** *Let $G = (V, E, w)$ be a weighted graph and $d_w$ is the $w$-distance. Then, $d_w$ is a metric on V.*

***Proof*** Let $P = u_0, u_1, u_2, \ldots, u_n$ be any path in $G$. Then $w(u_{i-1}u_i) > 0$ for $i = 1, 2, 3, \ldots, n$. So $w$-length of $P > 0$. Therefore $w$-distance, $d_w(u, v) \geq 0$ for every pair of vertices $u$ and $v$ in $G$. Also, from the definition of $d_w$, we get $d_w(u, v) = 0$ if and only if $u = v$. The reversal of a path from $u$ to $v$ is a path from $v$ to $u$ and vice versa. So $d_w(u, v) = d_w(v, u)$ for all $u, v \in V$.

Suppose $d_w(u, v) > d_w(u, w) + d_w(w, v)$. Then there exist a path $P$ from $u$ to $w$ and a path $Q$ from $w$ to $v$ such that the $u - v$ path contained in $P \cup Q$ has $w$-length strictly less than the minimum $w$-length of all $u - v$ paths, which is a contradiction. Therefore $d_w(u, v) \leq d_w(u, w) + d_w(w, v)$, for all $u, v, w \in V$. Hence $d_w$ is a metric. ∎



S. Mathew et al., *Weighted and Fuzzy Graph Theory*, Studies in Fuzziness and Soft Computing 429, https://doi.org/10.1007/978-3-031-39756-1_7

Followed by the introduction of $w$-distance in weighted graphs, here we introduce a new kind of distance called strong geodesic distance.

**Definition 7.1.3** The **strong geodesic distance** or **$sg$-distance** between two vertices $u$ and $v$ in a weighted graph $G = (V, E, w)$ denoted by $d_{sg}(u, v)$, is defined as the length of the shortest $u - v$ strong path. If $u$ and $v$ are not connected by a path, then $d_{sg}(u, v) = \infty$.

In the following theorem, we prove that $d_{sg}$ is a metric.

**Theorem 7.1.4** *Let $G = (V, E, w)$ be a weighted graph and $d_{sg}$ is the strong geodesic distance. Then, $d_{sg}$ is a metric on $V$.*

***Proof*** Clearly $d_{sg}(u, v) \geq 0$, for all $u, v \in V$. Also $d_{sg}(u, v) = 0$ if and only if $u = v$. Since, the reversal of a path from $u$ to $v$ is a path from $v$ to $u$ and vice versa. So $d_{sg}(u, v) = d_{sg}(v, u)$ for all $u, v \in V$.

Suppose $d_{sg}(u, v) > d_{sg}(u, w) + d_{sg}(w, v)$. Let $P$ be shortest $u - w$ strong path, $Q$ be a shortest $w - v$ strong path and $R$ be a shortest $u - v$ strong path. Then above inequality implies length of $R >$ length of $P$ + length of $Q$ = length of $P \cup Q$. Since $P \cup Q$ contains a $u - v$ path, it leads to a contradiction. So $d_{sg}(u, v) \leq d_{sg}(u, w) + d_{sg}(w, v)$ for all $u, v, w \in V$. Hence $d_{sg}(u, v)$ is a metric on $V$. ■

We will introduce two more distances in weighted graphs namely strongest strong distance and $\delta$ distance. Each of them is a metric in $G = (V, E, w)$.

**Definition 7.1.5** Let $G = (V, E, w)$ be a weighted graph. The **strongest strong distance** between two vertices $u$ and $v$ in $G$, denoted by $d_{ss}(u, v)$ is defined as $d_{ss}(u, v) = \frac{1}{CONN_G(u,v)}$ and $d_{ss}(u, u) = 0$, for all $u \in V$. If $G$ is disconnected and two vertices (say) $u$ and $v$ of $G$ are not connected by a path, then $CONN_G(u, v) = 0$ and $d_{ss}(u, v) = \infty$.

**Theorem 7.1.6** *Let $G = (V, E, w)$ be a weighted graph and $d_{ss}$ is the strongest strong distance. Then $d_{ss}$ is a metric on $V$.*

***Proof*** Between any two distinct vertices $u$ and $v$, $CONN_G(u, v) \geq 0$. So, $d_{ss}(u, v) \geq 0$ for all $u, v \in V$. Since, reversal of a path from $u$ to $v$ is a path from $v$ to $u$ and vice versa, $d_{ss}(u, v) = d_{ss}(v, u)$. For any three vertices $u, v, w \in V$, $CONN_G(u, v) \geq CONN_G(u, w) \wedge CONN_G(w, v)$, where $\wedge$ represents the minimum. This gives,

$$\frac{1}{CONN_G(u,v)} \leq \frac{1}{CONN_G(u,w) \wedge CONN_G(w,v)}$$
$$\leq \frac{1}{CONN_G(u,w)} + \frac{1}{CONN_G(w,v)}.$$

That is, $d_{ss}(u, v) \leq d_{ss}(u, w) + d_{ss}(w, v)$, for all $u, v, w \in V$.

Since $d_{ss}$ satisfies all the conditions for a metric, $d_{ss}$ is a metric on $V$. ■

**Definition 7.1.7** The **$\delta$-distance between two vertices** $u$ and $v$ in a connected weighted graph $G = (V, E, w)$ denoted by $\delta(u, v)$, is defined as $\delta(u, v) = 1 + \Delta_w - CONN_G(u, v)$, where $\Delta(w)$ is the maximum of the weights of all edges in $E$ and $\delta(u, u) = 0$, for every vertex $u$ in $V$.

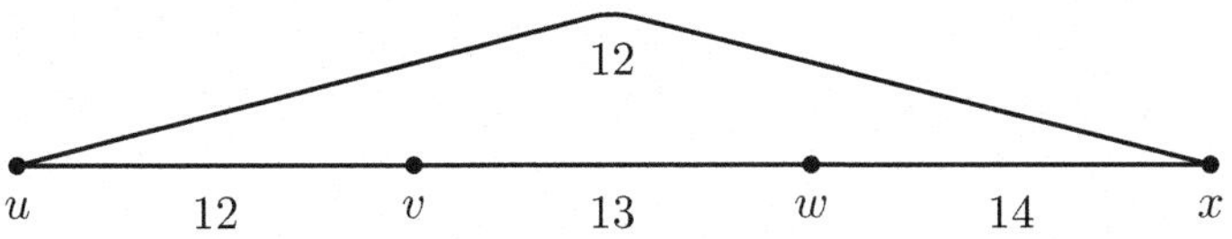


**Fig. 7.1** Illustration of distance

**Theorem 7.1.8** *$\delta$-distance in a weighted graph $G = (V, E, w)$ is a metric on $V$.*

***Proof*** $CONN_G(u, v) \leq \Delta_w$, for every pair of vertices $u, v \in V$. Also, $\Delta_w + 1 \geq 0$. Therefore, $\delta(u, v) \geq 0$ for all $u, v \in V$. Also $\delta(u, v) = 0$ if and only if $u = v$. Since, $CONN_G(u, v) = CONN_G(v, u)$ for all $u, v \in V$, $\delta(u, v) = \delta(v, u)$ for all $u, v \in V$. Also for any three vertices $u, v, w \in V$,

$$CONN_G(u, v) \geq CONN_G(u, w) \wedge CONN_G(w, v),$$

where $\wedge$ represents the minimum. So, $\Delta_w + 1 - CONN_G(u, v) \leq \Delta_w + 1 - [CONN_G(u, w) \wedge CONN_G(w, v)]$. That is, $\delta(u, v) \leq \delta(u, w) \vee \delta(u, v)$. Therefore $\delta(u, v) \leq \delta(u, w) + \delta(u, v)$ for all $u, v, w \in V$. Hence $\delta$ is a metric on $V$. ■

**Example 7.1.9** Consider the weighted graph given in Fig. 7.1.

$$d_w(u, v) = \frac{1}{12}, d_{sg}(u, v) = 1, d_{ss}(u, v) = \frac{1}{12}, \delta(u, v) = 3.$$

From this example, we can see that the distance between same pair vertices can be different according to the metric chosen.

**Proposition 7.1.10** *Let $G = (V, E, w)$ be a partial tree and $F$ be the maximum spanning tree of $G$. Then $d_{sg}$, $d_{ss}$ in $G$ are equivalent to $d_{sg}$, $d_{ss}$ respectively in $F$.*

Since edges of $F$ are the only strong edges in $G$, proof of Proposition 7.1.10 is straight forward. In the next theorem, we show that center of a partial tree $G$ and that of the maximum spanning tree associated with $G$ are isomorphic with respect to the metric $d_{sg}$.

**Theorem 7.1.11** *Let $G = (V, E, w)$ be a partial tree and $F$ be the corresponding maximum spanning tree of $G$. Then $< C_{sg}(G) >=< C_{sg}(F) >$.*

***Proof*** Consider a vertex $v$ in $G$. Let $e_{sg}(v) = k$ in $G$. We want to prove that $e_{sg}(v) = k$ in $F$. $e_{sg}(v) = k$ in G implies there exists a vertex $u = v^*$ such that there is a strong $v - u$ path $P$ of length $k$ in $G$. Since $G$ is a partial tree, $P$ is the unique strong $v - u$ path in $G$ and $F$ contains all strong edges in $G$. Thus $F$ contains the path $P$. So $e_{sg}(v) = k$ in $F$.

That is, for any vertex $v$ in $G$ eccentricities in $G$ and $F$ are the same. Therefore by the definition of center of a weighted graph $< C_{sg}(G) =< C_{sg}(F) >$. ■

Let $G$ be a partial tree and $F$ be the maximum spanning tree of $G$. Then, the set of $g$-eccentric vertices, the set of $ss$-eccentric vertices and the set of $\delta$-eccentric vertices related to $G$ and $F$ are the same.

**Proposition 7.1.12** *Let $G = (V, E, w)$ be a partial tree and $F$ be the maximum spanning tree of $G$. Then,*

*(1)* $< c_{ss}(G) >=< c_{ss}(F) >$
*(2)* $< c_{\delta}(G) >=< c_{\delta}(F) >$

***Proof*** For any pair of vertices $u$ and $v$ in $G$, $CONN_G(u, v) = CONN_F(u, v)$. So eccentricity of $u$ in both $G$ and $F$ are the same with respect to metrics $d_{ss}$ and $\delta$. ■

If each vertex of a connected weighted graph $G$ is a central vertex with respect to a metric $d$ then, $G$ is called a **self-centered** graph with respect to $d$.

**Theorem 7.1.13** *Let $G = (V, E, w)$ be connected weighted graph and $d$ be any one of the metrics $d_w, d_{sg}, d_{ss}$ or $\delta$, then $G$ is self-centered with respect to $d$ if $CONN_G(u, v) = w(uv), \forall u, v \in V$ and*

*(1)* $r_{dw}(G) = \frac{1}{w(u,v)}$, *where $w(uv)$ is the minimum weight of all edges in $G$*
*(2)* $r_{sg}(G) = 1.$
*(3)* $r_{ss}(G) = \frac{1}{w(u,v)}$, *where $w(uv)$ is the least.*
*(4)* $r_{\delta}(G) = 1 + \Delta_w - w(uv)$, *where $w(uv)$ is the least.*

***Proof*** By assumption, since $CONN_G(u, v) = w(uv)$ the underlying graph is complete. Also every edge $uv$ is a strongest $u - v$ path and every edges is strong.

(1) $CONN_G(u, v) = w(uv)$, for all $u, v \in V$. This gives that the weight of the weakest edge in any other strongest $u - v$ path is $w(uv)$. Hence the $w$-length of a strongest $u - v$ path is at least $\frac{1}{w(uv)}$. Let $\rho : u = u_0, u_1, u_2, \ldots, u_n = v$ be any $u - v$ path which is not strongest. Then the strength of $\rho$ is strictly less than $w(uv)$. So $w$-length of $\rho$ is strictly greater than $\frac{1}{w(uv)}$, and hence $d_w(u, v) = \frac{1}{w(uv)}$. Also,

$$e_w(u, v) = \max_v d_w(u, v) = \max_v \frac{1}{w(uv)} = \frac{1}{\min_v w(uv)}. \ldots\ldots\ldots\ldots(i)$$

**Claim** $e_w(v_i) = e_w(v_j)$ for all $v_i, v_j \in V$.

If not, let $e_w(v_i) < e_w(v_j)$ and let $u_i$ and $u_j$ are two vertices in $G$ such that

$$e_w(v_i) = \frac{1}{w(v_i u_i)} \text{ and } e_w(v_j) = \frac{1}{w(v_j u_j)}.$$

$$e_w(v_i) < e_w(v_j) \Rightarrow \frac{1}{w(v_i u_i)} < \frac{1}{w(v_j u_i)} \Rightarrow w(v_i u_i) > w(v_j u_j) \ldots\ldots\ldots (ii)$$

Consider the path $\rho : v_j, v_{i,} u_j$, then $w(v_j, v_i) \geq w(v_i, u_i)$, since $u_i = v_i^*$ and by $(i)$.

I.e., $w(v_i u_j) \geq w(v_i u_i)$.

So, $w(v_j v_i) \wedge w(v_i u_j) \geq w(v_i u_i) > w(v_j u_j)$, by $(ii)$.

That is, strength of a $v_j - u_j$ path exceeds $w(v_j u_j)$, which contradicts our assumption that every edge is a strongest path. Interchanging $i$ and $j$, a similar argument holds.

Thus $e_w(v_i) = e_w(v_j)$ for all $v_i, v_j \in V$.

(2) By assumption every edge is strong and the underlying graph is complete.

So, $d_{sg}(u, v) = 1$ for all $u, v \in V$. Therefore $e_{sg}(v) = 1$, for all $v \in V$. So $G$ is self centered and $r_{sg}(v) = \min e_{sg}(v) = 1$.

(3) $d_{ss}(u, v) = \frac{1}{CONN(uv)} = \frac{1}{w(uv)}$, by assumption.

**Claim** $e_{ss}(v_i) = e_{ss}(v_j)$, for $v_i, v_j \in V$.

The proof is same as in (1). Hence $G$ is self-centered and $r_{ss}(G) = \frac{1}{w(uv)}$, where $w(uv)$ is the least.

(4) $\delta(u, v) = 1 + \Delta_w - CONN_G(u, v)$, where $\Delta_w$ is the maximum weight of all edges. So, $\delta(u, v) = 1 + \Delta_w - w(uv)$.

**Claim** $e_\delta(v_i) = e_\delta(v_j)$ .

Suppose not, that is, $e_\delta(v_i) < e_\delta(v_j)$. Let $u_i = v_i^*$ and $u_j = v_j^*$. Then $e_\delta(v_i) < e_\delta(v_j)$ $\Rightarrow 1 + \Delta_w - w(u_i, v_i) < 1 + \Delta_w - w(u_j, v_j)$.

That is $CONN_G(u_i, v_i) > CONN_G(u_j, v_j)$.

That is, $w(u_i v_i) > w(u_j v_j)$. ..........................$(iii)$

$u_i = v_i^*$ and $u_j = v_j^*$ implies that $\delta(u_i, v_i) = \min_v \delta(u_i, v)$ and $\delta(u_j, v_j) = \min_v \delta(u_j, v)$................. $(iv)$

Consider the path $\rho : v_j v_i u_j$ then $(iv)$ gives $w(v_j v_i) \geq w(v_i u_i)$ and $w(v_i u_j) \geq w(v_i u_i)$.

That is, $w(v_j v_i) \wedge w(v_i u_j) \geq w(v_i u_i) > w(v_j u_j)$.

That is, $w(v_j v_i) \wedge w(v_i u_j) \geq w(v_j u_j)$. This is a contradiction to the assumption that $CONN_G(u, v) = w(uv)$ for all $u, v \in V$.

So $e_\delta(v_i) = e_\delta(v_j)$, for all $v_i, v_j \in V$.

So $G$ is self centered and $r(G) = 1 + \Delta_w - w(uv)$, Where $w(uv)$ is the minimum of all edge weights in $G$. ■

Theorem 7.1.13 says that irrespective of the metric, each vertex in a self centered graph is eccentric.

**Proposition 7.1.14** *If $G = (V, E, w)$ is a self-centered graph. Then each vertex of $G$ is eccentric. This property is independent of the metric defined on it.*

When the connectivity between $u$ and $v$ is equal to the edge weight of $uv$, that is $CONN_G(u, v) = w(uv)$ for all $u, v \in V$, the metrics $d_w$ and $d_{ss}$ coincide.

The condition in Proposition 7.1.14 is not necessary for a weighted graph to be self-centered.

**Example 7.1.15** Consider Fig. 7.2.

There is no edge connecting $a$ and $c$ and hence $w(ac) = 0$ and $CONN_G(a, c) = 1$. Note that $CONN_G(a, c) \neq w(ac)$. We can see that,

$$e_w(a) = e_w(b) = e_w(c) = e_w(d) = \frac{3}{2}.$$

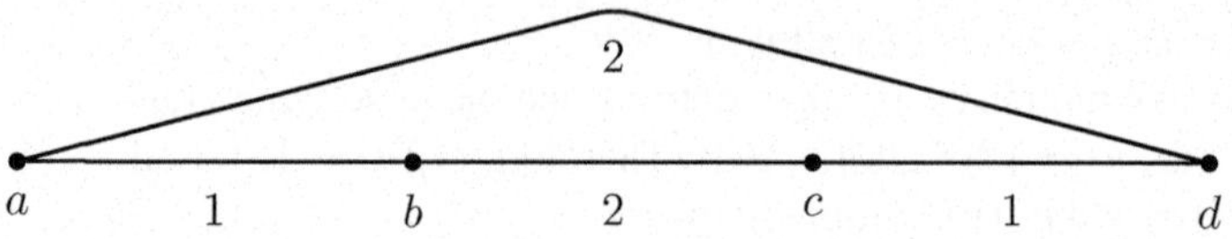


**Fig. 7.2** A $d_w$ self centered graph

Thus the graph is self-centered with respect to the metric $d_w$. Since every edge is strong, $e_{sg}(a) = e_{sg}(b) = e_{sg}(c) = e_{sg}(d) = 2$. So $G$ is self-centered with respect to the metric $d_{sg}$. Also, $e_{ss}(a) = e_{ss}(b) = e_{ss}(c) = e_{ss}(d) = 1$. That is, G is self-centered with respect to the metric $d_{ss}$. $e_\delta(a) = \Delta_(w) - \min_{v \in G} CONN(a, v)$ $\Rightarrow e_\delta(a) = e_\delta(b) = e_\delta(c) = e_\delta(d) = 1$. So $G$ is self-centered with respect to the metric $\delta$.

If $G$ is a weighted cycle in which there exist at least 2 edges having weakest weight $w_0$, Then $e_{ss}(a) = \frac{1}{w_o}, \forall a \in V$. Moreover all edges of $G$ are strong and hence $e_{ss}(a) = \frac{n}{2}$, if $n$ is even, and $e_{ss}(a) = \frac{n-1}{2}$, if $n$ is odd. This is true for every vertex $a$ in $G$. And $e_\delta(a) = 1 + \Delta_w - w_0, \forall a \in V$. So $G$ is self-centered with respect to $d_w, d_{sg}, d_{ss}$ and $\delta$.

**Theorem 7.1.16** *Every connected weighted graph $G = (V, E, w)$ is ss-self centered, as well as $\delta$-self centered.*

***Proof*** Let $P$ be a strongest $u - v$ path in $G$ having strength $s$ and let $s$ be the least strength of connectivity between any pair of vertices in $G$. Let $w(xy) = s$, where $xy$ is an edge in $P$. Now consider any vertex $z$ in $V$.

**Claim** $CONN_G(z, x) = s$, or $CONN_G(z, y) = s$.

Suppose not, that is there is a strongest $z - x$ path $P_1$ with strength $s_1$ and a strongest $z - y$ path $P_2$ with strength $s_2$ such that $s_1, s_2 > s$. Clearly both $P_1$ and$P_2$ do not contain the edge $xy$ and every edge in $P_1$ and $P_2$ has weight greater than $s$. $P_1 \cup P_2$ contains an x-y path and it does not contain the edge $xy$. So this path has strength greater than $s$. Then there exists a $u - v$ path $P_3$, which has strength greater than $s$. Which contradicts our assumption. Therefore $CONN_G(z, x) = s$, or $CONN_G(z, y) = s$ for all $z \in V$. Since $s$ is least, $min\{ CONN_G(z, x), CONN_G(z, y)\} \leq CONN_G(z, a)$, where $a$ is any vertex in $G$. This gives,

$d_{ss}(z, x) \wedge d_{ss}(z, y) \geq d_{ss}(z, a)$, where $\wedge$ denotes the minimum.

That is, $e_{ss}(z) = d_{ss}(z, x) \wedge d_{ss}(z, y) = \frac{1}{s}, \forall z \in V$. Therefore $G$ is $ss$-self-centered.

To prove that $G$ is $\delta$- self-centered, we have $min\{CONN_G(z, x), CONN_G(z, y)\} \leq CONN_G(z, a)$. This gives $1 + \Delta_w - min\{CONN_G(z, x), CONN_G(z, y)\} \geq 1 + \Delta_w - CONN_G(z, a)$.

$\delta(z, x) \vee \delta(z, y) \geq \delta(z, a)$, where $a$ is any vertex in $G$. That is, $e_\delta(z) = \delta(z, x) \vee \delta(z, y)$, where $\vee$ denotes the maximum. $e_\delta(z) = 1 + \Delta_w - s$ for all $z \in V$. Therefore $G$ is $\delta$-self-centered. ■

## 7.2 Intervals in Weighted Graphs

The contents of this section are from Dhanyamol and Sunil [3].

Interval function is a widely used concept in graph theory and it was introduced by Henry Martin Mulder. This section discusses the concept of intervals in weighted graphs.

Let us recall the definition of interval functions in graph theory. A similar definition is possible in weighted graph theory also.

**Definition 7.2.1** ([1]) Let $G(V, E)$ be a finite, connected, simple, loop less graph with distance function $d$ where $d(u, v)$ is the length of the shortest $u - v$ path. The **interval function** $I$ on $G$ is defined as $I(u, v) = \{w \in V : d(u, v) = d(u, w) + d(w, v)\}$, for every $u$ and $v$ in $V$. The set $I(u, v)$ is the interval between $u$ and $v$.

Figure 7.3 is a simple and connected graph in which $I(a, d) = \{a, d\}$, $I(d, b) = \{d, b\}$, $I(d, c) = \{d, c\}$ and $I(a, c) = \{a, b, c, d\}$. Now, we define Mulder interval in weighted graphs as follows.

**Definition 7.2.2** **Mulder's interval** $I_m(u, v)$ in a weighted graph $G = (V, E, w)$, is defined as $I_m(u, v) = \{w \in V \mid d(u, v) = d(u, w) + d(w, v)\}$, where $u, v \in V$ and $d(u, v)$ denotes the usual metric in $G$. That is, $d(u, v)$ is the minimum weight of all $u - v$ paths in $G$. Here, **weight of a path** $P$ is the sum of weights of all its edges.

**Example 7.2.3** Consider the weighted graph in Fig. 7.4.

This graph contains five vertices. Mulder's intervals relating all pairs of vertices are shown in Table 7.1.

Since there exist different types of distances in weighted graphs, we can have different types of intervals. Now we focus the strong weighted intervals arising from the strong geodesic distance $d_{sg}$.

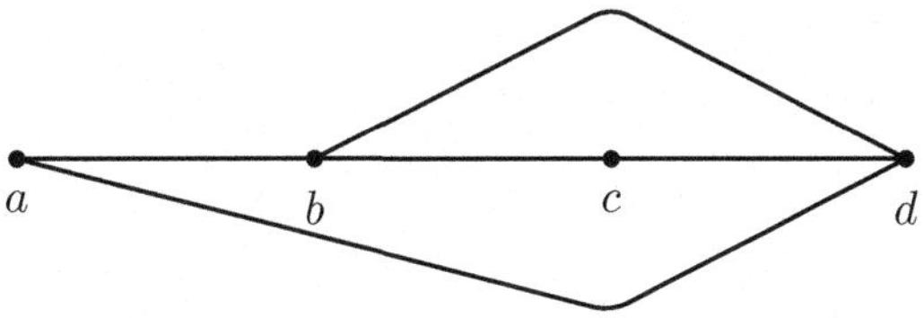


**Fig. 7.3** A simple graph

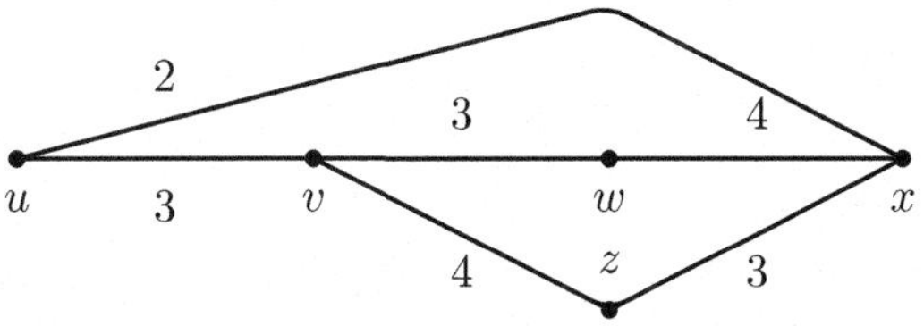


**Fig. 7.4** Illustration of interval

**Table 7.1** Intervals in Fig. 7.4

| *vertices* $a, b$ | $d(a, b)$ | $I_m(a, b)$ |
|---|---|---|
| $x, z$ | 3 | $\{x, z\}$ |
| $x, u$ | 2 | $\{x, u\}$ |
| $x, v$ | 5 | $\{x, v, u\}$ |
| $x, w$ | 4 | $\{x, w\}$ |
| $u, z$ | 5 | $\{u, z\}$ |
| $u, v$ | 3 | $\{u, v\}$ |
| $u, w$ | 6 | $\{u, w, v, x\}$ |
| $v, z$ | 4 | $\{v, z\}$ |
| $v, w$ | 3 | $\{v, w\}$ |
| $z, w$ | 7 | $\{z, w, v, x\}$ |

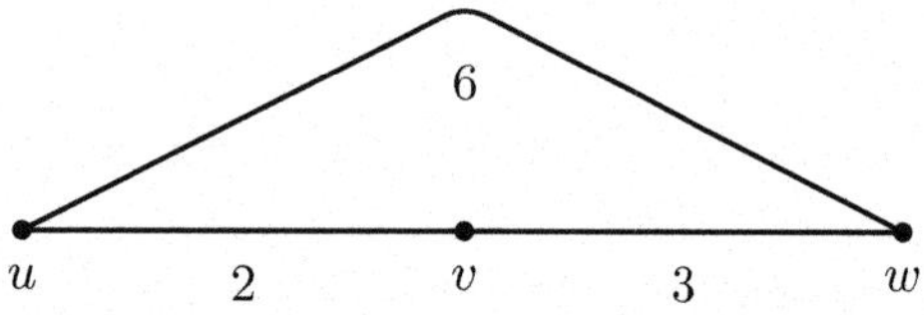


**Fig. 7.5** A partial tree

Recall that the strong geodesic distance or $sg$-distance between two vertices $u$ and $v$ in a weighted graph $G$ denoted by $d_{sg}(u, v)$, is defined as the length of the shortest $u - v$ strong path in $G$. If $u$ and $v$ are not connected by a path, then $d_{sg}(u, v) = \infty$.

**Definition 7.2.4 Strong weighted interval on a weighted graph** $G = (V, E, w)$ is defined as $I_s(u, v) = \{x \in V \ / \ d_{sg}(u, v) = d_{sg}(u, x) + d_{sg}(x, v)\}$, where $d_{sg}$ denotes strong geodesic distance ($sg$-distance). That is, strong interval corresponding to the pair $(u, v)$ is the set of all vertices which are lying on any strong $u - v$ geodesic.

In Fig. 7.4, $I_s(u, x) = \{u, v, w, x, z\}$ because $ux$ is not strong. All edges other than $ux$ are strong. Thus for all other edges $mp$, $I_s(m, p) = \{m, p\}$.
Also, $I_s(x, v) = \{x, w, z, v\}$, $I_s(x, z) = \{x, z\}$, $I_s(a, e) = \{a, e\}$.

The relation $I_m(u, v) = I_s(u, v)$ for all $u, v \in V(G)$ holds in any unweighted graph $G$. Generally the said relation does not hold in a weighted graph.

In Fig. 7.5, $I_m(u, v) = \{u, v\}$ and $I_s(u, v) = \{u, w, v\}$. Also, $I_m(u, w) = \{u, v, w\}$ and $I_s(u, w) = \{u, w\}$. It is difficult to find a relation between Mulders interval and strong interval in a weighted graph.

If there exists a unique path joining any two vertices in a weighted graph $G = (V, E, w)$, then $I_s(u, v) = I_m(u, v)$ for all $u, v \in V$. This relation trivially holds for a weighted tree.

Existence of unique strong path between pairs of vertices does not assure the equality of Mulders interval and strong interval. For, consider the partial tree in

Fig. 7.5. Edge $uw$ is the only strong path connecting vertices $u$ and $w$. But $I_m(u, w) \neq I_s(u, w)$. Both the intervals are equal only for the pair of vertices $v$ and $w$.

**Theorem 7.2.5** *For every weighted graph $G = (V, E, w)$ with at least one edge, there exist vertices $u, v, x, y$ such that $I_s(u, v) \subseteq I_m(u, v)$ and $I_m(x, y) \subseteq I_s(x, y)$.*

***Proof*** Every weighted graph $G$, which is not totally disconnected, has at least one strong edge. Let $uv$ be a strong edge of $G$, which has minimum weight and $xy$ be the edge in $G$, which has minimum weight (need not be strong) in $G$, then $I_s(u, v) = \{u, v\}$ and $I_m(x, y) = \{x, y\}$. Also, $a, b \in I_s(a, b) \cap I_m(a, b)$ for any pair of vertices $a, b$. Therefore $I_s(u, v) \subseteq I_m(u, v)$ and $I_m(x, y) \subseteq I_s(x, y)$. ■

Next theorem proves that $I_s(u, v)$ is always a subset of $I_m(u, v)$ if the edge $uv$ is strong.

**Theorem 7.2.6** *If $uv$ is a strong edge in a weighted graph $G = (V, E, w)$, then, $I_s(u, v) \subseteq I_m(u, v)$.*

***Proof*** If $uv$ is a strong edge, then, $I_s(u, v) = \{u, v\}$. Also, for any pair of vertices $(x, y)$, $x, y \in I_m(x, y)$. So $I_s(u, v) \subseteq I_m(u, v)$. ■

Consider a cycle $G$. If it is a partial tree, then for all $x, y \in V(G)$ there exists a unique strong path in $G$. So $I_s(u, v)$ is the set of all vertices on the unique strong $u - v$ path. Moreover $I_s(u, v) = V(G)$ if and only if $uv$ is the unique weakest edge in $G$. If $G$ is a strong cycle with $n$ vertices, then for all $x, y \in V(G)$, $I_s(x, y)$ is the set of all vertices on the shortest $x - y$ paths. If $n$ is even, then there exists $\frac{n}{2}$ pairs of diametrically opposite vertices $x, y$ in $G$ such that $I_s(x, y) = V(G)$. And for all other pair of vertices $u, v$, $I_s(u, v) \subset V(G)$. If $n$ is odd, then there does not exist any pair of vertices $x, y$ such that $I_s(x, y) = V(G)$. If an edge $xy$ of $G$ is either $\alpha$-strong or $\beta$-strong, then $I_s(x, y) = \{x, y\}$. Other wise (the case where $xy$ is a $\delta$-edge), $I_s(x, y) = V(G)$.

Following result proves certain properties of strong weighted intervals.

**Theorem 7.2.7** *Let $G = (V, E, w)$ be a weighted graph. Then, for all $u, v, w, x \in V$,*

1. *$I_s(u, v) = \{u, v\}$ if and only if $uv$ is a strong edge.*
2. *$I_s(u, v) = I_s(v, u)$.*
3. *If $w \in I_s(u, v)$ then $I_s(u, w) \subseteq I_s(u, v)$ and $I_s(w, v) \subseteq I_s(u, v)$.*
4. *If $w \in I_s(u, v)$ and If $x \in I_s(w, v)$ then $x \in I_s(u, v)$ and $w \in I_s(u, x)$.*
5. *If $x \in I_s(u, v)$ then $I_s(u, x) \cap I_s(x, v) = \{x\}$ and hence $| I_s(u, x) \cap I_s(x, v) |= 1$.*
6. *If $x \in I_s(u, v)$ then $I_s(u, x) \cup I_s(x, v) \subseteq I_s(u, v)$.*
7. *If $| I_s(u, v) |= 2 =| I_s(x, y) |$, $v \in I_s(u, x)$ and $u \in I_s(v, y)$, then $x \in I_s(v, y)$.*

***Proof*** Results in $1 - 5$ are obvious.

6. Let $x \in I_s(u, v)$. Then, $x$ is on a strong geodesic joining $u$ and $v$. So every strong geodesic $P$ joining $u$ to $x$ followed by a strong geodesic $Q$ joining $x$ to $v$ is again a

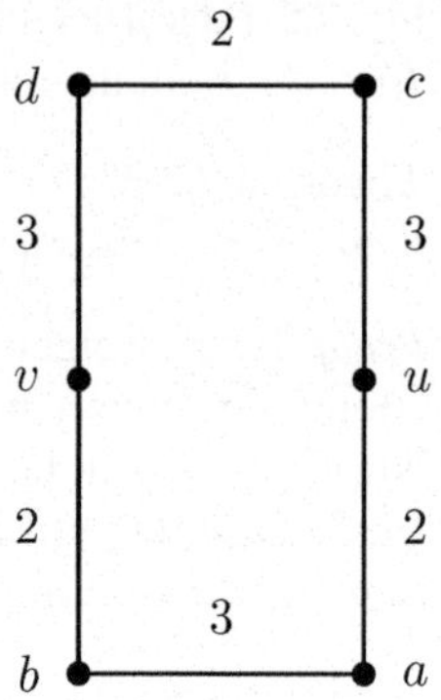


**Fig. 7.6** A weighted cycle

strong geodesic joining $u$ to $v$. So every vertex on $P \cup Q$ is in $I_s(u, v)$. Therefore, $I_s(u, x) \cup I_s(x, v) \subseteq I_s(u, v)$.

7. Let the assumption holds. Then $d_{sg}(v, y) \leq d_{sg}(v, x) + d_{sg}(x, y) = d_{sg}(v, x) + 1$ since $| I_s(x, y) |= 2$

$$
\begin{aligned}
&= d_{sg}(u, x) \quad \text{since } v \in I_s(u, x) \\
&\leq d_{sg}(u, y) + d_{sg}(y, x) \\
&= d_{sg}(u, y) + 1 \quad \text{since } | I_s(x, y) |= 2 \\
&= d_{sg}(v, y) \quad \text{since } u \in I_s(v, y) \text{ and } | I_s(u, v) | = 2.
\end{aligned}
$$

That is, $x \in I_s(v, y)$.

■

Figure 7.6 gives an illustration for the case where the equality of condition 6 of Theorem 7.2.7 does not hold.

In this graph, $I_s(u, v) = V(G)$ and $a \in I_s(u, v)$. But $I_s(u, a) \cup I_s(a, v) = \{u, a, b, v\} \neq I_s(u, v)$.

The condition $I_s(u, a) \cup I_s(a, v) = I_s(u, v)$ strictly holds for any $a \in I_s(u, v)$ if the vertices $u$ and $v$ are joined by a unique strong path. Since there exists a unique strong path between any two vertices in a partial tree, we have the following result.

**Theorem 7.2.8** *In a partial tree $G = (V, E, w)$, $I_s(u, v) = I_s(u, x) \cup I_s(x, v)$ for all $x, u, v \in V(G)$ such that $x \in I_s(u, v)$.*

***Proof*** In a partial tree $G$ for all $u, v \in V(G)$, there exists a unique $u - v$ strong path. So $I_s(u, v) = I_s(u, x) \cup I_s(x, v)$ for all $x \in I_s(u, v)$. ■

**Theorem 7.2.9** *If $x$ is a cut vertex in a weighted graph $G = (V, E, w)$, then there exists a pair of vertices $u, v$ in $G$ such that $I_s(u, v) = I_s(u, x) \cup I_s(x, v)$.*

***Proof*** Let $x$ be a cut vertex of $G$ then, there exists a pair of vertices $u, v$ in $G$ such that the removal of $x$ from $G$ deletes all $u - v$ paths. This means every $u - v$ path contains $x$. That is, every strong $u - v$ path contains $x$. So vertices on $u - v$ paths can be partitioned into two sets such as vertices before $x$ and vertices after $x$. So we get $I_s(u, v) = I_s(u, x) \cup I_s(x, v)$. Hence the proof. ■

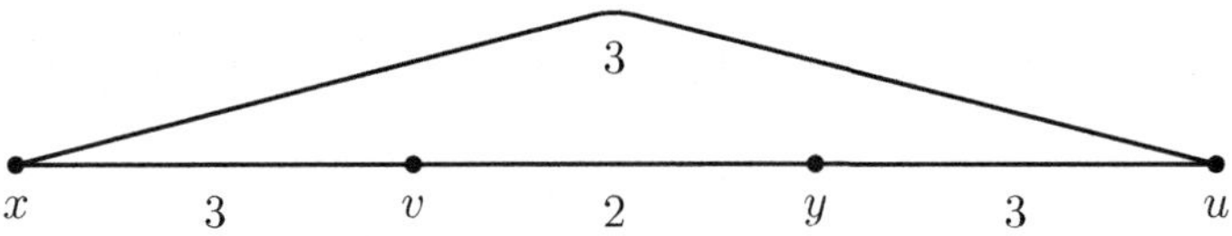


**Fig. 7.7** A partial tree

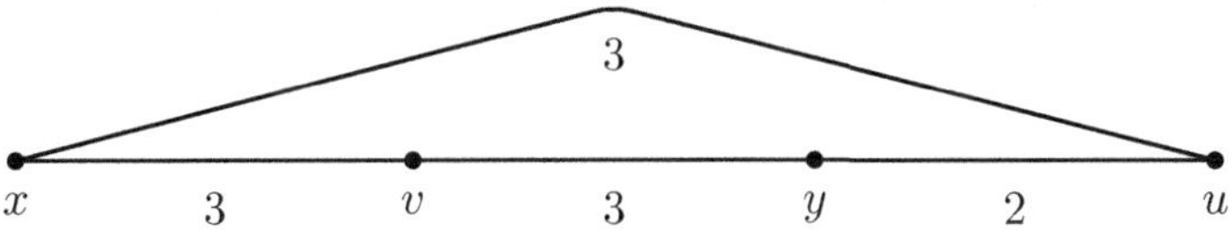


**Fig. 7.8** A weighted graph with a partial cut vertex

**Fig. 7.9** A weighted cycle

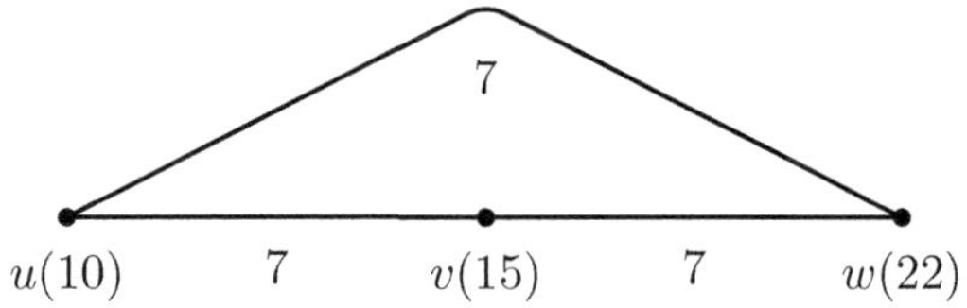


In Fig. 7.7, even if the condition $I_s(u, v) = I_s(u, x) \cup I_s(x, v)$ holds in $G$, $x$ is not a cutvertex in $G$. Note that $x$ is only a partial cut vertex.

Then there arises a question whether the condition in Theorem 7.2.9 holds if $x$ is a partial cut vertex. Consider the graph in Fig. 7.8, where $x$ is a partial cut vertex. But we can not find any pair of vertices $u$ and $v$ such that $I_s(u, v) = I_s(u, x) \cup I_s(x, v)$.

**Theorem 7.2.10** *In a precisely weighted graph $G = (V, E, w)$, $I_s(u, v) = \{u, v\}$, $\forall\, u, v \in V(G)$.*

***Proof*** In a precisely weighted graph every two vertices are joined by a strong edge [2]. Thus $I_s(u, v) = \{u, v\}\ \forall u, v \in V(G)$. ■

Converse of the Theorem 7.2.10 is not true in general. To illustrate this, Consider the graph in Fig. 7.9.

Here, $I_s(u, v) = \{u, v\}$, $I_s(v, w) = \{v, w\}$ and $I_s(u, w) = \{u, w\}$ and $G$ is not precisely weighted.

**Theorem 7.2.11** *If a partial tree $G = (V, E, w)$ is a path or a cycle, then there exist two vertices $u$ and $v$ such that $I_s(u, v) = V(G)$.*

***Proof*** Consider a partial tree $G$. If $G$ is a path, then take $u$ and $v$ as the end vertices of $G$. Then $I_s(u, v) = V(G)$. If $G$ is a cycle, then it has exactly one weak edge (say) $uv$. So $I_s(u, v) = V(G)$. ■

The Converse of Theorem 7.2.11 is not true always. For consider an example in Fig. 7.10. It is a partial tree and $I_s(u, x) = V(G)$. But it is neither a path nor a cycle.

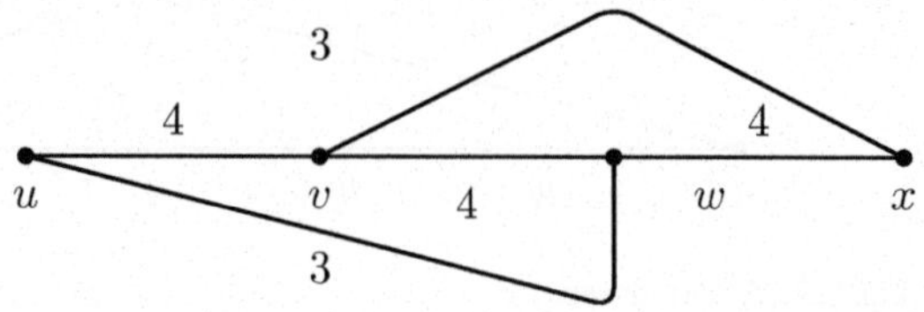


**Fig. 7.10** Converse of Theorem 7.2.11

**Theorem 7.2.12** *Let $G = (V, E, w)$ be a connected weighted graph and $u, v, x, y \in V(G)$. Then $I_s(u, x) \cup I_s(x, v) = I_s(u, v)$ if and only if $x$ is on every strong $u - v$ geodesic.*

***Proof*** If $x$ is on every strong $u - v$ geodesic, then every strong $u - v$ geodesic can be considered as the union of a $u - x$ strong geodesic and a strong $x - v$ geodesic. Then,

$$I_s(u, x) \cup I_s(x, v) = I_s(u, v) \tag{7.1}$$

Now consider the converse. We want to prove that $x$ is on every strong $u - v$ geodesic whenever

$$I_s(u, x) \cup I_s(x, v) = I_s(u, v). \tag{7.2}$$

Suppose this is not true. Then there exist at least one strong $u - v$ geodesic $P$ which does not contain $x$. Let $P$ be $y_0 y_1 y_2 \cdots y_{n-1} y_n$, where $y_0 = u$ and $y_n = v$. Since $u \in I_s(u, x)$ and $v \in I_s(x, v)$, there exist $i \in \{1, 2, 3, \ldots, n-1\}$ such that $y_i \in I_s(u, x)$ and $y_{i+1} \in I_s(x, v)$ and $y_i y_{i+1}$ is a strong edge (since it is an edge of the strong path $P$).

Now, $y_i \in I_s(u, x)$ implies

$$d_{sg}(u, y_i) + d_{sg}(y_i, x) = d_{sg}(u, x). \text{ So we get } d_{sg}(u, y_i) + 1 \leq d_{sg}(u, x) \tag{7.3}$$

Also, $y_{i+1} \in I_s(x, v)$ implies

$$d_{sg}(x, y_{i+1}) + d_{sg}(y_{i+1}, v) = d_{sg}(x, v). \text{ So we get } 1 + d_{sg}(y_{i+1}, v) \leq d_{sg}(x, v) \tag{7.4}$$

Equation (7.2) and $x \in I_s(u, v)$ together implies

$$d_{sg}(u, x) + d_{sg}(x, v) = d_{sg}(u, v) \tag{7.5}$$

Let $P_1$ be the $u - x$ strong geodesic containing $y_i$ and $P_2$, the $x - v$ strong geodesic containing $y_{i+1}$. Now consider the strong $u - v$ path, $u - y_i$ part of $P_1$ followed by strong edge $y_i y_{i+1}$ along with $y_{i+1} - v$ part of $P_2$ and which has length

$$\begin{aligned} &= d_{sg}(u, y_i) + d_{sg}(y_i, y_{i+1}) + d_{sg}(y_{i+1}, v) \\ &\leq d_{sg}(u, x) - 1 + 1 + d_{sg}(x, v) - 1 = d_{sg}(u, v) - 1 \end{aligned} \tag{7.6}$$

by Eqs. (7.1), (7.2) and (7.3). Which is a contradiction. That is, $x$ is on every strong $u - v$ geodesic. ■

## 7.3 Operations on Strong Intervals

It is sensible to define the union and intersection of strong intervals with respect to the conventional union and intersection of sets. Because strong intervals are basically subsets of $V$ in a weighted graph $G = (V, E, w)$. Union and intersection of two strong intervals say $I_s(u, v)$ and $I_s(x, y)$ are defined by $I_s(u, v) \cup I_s(x, y)$ and $I_s(u, v) \cap I_s(x, y)$ respectively.

**Theorem 7.3.1** *Union of two or more strong intervals need not be a strong interval.*

**Example 7.3.2** Figure 7.11a is a star graph in which $I_s(u, v) = \{u, x, v\}$, and $I_s(u, y) = \{u, y, x\}$. Consider $I_s(u, v) \cup I_s(u, y) = \{u, x, v\} \cup \{u, y, x\} = \{u, v, x, y\}$. We can see that $\{u, v, x, y\}$ is not a strong interval between any two vertices in Fig. 7.11a.

Consider a bipartite graph $G = (V, E, w)$ on five vertices with unit weight assigned to all the edges (Fig. 7.11b). See that $I_s(u, v) = \{u, x, v, y\}$ and $x, y \in I_s(u, v)$. Consider $I_s(u, v) \cup I_s(x, y)$. The resultant set is not equal to $I_s(u, v)$.

In Fig. 7.11b, $I_s(u, v) = \{u, x, y, v\}$ and $I_s(u, w) = \{u, x, y, w\}$. Consider $I_s(u, v) \cap I_s(u, w) = \{u, x, y\}$ and this is not a strong interval in $G$.

Figure 7.11c is a tree with weight 2 is assigned to all the edges. Here, $I_s(u, v) = \{u, v\}$ and $I_s(x, y) = \{x, y\}$. Their union $I_s(u, v) \cup I_s(x, y) = \{u, v, x, y\} = I_s(u, x)$. Note that union of two strong intervals can be a strong interval.

In general we can say that union and intersection of two strong intervals need not be a strong interval.

**Theorem 7.3.3** *In a partial tree $G = (V, E, w)$, if union of two disjoint strong intervals $I_s(u, v)$ and $I_s(x, y)$ is again a strong interval then the vertices $u, v, x, y$ lie on a path.*

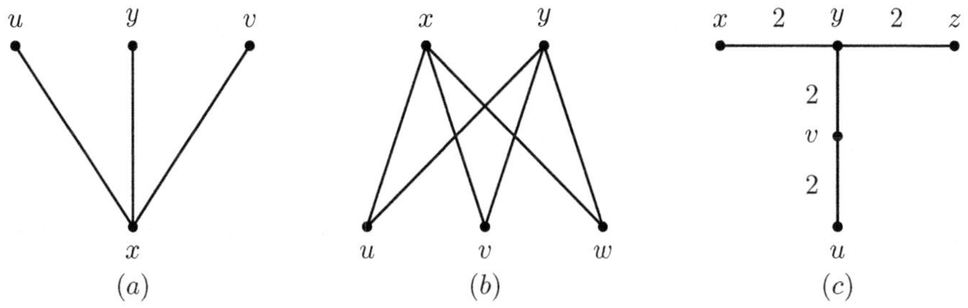


**Fig. 7.11** Operations on intervals

***Proof*** Given that $I_s(u, v) \cup I_s(x, y)$ is a strong interval. Let $I_s(u, v) \cup I_s(x, y) = I_s(a, b)$. Then clearly $u, v, x, y \in I_s(a, b)$. If possible assume that the vertices $u, v, x, y$ do not lie together on any $a - b$ strong geodesic. Consider two vertices $m, n \in \{u, v, x, y\}$ such that $m$ and $n$ lie on different strong $a - b$ geodesics (say) $P_1$ and $P_2$. Then $P_1$ and $P_2$ together gives a strong cycle. This is a contradiction since $G$ is a partial tree. ■

**Definition 7.3.4** In a weighted graph $G = (V, E, w)$, a set $X \subseteq V(G)$, is said to be $I_s$**-convex** $I_s$-, if for all $u, v \in X$, $I_s(u, v) \subseteq X$.

Corresponding to every $I_s$-convex set $X$ in a weighted graph $G$, we can find a pair of vertices $u, v$ so that $I_s(u, v) = X$.

**Theorem 7.3.5** *In a weighted graph $G = (V, E, w)$, $I_s$-convex strong intervals are closed under nonempty intersections.*

***Proof*** Consider two $I_s$-convex strong intervals $I_s(u, v)$ and $I_s(x, y)$ such that $I_s(u, v) \cap I_s(x, y) \neq \emptyset$. We want to prove that $I_s(u, v) \cap I_s(x, y)$ is again a $I_s$-convex strong interval. There are two cases.

**Case 1** $I_s(u, v) \cap I_s(x, y) = \{a\}$ for some $a \in V$.
In this case, since $I_s(a, a) = \{a\}$ for any vertex $a \in V$, we get $I_s(u, v) \cap I_s(x, y) = \{a\} = I_s(a, a)$. That is $I_s(u, v) \cap I_s(x, y)$ is again a strong interval.

**Case 2** $I_s(u, v) \cap I_s(x, y)$ contains at least 2 vertices.
Here for any two vertices $a, b \in I_s(u, v) \cap I_s(x, y)$, $I_s(a, b) \subset I_s(u, v) \cap I_s(x, y)$. Now choose $m, n \in I_s(u, v) \cap I_s(x, y)$ such that $d_s(m, n) = \max_{p,q \in I_s(u,v) \cap I_s(x,y)} d_s(p, q)$. Then, $I_s(u, v) \cap I_s(x, y) = I_s(m, n)$. That is, $I_s(u, v) \cap I_s(x, y) = I_s(m, n)$. That is, $I_s(u, v) \cap I_s(x, y)$ is again a strong interval. From both these cases it is clear that every nonempty intersection of $I_x$-convex strong intervals is again a $I_s$- convex strong interval. Hence the proof. ■

Next we have an obvious result.

**Theorem 7.3.6** *In a weighted graph $G = (V, E, w)$, $I_s$-convex strong intervals are closed under nested unions.*

## 7.4 Strong Gates in Weighted Graphs

In the computer and internet networks, all the communication to a certain locality is monitored by a particular server. The traffic of data packets from outside world to that locality is entered in to this server and the server takes decision to route them to various receiving points. Following definition of gate and the related concepts can be applied in communication networks effectively.

**Definition 7.4.1** A subgraph $H = (V(H), E(H), w_1)$ of a weighted graph $G = (V(G), E(G), w)$ is said to be **gated** with respect to a metric $d$, if for all $v \in V(G) \setminus V(H)$, there exists a vertex $x \in V(H)$ such that $d(u, v) = d(u, x) + d(x, v)$, for all $u \in V(H)$. $x$ is said to be a **gate** of $H$ in $G$.

**Definition 7.4.2** A nontrivial subgraph $H = (V(H), E(H), w_1)$ of a weighted graph $G = (V(G), E(G), w)$ is said to be **gated with respect to an interval function** $I$ if for all $v \in V(G) \setminus V(H)$, there exists a vertex $x \in V(H)$ such that $x \in I(u, v)$ for all $u \in V(H)$. Also, $x$ is called a gate of $H$ in $G$ if and only if $x \in \cap_{u \in V(H)} I(u, v)$, for all $v \in V(G) \setminus V(H)$.

From the definition itself it is clear that gates are interior vertices but all interior vertices need not be gates.

**Definition 7.4.3** A subset of $V(G)$ which induces a gated subgraph of $G$, is known as a **gated set** in $G$.

If a nontrivial subgraph $H$ of $G$ is gated with respect to the strong geodesic distance $d_{sg}$ (or strong interval function $I_{sg}$), then $H$ is called **strongly gated** and if $x$ is a gate of $H$ in $G$ with respect to the strong geodesic distance $d_{sg}$(or strong interval function $I_{sg}$), then $x$ is known as a **strong gate** of $H$ in $G$. Through out this section, we consider the strong geodesic distance only. All internal vertices of a weighted path are strong gates. A vertex $x$ is called as a gate of a graph $G$ if it is the gate of some nontrivial subgraph $H$ in $G$. Also, it is clear that only connected weighted graphs can have strong gates.

**Proposition 7.4.4** *A nontrivial induced subgraph $H$ of a weighted graph $G = (V, E, w)$ has at most one strong gate in $G$.*

***Proof*** Suppose $x$ and $y$ are strong gates of a subgraph $H$ in $G$. $x$ is a strong gate of $H$ implies $x \in I_s(y, v)$ for all $v \in V(G \setminus H)$. This gives,

$$d_{sg}(y, v) = d_{sg}(y, x) + d_{sg}(x, v) \tag{7.7}$$

$y$ is a strong gate of $H$ implies $y \in I_s(x, v)$ for all $v \in V(G \setminus H)$
This gives,

$$d_{sg}(x, v) = d_{sg}(x, y) + d_{sg}(y, v) \tag{7.8}$$

Equations (7.7) and (7.8) implies $d_{sg}(x, y) = 0$ and hence $x = y$, since $d_{sg}$ is a metric by Theorem 7.1.4. ■

Consider a weighted cycle $G$ with four vertices in Fig. 7.12, where $c$ and $d$ are the strong gates.

**Theorem 7.4.5** *Let $G = (V, E, w)$ be a connected weighted graph and $H$ be a nontrivial induced subgraph of $G$. Then, $x$ is a strong gate of $H$ in $G$ if and only if every strong path from $H$ to $G \setminus H$ passes through $x$.*

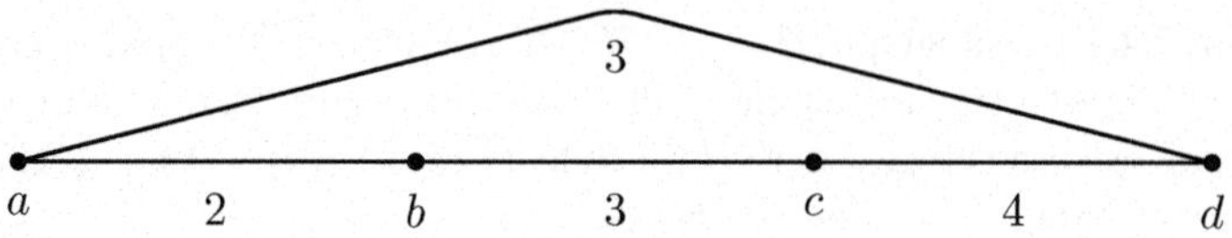


**Fig. 7.12** A weighted cycle with two strong gates

**Fig. 7.13** Weighted graph with a unique strong gate

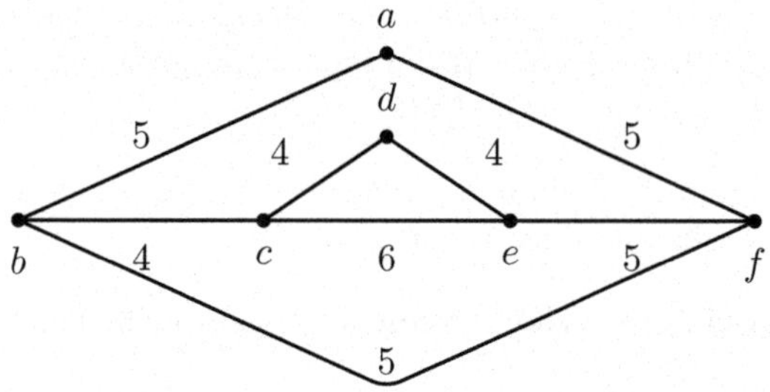


Proof of the above theorem is trivial and may be proved by the method of contradiction.

A vertex $x$ is a strong gate of $H$ in $G$ does not mean that all strongest paths from $H$ to $G \setminus H$ passes through $x$. To illustrate this, consider Fig. 7.13.

In Fig. 7.13, consider the gate of the subgraph induced by $H = \{a, b, f\}$. See that $bc$ is the only weak edge in $G$ and hence $e$ is the gate of the subgraph induced by $H$. The strongest $a - d$ path $abcd$ does not contain $e$.

The number of strong gates in a weighted cycle can be predicted based on the number of vertices in it.

**Theorem 7.4.6** *Let $G = (V, E, w)$ be a weighted cycle with n vertices. Then,*
$$\zeta(G) = \begin{cases} 0 & \text{if } G \text{ is a strong cycle} \\ n-2 & \text{if } G \text{ is not a strong cycle} \end{cases}$$

Here $\zeta(G)$ denotes the number of strong gates in $G$.

***Proof*** Disconnected subgraphs of $G$ has no strong gates. So, it is enough to consider connected subgraphs of $G$. Consider a strong cycle $G$ and a nontrivial connected induced subgraph $H$ of $G$. Then $H$ has at least two vertices. Clearly, no interior vertex of $H$ can be a strong gate of $H$ in $G$. Let $u$ and $v$ be pendent vertices of $H$ and let $w_1$ be the vertex in $G \setminus H$ which is adjacent to $u$. Then $v \notin I_s(u, w_1)$ and $I_s(u, w_1) = \{u, w_1\}$. Also, let $w_2$ be the vertex in $G \setminus H$ which is adjacent to $v$. Then, $u \notin I_s(v, w_2)$ and $I_s(v, w_2) = \{v, w_2\}$. So $H$ has no strong gates in $G$. Since H is arbitrary, it leads to the conclusion that, $G$ has no strong gates. Now consider the case where $G$ is not a strong cycle. Then $G$ has exactly one weakest edge. Let it be $xy$. Then any nontrivial connected induced subgraph $H$ of $G$ containing $xy$ has no strong gates. Also any nontrivial connected induced subgraph, not containing both $x$ and $y$, has no strong gates. Now consider a nontrivial connected subgraph $H$ in $G$ containing either $x$ or $y$ (not both). Then exactly one of the pendent vertices of

$H$ is, either $x$ or $y$. Let $w$ be the other pendent vertex of $H$, different from $x$ and $y$. Then $w$ is a strong gate of $H$ in $G$. Also, $x$ and $y$ cannot be strong gates. That is, all vertices of $G$, except $x$ and $y$, are strong gates of $G$. Therefore, $\zeta(G) = n - 2$. ■

It should be noted that no end vertex in a partial tree is a strong gate. Now we define $I$-gate free and $d$-gate free weighted graphs which are useful to categorize weighted graphs to recognize their connectivity.

**Definition 7.4.7** A subgraph of a weighted graph $G = (V, E, w)$ is called **$I$-gate free** ($d$-gate free) if it has no $I$-gates($d$-gates). Here $I - gate$ means gate with respect to the interval function $I$. Similarly $d$-gate means a gate with respect to the metric $d$.

Another categorization of weighted graphs in terms of gates can be seen in Definition 7.4.8.

**Definition 7.4.8** A weighted graph $G = (V, E, w)$ is said to be **partially strong gated** if some connected induced subgraphs of $G$ are gated. $G$ is said to be **fully strong gated** if every nontrivial connected induced subgraphs of $G$ has a gate. $G$ is said to be **strongly restricted** if no connected induced subgraph of $G$ is gated.

Note that all precisely weighted graphs are strongly restricted.

**Theorem 7.4.9** *Let $G = (V, E, w)$ be any weighted graph and $H$, an induced connected subgraph of $G$, which is strongly gated in $G$. Then,*

*(1) If $x$ is a strong gate of $H$ in $G$, then for any nontrivial connected induced subgraph $K$ of $H$ containing $x$, having a strong gate in $G$, $x$ is a strong gate of $K$ in $G$.*
*(2) If $x \in H$, is not a strong gate of $H$ in $G$, then there does not exist any nontrivial induced subgraph $M$ of $G$ such that $M$ contains $H$ and $x$ is a strong gate of $M$.*

***Proof***

(1) Using Proposition 7.4.4, assume that $y$ is a strong gate of $K$. Consider the vertices $u$ and $v$ such that $u \in V(K)$ and $v \in V(G \setminus H)$. Then,

$$d_{sg}(u, v) = d_{sg}(u, x) + d_{sg}(x, v), \quad \text{since } x \text{ is a strong gate of } H \tag{7.9}$$

$$d_{sg}(u, v) = d_{sg}(u, y) + d_{sg}(y, v), \quad \text{since } y \text{ is a strong gate of } K \tag{7.10}$$

$$\text{Equation (7.9) implies } d_{sg}(u, v) = d_{sg}(u, x) + d_{sg}(x, y) + d_{sg}(y, v) \tag{7.11}$$

This is because $y$ is a strong gate of $K$ and $x$ is a strong gate of $H$ in $G$. Also we get,

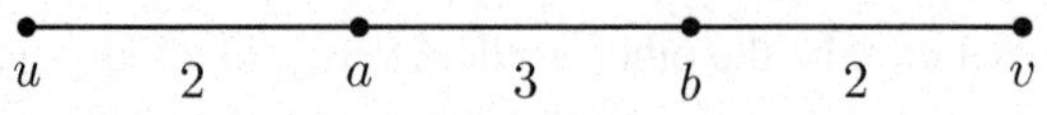


**Fig. 7.14** A weighted path

$$d_{sg}(u,v) = d_{sg}(u,y) + d_{sg}(y,x) + d_{sg}(x,v), \text{ using Equation (7.10)} \quad (7.12)$$

$$\text{Equations (7.11)–(7.12) implies } d_{sg}(u,x) + d_{sg}(y,v) = d_{sg}(u,y) + d_{sg}(x,v) \quad (7.13)$$

$$\text{Equations (7.9)–(7.10) implies } d_{sg}(u,x) + d_{sg}(x,v) = d_{sg}(u,y) + d_{sg}(y,v) \quad (7.14)$$

$$\text{Equations (7.13) and (7.14)implies } d_{sg}(u,x) = d_{sg}(u,y) \text{ and } d_{sg}(y,v) = d_{sg}(x,v) \quad (7.15)$$

$u$ is any vertex in $K$. Also, $x \in K$. So $u = x$ in (7.15) implies $x = y$, Since $d_{sg}$ is a metric.

(2) Follows from (1). ■

The condition $x \in H$ is necessary in Theorem 7.4.10. It is clear from Fig. 7.14. $b$ is a strong gate of the subgraph $H$ induced by $\{u, a, b\}$ and its subgraph $K$ induced by $\{u, a\}$ is strongly gated. But, $b$ is not a strong gate of $K$. For Theorem 7.4.9(2) to be valid, $H$ should be strongly gated in $G$. That is, even if $x \in H$ is not a strong gate of $H$, there exists subgraphs $M$ of $G$ such that $M$ contains $H$ and $x$ is a strong gate of $M$.

**Theorem 7.4.10** *Every strong internal vertex (vertex whose strong degree greater than or equal to 2) of a partial tree $G = (V, E, w)$ is a strong gate.*

***Proof*** Between any two vertices of a partial tree, there exists a unique strong path and every interior vertex lies in a strong path joining a pair of its strong neighbors. So every strong interior vertex is a strong gate. ■

**Theorem 7.4.11** *Every strong gate in a weighted graph $G = (V, E, w)$ is a partial cut vertex.*

***Proof*** Assume that $x$ is a strong gate in $G$. We want to prove that $x$ is a partial cut vertex. Let $x$ be the strong gate of a nontrivial connected induced subgraph $H$ in $G$. Consider $u \in V(H)$ and $v \in V(G) \setminus V(H)$ such that $u, v$ has the maximum strength of connectedness among all pair of vertices $(m, n) \in V(H) \times V(G) \setminus V(H)$. Assume that $CONN_G(u, v) = m$. Suppose there exists a strongest $u - v$ path $P$ which does not pass through $x$. Then strength of $P$, $s(P) = m$.

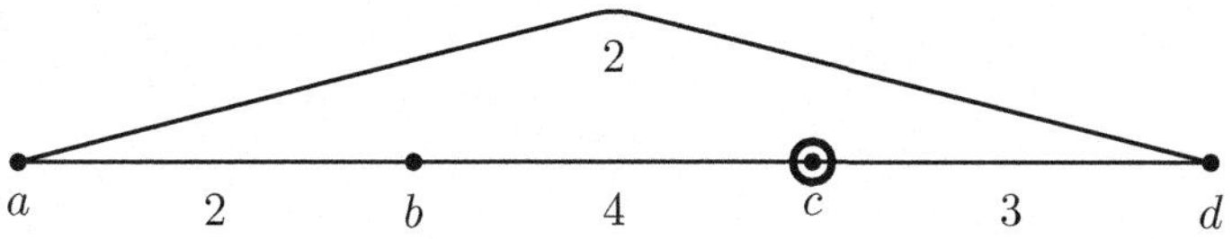


**Fig. 7.15** Strong cycle $H$ with a partial cut vertex $c$

**Claim** $P$ is not a strong path.
Assume, $P : a_1a_2a_3 \cdots a_k$ is a strong path, where $a_1 = u$ and $a_k = v$. Since $u \in V(H)$ and $v \in V(G) \setminus V(H)$ there exists $i \in \{1, 2, 3, \ldots, k-1\}$ such that $a_i \in V(H)$ and $a_{i+1} \in V(G) \setminus V(H)$. Clearly $a_ia_{i+1}$ is a strong edge in $G$, since it is an edge in the strong path $P$. Therefore $I_s(a_i, a_{i+1}) = \{a_i, a_{i+1}\}$. So $x \notin I_s(a_i, a_{i+1})$. This is a contradiction to the assumption that $x$ is a strong gate of $H$ in $G$. Therefore, $P$ is not a strong path. Hence the claim.

Since $a_ia_{i+1}$ is a weak edge, $CONN_G(a_i, a_{i+1}) > w(a_i, a_{i+1})$, weight of the edge $a_ia_{i+1}$. Since $a_ia_{i+1} \in P$, we get $m \leq w(a_ia_{i+1})$. Both these equations together gives, $CONN_G(u, v) = m < CONN_G(a_i, a_{i+1})$. This is a contradiction to the assumption that $uv$ has the maximum strength of connectivity among all pair of vertices $(m, n) \in V(H) \times V(G) \setminus V(H)$. That is, every strongest $u - v$ path is passing through $x$. So, deletion of $x$ makes a reduction in the strength of connectedness between $u$ and $v$. Therefore, $x$ is a partial cut vertex. ■

Converse of the Theorem 7.4.11 is not true always. For consider Fig. 7.15.

Here, $c$ is a partial cut vertex and it does not act as a strong gate. Since cut vertices are not part of a partial block, the corollary follows.

**Corollary 7.4.12** Every partial block is strong gate free.

We can have some sufficient conditions for a partial cut vertex to be a strong gate.

**Theorem 7.4.13** *Let $G = (V, E, w)$ be a weighted graph. A partial cut vertex $w$ of $G$ is a strong gate if either one of the following statement are true.*

*(1) $w$ is a cut vertex of $G$*
*(2) $w$ do not lie on any strong cycle.*

***Proof*** Let $G$ be a weighted graph and $w$ a partial cut vertex of $G$.

(1). Consider the case where $w$ is a cut vertex. $w$ is a cut vertex of $G$ implies $G - w$ is disconnected. Let $H$ be any component of $G - w$. Then, $w$ is a strong gate of the subgraph induced by $H$ in $G$.

(2). Assume that $w$ do not lie on any strong cycle. Since $w$ is a partial cut vertex, we can find two vertices $u$ and $v$ such that, $u$ and $v$ are strongly adjacent to $w$ and $CONN_{G-w}(u, v) < CONN_G(u, v)$. Also $G$ does not have a strong $(u, v)$ edge. Because, $w$ do not lie on any strong cycle. Now consider two sets $G_u$ and $G_v$ such that

$G_u = \{x \in V(G)\ /\ x\, is\, strongly\, connected\, to\, u\, in\, G - w\}$ and $G_v = \{y \in V(G)\ /\ y\, is\, strongly\, connected\, to\, v\, in\, G - w\}$. Here, two vertices are strongly connected means, there exists a strong path between them.
Then, clearly $G_u \cap G_v = \emptyset$. If not; that is if, $a \in G_u \cap G_v$, then, there exists a strong $u - a$ path $P_1$ (say) and a strong $a - v$ path $P_2$ (say). Then, the strong walk $P_1$ followed by $P_2$ contains a strong $u - v$ path $P$ (say) which does not contain the vertex $w$. $P$ followed by the strong path $vwu$ is a strong cycle which contains $w$. This gives a contradiction to our assumption that $w$ do not lie on any strong cycle. So, $G_u \cap G_v = \emptyset$. That is, every strong path from $G_u$ to $G \setminus G_u$ passes through $w$. So $w$ is a strong gate in $G$. ■

Following definitions are helpful in the characterization of strong gates in a weighted graph.

**Definition 7.4.14** A weighted graph $G = (V, E, w)$ is said to be **strongly connected** if, there exists a strong path between any two vertices in $G$. Maximal strongly connected subgraphs of $G$ are called **strong components** of $G$. A weighted graph is said to be **strongly disconnected** if it has at least two strong components. $\omega_s(G)$ denotes the number of strong components of $G$.

Theorem 7.4.15 is a characterization of strong gates in weighted graphs in terms of strong disconnection. Theorem 7.4.16 is also a characterization related to strongly restricted weighted graphs.

**Theorem 7.4.15** *In a connected weighted graph $G = (V, E, w)$, a vertex $x$ is a strong gate if and only if $G - x$ is strongly disconnected.*

***Proof*** Let $x$ be a strong gate of $H$ in $G$. Then every strong path from $H$ to $G \setminus H$ passes through $x$. So $G \setminus H$ is strongly disconnected.

Now consider the converse. $G$ is strongly connected but $G - x$ is strongly disconnected. Let $H$ be any strong component of $G - x$ and let $K$ be the subgraph of $G$ induced by $H$ and $x$. Then $x$ is a strong gate of $K$ in $G$. ■

**Theorem 7.4.16** *A nonempty connected weighted graph $G = (V, E, w)$ is strongly restricted if and only if any two vertices lie on a strong cycle.*

***Proof*** Consider a nonempty connected weighted graph $G = (V, E, w)$ such that $G$ is strongly restricted. Now, we want to prove that any two vertices lie on a strong cycle. We will prove this result by mathematical induction on $d_{sg}(u, v)$, where $u$ and $v$ are any two vertices in $V(G)$. Let $d_{sg}(u, v) = 1$. That is, $uv$ is a strong edge in $G$. Since, $G$ is strongly restricted it has no strong gates. So, it has no strong pendent vertices. Therefore, $d_{sg}(u) \geq 2$ and $d_{sg}(v) \geq 2$. So, consider a vertex $w \neq u$ such that $w$ is strongly adjacent to $v$. Then, $uvw$ is a strong $u - w$ path.

**Claim** There exists a strong $u - w$ path $P$ such that $P$ is internally disjoint from $uvw$.
Assume that all strong $u - w$ paths pass through $v$. Then $G - v$ is strongly disconnected. Let $\{V_u, V_w, V_*\}$ be a partition of $V(G - v)$, defined as follows. $V_u = \{x \in V(G - v)/x$ is strongly connected to $u\}$,
$V_w = \{x \in V(G - v)/x$ is strongly connected to $w\}$ and $V_* = V(G - v) \setminus (V_u \cup V_w)$. Then, $u \in V_u$ and $w \in V_w$. From the construction, it is clear that, all three, $V_u$, $V_w$ and $V_*$ are pairwise disjoint and their union is $V(G - v)$. Now, consider the subgraph $H$ of $G$, induced by $V_u \cup v$. Then $v$ is a strong gate of $H$ in $G$. It is a contradiction, since $G$ is strongly restricted. So, there exists a strong $u - w$ path $P$ such that $P$ is internally disjoint to $uvw$. Then $P$ followed by $wvu$ is a strong cycle containing $u$ and $v$.

Now assume that $\forall u, v$ with $d_{sg}(u, v) \leq n$, $u$ and $v$ lie on a strong cycle, where $n \geq 2$. Now consider two vertices $x, y$ of $G$ such that $d_{sg}(x, y) = n + 1$. Consider any $x - y$ strong geodesic. Let it be $u_0, u_1, u_2, \ldots, u_n, u_{n+1}$, where $u_0 = x$ and $u_{n+1} = y$. Since, $d_{sg}(x, u_n) = n$, by assumption there exist two internally disjoint $x - u_n$ strong paths, say $P_1$ and $P_2$. Let $w$ be the last vertex in $P_1$, strongly adjacent to $u_n$. Clearly $d_{sg}(w, y) \leq 2$. Since $u_n$ is not a strong gate, there exists a strong $w - y$ path internally disjoint from $wu_ny$. Let it be $P_3$. $P_3$ can have common vertices with $P_1$ and $P_2$. Let $a$ be the last such vertex in $P_3$. Then there are two cases.

**Case 1** $a$ lies on $P_1$.
$x - a$ sub path of $P_1$ followed by $a - y$ sub path of $P_3$ is a strong $x - y$ path. This is internally disjoint from the strong $x - y$ path formed by $P_2$ followed by the strong edge $u_ny$. Now we get two internally disjoint $x - y$ strong paths. They together form a strong cycle containing $x$ and $y$.

**Case 2** $a$ lies on $P_2$.
$P_1$ followed by the strong edge $u_ny$ is a strong $x - y$ path. Another $x - y$ strong path can be constructed as $x - a$ sub path of $P_2$ followed by $a - y$ sub path of $P_3$. These two strong paths are internally disjoint and they together form a strong cycle containing $x$ and $y$.

By mathematical induction we proved that any two vertices lie on a strong cycle, if the graph is strongly restricted. Now consider the converse. That is, we want to prove that $G$ is strongly restricted, if any two vertices lie on a strong cycle. Now consider a graph $G$, which has the property that any two vertices lie on a strong cycle. To prove that $G$ is strongly restricted, it is enough to prove that $G$ has no strong gates. since any two vertices lie on a strong cycle, for any vertex $x$ and for any pair of vertices $u, v$, there is a strong $u - v$ path not passing through $x$. So $x$ can not be a strong gate of $G$. Since $x$ is arbitrary vertex of $G$, it cannot have any strong gates. So, $G$ is strongly restricted. Hence the proof. ■

Let $x$ be a strong pendant vertex of a connected weighted graph $G$. Consider a vertex $y$ which is strongly adjacent to $x$. Let $H$ be a subgraph of $G$ induced by $x$ and $y$. Then $y$ will be a strong gate of $H$ in $G$.

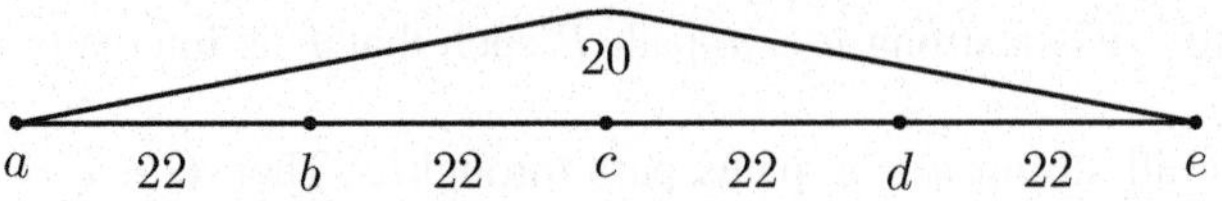


**Fig. 7.16** A weighted cycle with unique weak edge

**Fig. 7.17** A weighted graph with two triangles

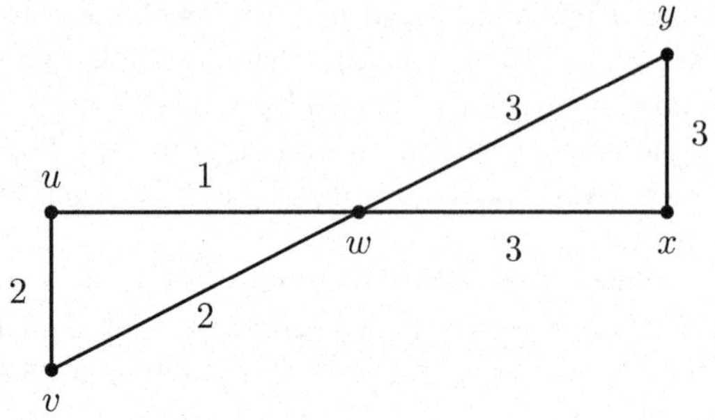


We may feel curious that whether all strong gates are cut vertices in a weighted graph. Figure 7.16 shows that it is not true generally. In this graph, $b$, $c$ and $d$ are strong gates but they are not cut vertices.

In a partial tree, not all nontrivial connected induced subgraphs are strongly gated. Star graphs and paths of length two are fully strong gated. All other partial trees are partially strong gated. In other words, all weighted graphs containing strong paths of length greater than or equal to three are not fully strong gated. They are only partially strong gated.

In a weighted graph $G$, the set of cut vertices of $G$ is a subset of the set of strong gates of $G$. The latter set is again a subset of the set of partial cut vertices of $G$. In a star graph $S_{1,n}$, all nontrivial connected induced subgraphs are strongly gated. This means that star graphs are fully strong gated. But it has only one strong gate precisely the vertex of degree $n$. In a partial tree, all vertices of strong degree more than one are strong gates. But this is not the case in a general weighted graph. In other words, all vertices of strong degree greater than one cannot be strong gates. Consider the Fig. 7.17.

$x$ and $y$ are two vertices of strong degree two. But they are not strong gates in $G$. If we take $H$ as the edge $xy$ then $b \in \bigcap_{u \in V(G) \setminus V(H)} \bigcap_{v \in V(H)} I_s(u, v)$. But $b$ is not a strong gate of $H$, since $b$ not in $V(H)$.

Strong pendant vertices cannot be strong gates. There are no strong gates in a precisely weighted graph. So minimum number of strongly gated subgraphs in a weighted graph is zero. In a star graph $S_{1,n-1}$, all nontrivial connected induced subgraphs are strongly gated.

# References

1. Mulder, H.: The interval function of graphs, MCT132. Math. Centrum, Amsterdam (1980)
2. Mathew, S., On totally weighted interconnection networks, J. Interconnection Networks, 14 (2013) 1350004(1- 16)
3. Dhanyamol, M.V., Suni, M.: On transit functions in weighted graphs, Discrete mathematics, algorithms and applications. **9**(3), 1750036 (23 Pages) (2017). https://doi.org/10.1142/S1793830917500367

# Chapter 8
# Weighted Graphs and Fuzzy Graphs

One of the most important papers concerning fuzzy set theory in recent years is one by Klement and Mesiar, [1]. In this paper, it is shown that differently defined families of fuzzy sets have lattice structures that are actually isomorphic and so theoretical results for one family can be carried over to another family.

One of the purposes of this chapter is to show that there are isomorphisms and other methods in fuzzy set theory to obtain results from one family for another. We also show by using a real world problem with real world data that even though theoretical results can be obtained for one family from another, the two families may arise naturally in an application. We wish to focus in particular on the connection between weighted graphs and fuzzy graphs.

We show the connection between weighted graphs, fuzzy graphs, $t$-norm fuzzy graphs, $t$-conorm fuzzy graphs, and a new definition of a fuzzy set proposed in [2]. We use the concepts of vulnerability and government response to modern slavery to illustrate our findings. In [2], it is stated that the departing point is the fact that not only fuzzy sets originate in Language, but that they are just 'linguistic entities' genetically different from the concept of 'crisp sets' whose origin is either in a physical collection of objects, or in a list of them. A new definition of a fuzzy set is presented by means of two magnitudes: A qualitative one, called a graph, the basic magnitude, and a quantitative one, a scalar magnitude. If the first reflects the language's relational ground of the fuzzy set, the second reflects the (numerical) extensional state in which it currently appears.

We next illustrate these ideas using the concepts of vulnerability and government response with respect to modern slavery, [3].

**Vulnerability Measures**

(1) Government issues
(2) Nourishment and access
(3) Inequality
(4) Disenfranchised groups
(5) Effect of conflicts



S. Mathew et al., *Weighted and Fuzzy Graph Theory*, Studies in Fuzziness and Soft Computing 429, https://doi.org/10.1007/978-3-031-39756-1_8

Countries are scored with respect to these five measures. Then a weighted average is of these scores is taken to provide a single score for each number. For example, the final score for Brazil is 36.4. The countries are placed into regions. Brazil is in the Americas. For this region, the highest score was 69.6 and the smallest was 10.2. The country scores were normalized using the formula (number - minimum)/(maximum - minimum) to obtain $(36.4 - 10.2)/(69.6 - 10.2) = 0.443$

**Government Response**

(1) Support for survivors
(2) Criminal justice
(3) Coordination
(4) Response
(5) Supply chains

Similarly, as for the vulnerability measures a final score is determined for each country. For example the final score for Brazil is 55.6. For the Americas, the maximum score was 71.7 and the minimum was 20.8. Hence the normalized value for Brazil was $(55.6 - 20.8)/(71.6 - 20.8) = 0.684$.

In [2], it is stated that shortening the statement $x$ is less $P$, where $P$ is a predicate, by $x \prec_P y$ facilitates the basic magnitude. That is, $x \prec_P y \subseteq X \times X$.

For our illustration, we let $P$ denote the predicate vulnerable and $X$ denote the set of countries under consideration. Now the final vulnerable score for Mexico was 57.3. Brazil's was 36.4. Hence Brazil $\prec_P$Mexico. The final value for government response for Mexico was 52.4 and for Brazil 55.6. In the case, we have Mexico $\prec_P$ Brazil if $P$ denotes government response and $\prec_P$is the linguistic relation $x$ has less government response than $y$.

In [2], a membership function $m_P : X \times X \to [0, 1]$ was introduced. It provides a numerical value for measuring the degree to which $x$ is $P$. The membership function is required to satisfy the following three properties:

(i) $x \prec_P y$ implies $m_P(x) \leq m_P(y)$.
(ii) If $z$ is minimal, then $m_P(z) = 0$.
(iii) If $w$ is maximal, then $m_P(w) = 1$.

We see that our membership function $m_P(x) = \frac{\#(x)-\min}{\max - \min}$, where $\#(x)$ denotes the final score of $x$, satisfies these three properties. Thus $m_V(Brazil) = 0.443$, where $V$ denotes vulnerable and $m_G(Brazil) = 0.684$, where $G$ denotes government response. For Mexico, we have $m_V(Mexico) = 0.796$ and $m_G(Mexico) = 0.621$.

In Sect. 8.1, we present some isomorphisms and other methods in fuzzy set theory to obtain results from one family for another. Section 8.2. provides results that are the most pertinent to the book. These results deal with weighted and fuzzy graphs. Other sections consider some of these isomorphisms in a little less detail than Sect. 8.2.

## 8.1 Isomorphisms of Different Families of Fuzzy Sets

In this section, we show that there are isomorphisms and other methods in fuzzy set theory to obtain results from one family for another.

*Neutrosophic fuzzy sets and Pythagorean fuzzy sets:* Recall that a neutrosophic fuzzy set is a triple $(\sigma, \tau, \mu)$ of fuzzy subsets of a set, [4]. It is based on the lattice of elements $(x_1, x_2, x_3) \in [0, 1]^3$,where $(x_1, x_2, x_3) \leq (y_1, y_2, y_3)$ if and only if $x_1 \leq y_1, x_2 \leq y_2$, and $x_3 \geq y_3$. Also, a Pythagorean fuzzy set is a pair of fuzzy subsets $(\sigma, \tau)$ of a set $X$ such that for all $x \in X, \sigma(x)^2 + \tau(x)^2 \leq 1$,[5] We can see that vulnerability and government response corresponding to modern slavery are opposites, [6]. That is, an increase in government response by a country would lower the country's vulnerability. However, $m_V(Brazil) + m_G(Brazil) = 0.443 + 0.684 > 1$. This gives meaning to neutrosophic fuzzy sets even though certain theoretical results can following immediately from other types of fuzzy sets. Also, $(0.443)^2 + (0.684)^2 = 0.096 + 0.468 < 1$. Consequently, similar comments might be able to be made here even though Pythagorean fuzzy sets and intuitionistic fuzzy sets, [7], have corresponding isomorphic lattices. However, this isomorphism may make the situation different to the neutrosophic case since it is so straight forward. The lattice isomorphism $f$ involved here is $f : P^* \to L^*$ defined by $f((x_1, x_2)) = (x_1^2, x_2^2)$, where $L^* = \{(x_1, x_2) \in [0, 1] | x_1 + x_2 \leq 1\}$ and $P^* = \{(x_1, x_2) \in [0, 1] | x_1^2 + x_2^2 \leq 1\}$. The paper by Klement and Mesiar contains many other cases, where various families of fuzzy sets have isomorphic lattices.

*Linguistic variables*: The size of flow of trafficked people from country to country is given in [8]. It is reported in linguistic terms since accurate data concerning the size of the flow is impossible to obtain. Information is provided with respect to the reported human trafficking in terms of origin, transit, and/or destination according to the citation index. The data is provided in two columns. Information in the left column as to whether a country ranks(very) low, medium (very) high depends upon the total number of sources which made reference to this country as one of origin, transit, or destination. Information provided in then the right column provides further detail to the information provided in the left column. If a country in the right column was mentioned by one or two sources, the related country was ranked low. If linkage between the countries in the two columns was reported by 3-5 sources, the related country was ranked medium. If 5 or more sources linked the two countries, the country in the right was ranked high. This method of combining linguistic data provides an ideal reason for the use of mathematics of uncertainty to study the problem of trafficking by persons. For example, by assigning numbers in the interval $[0, 1]$ to the linguistic data, the data can be combined in a mathematical way. In [9], the notions of $t$-norms and $t$-conorms were used. The number 0.1 can be assigned very low, 0.3 to low. 0.5 to medium, 0.7 to high and 0.9 to very high. Using the notation and ideas from [2]], we have $x \prec_P y$ if and only if country $x$'s linguistic rank is less than country $y$'s linguistic rank. We have $m_P(x) = 0.1, 0.3, 0.5, 0.7$, or $0.9$ if $x$ is assigned very low, low, medium, high, or very high, respectively. We note that here $m_P$ does not satisfy $(ii)$ and $(iii)$.

*$t$-norms and $t$-conorms*: Suppose $X$ denotes a set of countries involved in human trafficking. Suppose also that $x, y \in X$ represent vertices of a graph and suppose that $xy$ is an edge in the graph such that there is trafficking between $x$ and $y$. Then $m_V(x)$ and $m_V(y)$ denotes the measure of vulnerability for $x$ and $y$, respectively. Suppose $m_V(x) = 0.6$ and $m_V(y) = 0.8$. If we wish to determine a joint vulnerability, we might use a $t$-conorm, say maximum. Then $m_V(x) \vee m_V(y) = 0.8$. However if $m_V(x) = 0.1$, then $m_V(x) \vee m_V(y) = 0.8$ also. It seems that the latter result should be smaller since $0.1 < 0.6$. Thus it seems more realistic to use another $t$-conorm, say algebraic sum $\oplus$. Then $m_V(x) \oplus m_V(y) = 0.82$ (compared with $0.92$). Now consider $m_G(x)$ and $m_G(y)$ as the measure of government response for $x$ and $y$, respectively. Suppose that $m_G(x) = 0.7$ and $m_G(y) = 0.4$. If we wish to determine a joint government response, we might use a $t$-norm, say minimum. Then $m_G(x) \wedge m_G(y) = 0.4$. However if $m_G(x) = 0.5$, then $m_G(x) \wedge m_G(y) = 0.4$ also. It seems that the latter result should be smaller since $0.5 < 0.7$. Hence it seems more realistic to use a different $t$-norm, say product $\bullet$. Then $m_G(x) \bullet m_G(y) = 0.2$ (compared with $0.28$).

In Sect. 8.3, we consider a complementary relationship between $t$-norm fuzzy graphs and $t$-conorm fizzy graphs due to a certain isomorphism between two lattices.

*Fuzzy algebra*: In [10, 11], the reduction of fuzzy algebra to classical algebra is considered. In [10], the following three ideas are presented, the Metatheorem, the lattice embedding for sets, and the lattice embedding for algebras. The Metatheorem allows the conversion of existing theorems about classical subsets into corresponding theorems about fuzzy subsets. The concept of a fuzzifiable operation on a power set is defined. The main result states that any implication or identity which can be stated using fuzzifiable operations is true if and only if it is true about classical subsets. The lattice embedding theorem for sets shows that for any set $X$, there is a set $Y$ such that the lattice of fuzzy subsets of $X$ is isomorphic to a sublattice of the classical subsets of $Y$. In fact, if $X$ is infinite, then $Y$ can be chosen to be $X$. Thus the lattice of fuzzy subsets of $X$ is isomorphic to a sublattice of the classical subsets of $X$ itself. The lattice embedding theorem for algebras shows that under certain circumstances the lattice of fuzzy subalgebras of an algebra $A$ embeds into the lattice of classical subalgebras of a closely related algebra. We consider the situation here in detail in Chap. 9.

*Nonstandard analysis*: We next consider nonstandard analysis and the transfer principle, [12]. It is known for the real numbers $\mathbb{R}$, that the structure $(\mathbb{R}, +, *, 0, 1, <$ $)$ is an ordered field of real numbers. It has an extension to an ordered field $(\mathbb{R}^*, +, *, 0, 1, <)$, [12–15]. The field $\mathbb{R}^*$has a positive infinitesimal element $\in$, i.e, $\in > 0$, but $\in <$ every positive real number $r$. Since $\mathbb{R}^*$ is a field $\in^{-1}$ exists in $\mathbb{R}^*$ as does $-\in$. The development of $\mathbb{R}^*$ by Abraham Robinson provides a powerful mathematical tool. A complete discussion here is beyond our purpose. Our purpose is to mention the transfer principle. Let $\mathbb{R}_{fin} = \{x \in \mathbb{R}^* | \ |x| < n \text{ for some } n \in \mathbb{N}\}$. Then $\mathbb{R}_{fin}$ is a subring of $\mathbb{R}^*$. Let $M = \{x \in \mathbb{R}^* | \ |x| \leq \frac{1}{n} \text{ for all } n \in \mathbb{N} \text{ such that } n > 0\}$. Then $M$ is the unique maximal ideal of $\mathbb{R}_{fin}$. The transfer principle allows one to show that $\mathbb{R}^*$ has all the properties of $\mathbb{R}$, but it also allows one to prove theorems about $\mathbb{R}$ by first proving them in $\mathbb{R}^*$ and then transferring them back to $\mathbb{R}$. The situation here is discussed further in Sect. 2.4. [16] is pertinent to a discussion above.

**Weighted graphs and fuzzy graphs**: We next consider weighted graphs and fuzzy graphs. We consider the isomorphism of the two given in [17]. This isomorphism is in keeping with those in [1]. The concepts of complements, $t$-norms, $t$-conorms, and nonstandard analysis, [12], are examined in [17].

## 8.2 Weighted Graphs and Fuzzy Graphs

In this section, we present the main results of the chapter. It has been shown in [1] that in the two-dimensional case, the lattices of truth values considered are pairwise isomorphic, and so are the corresponding families of fuzzy sets. Therefore, each result for one of these types of fuzzy sets can be directly rewritten for each (isomorphic) type of fuzzy sets. In this chapter, we show that there is a strong connection between weighted graphs and fuzzy graphs. We accomplish this by using lattice isomorphisms. Consequently, under certain conditions, results for one area can be carried over immediately to the other. Many situations in fuzzy graph theory do not depend on the weights of the vertices. The situation of providing weights for the vertices of a weighted graph is also considered. We also consider lattice homomorphisms with an illustration involving nonstandard analysis. In particular, we consider a nonstandard weighted graph, i.e., a graph where the weights of the edges are from a nonstandard interval.

In [1], it is shown that many well-known generalizations of the concept of fuzzy sets with two-dimensional lattices of truth values are pairwise isomorphic and so are the corresponding families of fuzzy sets. Therefore, each result for one of these types of fuzzy sets can be directly rewritten for each isomorphic type of fuzzy set. In this chapter, it is shown that this also holds for weighted graphs and fuzzy graphs in certain circumstances. Many situations in fuzzy graph theory do not depend on the weights of the vertices. The situation of providing weights for the vertices of a weighted graph is also considered. We also develop some beginning results for lattice homomorphisms and illustrate the results using the nonstandard interval $[0, 1]^*$ in nonstandard analysis and the standard unit interval $[0, 1]$.

We begin by reviewing some results from [17]. We let $\mathbb{N}$ denote the set of positive integers.

For two partially ordered sets $(L_1, \leq_1)$ and $(L_2, \leq_2)$, a function $\varphi : L_1 \to L_2$ is called an **order homomorphism** if it preserves monotonicity, i.e., if $x \leq_1 y$ implies $\varphi(x) \leq_2 \varphi(y)$. If $(L_1, \leq_1)$ and $(L_2, \leq_2)$ are two lattices, then a function $\varphi : L_1 \to L_2$ is called a **lattice homomorphism** if it preserves finite meets and joins, i.e., if for all $x, y \in L_1$, $\varphi(x \wedge_1 y) = \varphi(x) \wedge_2 \varphi(y)$ and $\varphi(x \vee_1 y) = \varphi(x) \vee_2 \varphi(y)$. Each lattice homomorphism is an order homomorphism, but not conversely. A lattice homomorphism $\varphi : L_1 \to L_2$ is called an **embedding** (or **monomorphism**) if it is injective, an **epimorphism** if it is surjective, and an **isomorphism** if it is bijective, i.e., if it is both an embedding and an epimorphism.

Suppose that $(L_1, \leq_1)$ and $(L_2, \leq_2)$ are isomorphic lattices and that $\varphi : L_1 \to L_2$ is a lattice isomorphism of $L_1$ onto $L_2$. Let the bottom and top elements of $(L_1, \leq_1)$

be denoted by $0_1$ and $1_1$, respectively. Let $A_1 : L_1 \times L_1 \to L_1$ be an associative, commutative order homomorphism and define the function $A_2 : L_2 \times L_2 \to L_2$ by

$$A_2(x, y) = \varphi(A_1((\varphi^{-1}(x), \varphi^{-1}(y))).$$

If $A_1$ is a $t$-norm, then $A_2$ is a $t$-norm. If $A_1$ is a $t$-conorm, then $A_2$ is a $t$-conorm, [16], p. 5].

Note that $\varphi^{-1}(A_2(x, y)) = A_1((\varphi^{-1}(x), \varphi^{-1}(y))$ and so $\varphi^{-1}(A_2(\varphi(w), \varphi(z))) = A_1(w, z)$,where $w = \varphi^{-1}(x)$ and $z = \varphi^{-1}(y)$.

**Definition 8.2.1** Let $c$ be a function of $L$ into $L$. Here we are considering a lattice $(L, \vee, \wedge, 0, 1)$.

Consider the following conditions:

(1) $c(0) = 1$ and $c(1) = 0$ (boundary conditions).
(2) For all $a, b \in L$, if $a \leq b$, then $c(a) \geq c(b)$ (monotonicity).
(3)c is continuous.
(4)c is involutive, i.e., $c(c(a)) = a$ for all $a \in L$.

If $c$ satisfies conditions (1) and (2), we say that $c$ is a **complement** on $L$.

Suppose that $\varphi$ is a lattice isomorphism of $L_1$ onto $L_2$. Let $c_1$ be a complement on $L_1$. Define $c_2(x) = \varphi(c_1(\varphi^{-1}(x)))$ for all $x \in L_2$. Then $c_2(1_2) = \varphi(c_1(\varphi^{-1}(1_2))) = \varphi(c_1(1_1)) = \varphi(0_1) = 0_2$. Also, $c_2(0_2) = \varphi(c_1(\varphi^{-1}(0_2))) = \varphi(c_1(0_1)) = \varphi(1_1) = 1_2$.

Let $x, y \in L_2$ be such that $x \leq_2 y$. Then $\varphi^{-1}(x) \leq_1 \varphi^{-1}(y)$. Thus $c_2(x) = \varphi(c_1(\varphi^{-1}(x))) \geq_2 \varphi(c_1(\varphi^{-1}(y)) = c_2(y)$.

Let $c_2$ be a complement of $L_2$. Define $c_1(x) = \varphi^{-1}(c_2(\varphi(x)))$ for all $x \in L_1$. Then $c_1(1_1) = \varphi^{-1}(c_2(\varphi(1_1))) = \varphi^{-1}(c_2(1_2)) = \varphi^{-1}(0_2) = 0_1$. Also, $c_1(0_1) = \varphi^{-1}(c_2(\varphi(0_1))) = \varphi^{-1}(c_2(0_2)) = \varphi^{-1}(1_2) = 1_1$.

Let $x, y \in L_1$ be such that $x \leq_1 y$. Then $\varphi(x) \leq_2 \varphi(y)$. Thus $c_1(x) = \varphi^{-1}(c_2(\varphi(x))) \geq_2 \varphi^{-1}(c_2(\varphi(y))) = c_1(y)$.

We have just shown the following result.

**Theorem 8.2.2** *$c_1$ is a complement if and only if $c_2$ is a complement.*

*Note that if $c_2(x) = \varphi(c_1(\varphi^{-1}(x)))$, then $\varphi^{-1}(c_2(x)) = c_1(\varphi^{-1}(x))$ and so $\varphi^{-1}(c_2(\varphi(y))) = c_1(\varphi^{-1}(\varphi(y))) = c_1(y)$, where $y = \varphi^{-1}(x)$.*

**Theorem 8.2.3** *$c_1$ is involutive if and only if $c_2$ is involutive.*

***Proof*** Suppose $c_1$ is involutive. Let $x \in L_2$. Then

$$\begin{aligned} c_2(c_2(x)) &= c_2(\varphi(c_1(\varphi^{-1}(x))) = \varphi((c_1(\varphi^{-1}(\varphi(c_1(\varphi^{-1}(x)))) \\ &= \varphi(c_1(c_1(\varphi^{-1}(x))) = \varphi(\varphi^{-1}(x)) = x. \end{aligned}$$

The converse is now immediate.

We next apply the above results to weighted graphs.

Let $(V, E, w)$ be a weighted graph, where $V$ is a finite set of vertices and $w$ is a function of the set of edges $E$ into the positive real numbers. We assume no loops and at most one edge between two vertices. Let $m \geq \vee\{w(e) | e \in E\}$. We hold $m$ fixed throughout. We also assume $m \geq 1$. For all $e \in E$, define $\mu : E \to [0, 1]$ by $\mu(e) = \frac{1}{m}w(e)$. Then $G = (V, E, \sigma, \mu)$ is a fuzzy graph if for all $v \in V, \sigma(v) = 1$. Consider $(V, E, \mu)$ as a weighted graph. We note below that the lattices associated with $(V, E, w)$ and $(V, E, \mu)$ are isomorphic. (Clearly there one-to-one correspondence between their point sets which preserves adjacency, namely the identity map of $V$ onto $V$.)

Define $\varphi : [0, m] \to [0, 1]$ by for all $x \in [0, m], \varphi(x) = \frac{1}{m}x$. Then $\varphi$ is a lattice isomorphism of $L_1$ onto $L_2$, where $L_1 = [0, m]$ and $L_2 = [0, 1]$ and $\vee$ and $\wedge$ are the usual operations of maximum and minimum, respectively. Note that for all $x, y \in L_1, \varphi(x \wedge y) = \frac{x \wedge y}{m} = \frac{x}{m} \wedge \frac{y}{m} = \varphi(x) \wedge \varphi(y)$ and a similar result holds for $\vee$. Also, $\varphi$ is continuous and preserves $<$ .

We next consider Definition 8.2.1 for $[0, m]$.

**Theorem 8.2.4** *([18]) Let* $m = 1$. *If* $c : [0, 1] \to [0, 1]$ *satisfies* (2) *and* (4) *of Definition 8.1.1, then c satisfies* (1) *and* (3). *Also, c is bijective.*

Let $c$ satisfy (1) and (2) of Definition 8.2.1. Define $\widehat{c} : [0, m] \to [0, m]$ by for all $a \in [0, m], \widehat{c}(a) = \varphi^{-1}(c((\varphi(a))$. Then $\widehat{c}(0) = \varphi^{-1}(c(0)) = \varphi^{-1}(1) = m$ and $\widehat{c}(m) = \varphi^{-1}(c(1)) = \varphi^{-1}(0) = 0$.

Let $C = \{c | \ c$ satisfies (1) and (2) for $m = 1\}$ and $\widehat{C} = \{\widehat{c} | \ \widehat{c}$ satisfies (1) and (2) for $m > 1\}$. Define the function $f$ of $C$ into $\widehat{C}$ by for all $c \in C$, $f(c) = \widehat{c}$, where for all $a \in [0, m]$,

$$\widehat{c}(a) = mc(\frac{a}{m}).$$

We show that $f$ is a one-to-one function of $C$ onto $\widehat{C}$. We have that $\widehat{c}(0) = mc(0) = m$ and $\widehat{c}(m) = m(c(1)) = 0$. Since $m$ is fixed, that $\widehat{c}$ satisfies (2) holds since $c$ satisfies (2). Hence $f$ maps $C$ into $\widehat{C}$. Let $\widehat{c} \in \widehat{C}$. Define $c : [0, 1] \to [0, 1]$ by for all $a \in [0, 1], c(a) = \frac{1}{m}\widehat{c}(ma)$. Then $c(0) = \frac{1}{m}\widehat{c}(0) = 1$ and $c(1) = \frac{1}{m}\widehat{c}(m) = 0$. Thus $f$ maps $C$ onto $\widehat{C}$. Now

$$\begin{aligned} c_1 &= c_2 \Leftrightarrow \forall a \in [0, m], c_1(\frac{a}{m}) = c_2(\frac{a}{m}) \\ &\Leftrightarrow \forall a \in [0, m], mc_1(\frac{a}{m}) = mc_2(\frac{a}{m}) \\ &\Leftrightarrow \forall a \in [0, m], \widehat{c_1}(a) = \widehat{c_2}(a) \\ &\Leftrightarrow \widehat{c_1} = \widehat{c_2} \\ &\Leftrightarrow f(c_1) = f(c_2). \end{aligned}$$

**Theorem 8.2.5** *c is involutive if and only if* $\widehat{c}$ *is involutive, where* $f(c) = \widehat{c}$.

***Proof*** Suppose $c$ is involutive. Let $a \in [0, m]$. Then

$$\widehat{c}(\widehat{c}(a)) = \widehat{c}(m(c(\frac{a}{m})) = m(c(\frac{mc(\frac{a}{m})}{m}) = m(c(c(\frac{a}{m})) = m(\frac{a}{m}) = a.$$

Thus $\widehat{c}$ is involutive.

Conversely, suppose $\widehat{c}$ is involutive. Let $a \in [0, m]$. Then $c(c(\frac{a}{m})) = c(\frac{\widehat{c}(a)}{m}) = \frac{\widehat{c}(\widehat{c}(a))}{m} = \frac{a}{m}$. Hence $c$ is involutive.

**Definition 8.2.6** Let $\pi : [0, m] \times [0, m] \to [0, m]$. Then $\pi$ is called a ***t*-norm** on $[0, m]$ if the following conditions hold for all $a, b, d \in [0, m]$:

(1) $\pi(a, m) = a$ (boundary condition).
(2) $b \le d$ implies $\pi(a, b) \le \pi(a, d)$ (monotonicity).
(3) $\pi(a, b) = \pi(b, a)$ (commutativity).
(4) $\pi(a, \pi(b, d)) = \pi(\pi(a, b), d)$ (associativity).

Let $i$ be a $t$-norm on $[0, 1]$. Define $\pi : [0, m] \times [0, m] \to [0, m]$ by for all $a, b \in [0, m]$, $\pi(a, b) = mi(\frac{a}{m}, \frac{b}{m})$. Let $a \in [0, m]$. Then $\pi(a, m) = mi(\frac{a}{m}, \frac{m}{m}) = m(\frac{a}{m}) = a$. Note that $\pi(a, b) = \varphi(i(\varphi^{-1}(a), \varphi^{-1}(b)))$, where $\varphi(x) = \frac{x}{m}$ and $x \in [0, m]$. Note also that $\varphi^{-1}(y) = my$, where $y \in [0, 1]$. Check: $\varphi(i(\varphi^{-1}(y_1), \varphi^{-1}(y_2))) = \varphi(i(my_1, my_2)) = \frac{1}{m}i(my_1, my_2)$.

Let $\pi$ be a $t$-norm on $[0, m]$. Define $i : [0, 1] \times [0, 1] \to [0, 1]$ by for all $a, b \in [0, 1]$, $i(a, b) = \frac{1}{m}\pi(ma, mb)$. Let $a \in [0, 1]$. Then $i(a, 1) = \frac{1}{m}\pi(ma, m) = \frac{1}{m}ma = a$.

Suppose for example that $i$ is the $t$-norm product on $[0, 1]$. Let $a, b \in [0, m]$. Then $\pi(a, b) = mi(\frac{a}{m}, \frac{b}{m}) = m\frac{a}{m}\frac{b}{m} = \frac{ab}{m}$.

**Definition 8.2.7** Let $\rho : [0, m] \times [0, m] \to [0, m]$. Then $\pi$ is called a ***t*-conorm** on $[0, m]$ if the following conditions hold for all $a, b, d \in [0, m]$:

(1) $\rho(a, 0) = a$ (boundary condition).
(2) $b \le d$ implies $\rho(a, b) \le \rho(a, d)$ (monotonicity).
(3) $\rho(a, b) = \rho(b, a)$ (commutativity).
(4) $\rho(a, \rho(b, d)) = \rho(\rho(a, b), d)$ (associativity).

Let $u$ be a $t$-conorm on $[0, 1]$. Define $\rho : [0, m] \times [0, m] \to [0, m]$ by $a, b \in [0, m]$, $\rho(a, b) = mu(\frac{a}{m}, \frac{b}{m})$. Let $a \in [0, m]$. Then $\rho(a, 0) = mu(\frac{a}{m}, \frac{0}{m}) = m(\frac{a}{m}) = a$.

Let $\rho$ be a $t$-conorm on $[0, m]$. Define $u : [0, 1] \times [0, 1] \to [0, 1]$ by for all $a, b \in [0, 1]$, $u(a, b) = \frac{1}{m}\rho(ma, mb)$. Let $a \in [0, 1]$. Then $u(a, 0) = \frac{1}{m}\rho(ma, 0) = \frac{1}{m}ma = a$.

Suppose for example that $u$ is the $t$-conorm algebraic sum on $[0, 1]$. Let $a, b \in [0, m]$. Then $\rho(a, b) = mu(\frac{a}{m}, \frac{b}{m}) = m(\frac{a}{m} + \frac{b}{m} - \frac{a}{m}\frac{b}{m}) = a + b - \frac{ab}{m}$.

Recall that $\varphi : [0, m] \to [0, 1]$, where for all $a \in [0, m]$, $\varphi(a) = \frac{a}{m}$ is an isomorphism. Also if $e_1, e_2 \in E$, then $w(e_1) \le w(e_2)$ if and only if $\mu(e_1) \le \mu(e_2)$.

Define $w \circ w$ by for all $x, y \in V$, $(w \circ w)(x, y) = \vee\{w(xz) \wedge w(zy) | z \in V\}$. Let $w^2 = w \circ w$. Suppose $n$ is a positive integer and that $w^n$ has been defined. Define $w^{n+1}$ to be $w^n \circ w$. Define $w^\infty$ by $w^\infty(x, y) = \vee\{w^n(x, y) | n = 1, 2, ...\}$.

Define $\mu : E \to [0, 1]$ by $e \in E$, $\mu(e) = \frac{1}{m}w(e)$. Then

$$\begin{aligned}(\mu \circ \mu)(x, y) &= \vee\{\mu(xz) \wedge \mu(zy)|z \in V\}\\ &= \vee\{\frac{1}{m}w(xz) \wedge \frac{1}{m}w(zy)|z \in V\}\\ &= \frac{1}{m} \vee \{w(xz) \wedge w(zy)|z \in V\}\\ &= \frac{1}{m}(w \circ w)(x, y).\end{aligned}$$

It follows by induction that $\mu^n(x, y) = \frac{1}{m}w^n(x, y)$ for all positive integers $n$. Thus $\mu^\infty(x, y) = \frac{1}{m}w^\infty(x, y)$.

Let $xy \in E$. Let $G' = (V, E\backslash\{xy\}, w')$ be the weighted subgraph of $G = (V, E, w)$ obtained by deleting the edge $xy$ from $E$ and defining $w'$ on $E\backslash\{xy\}$ by $w'(uv) = w(uv)$ for all $uv \in E\backslash\{xy\}$. Then $xy$ is called a **bridge** in $G$ if $\omega'^\infty(uv) < w^\infty(uv)$ for some $uv \in E\backslash\{xy\}$. Clearly, $xy$ is a bridge in $G$ if and only if $xy$ is a bridge in the fuzzy graph $(V, E, \sigma, \mu)$, where $\mu(uv) = \frac{1}{m}w(uv)$ for all $uv \in E$.

It is now easy to see that the proof of the following result can be copied from the proof of Theorem 9.1, [[19], p. 90].

**Theorem 8.2.8** *Let $(V, E, w)$ be a weighted graph. Then the following statements are equivalent.*

*(1) $xy$ is a bridge;*
*(2) $w'^\infty(xy) < w(xy)$;*
*(3) $xy$ is not the weakest edge of any cycle.*

We next consider placing weights on the vertices of weighted graphs. Let $(V, E, w)$ be a weighted graph. Let $m \geq \vee\{w(e)|e \in E\}$. Hold $m$ fixed. Define $\tau : V \to [0, m]$ by for all $x \in V, \tau(x) = \vee\{w(xy)|y \in V\}$. Then for all $uv \in E, w(uv) \leq \tau(u) \wedge \tau(v)$. That is, $(V, E, w, \tau)$ is a weighted graph with a weight on the vertices. Define $\sigma : V \to [0, 1]$ by for all $x \in V, \sigma(x) = \frac{1}{m}\tau(x)$. Define $\mu : E \to [0, 1]$ by for all $xy \in E, \mu(xy) = \frac{1}{m}w(xy)$. Since $w(x) \leq \tau(x) \wedge \tau(y)$, it follows that $\mu(xy) \leq \sigma(x) \wedge \sigma(y)$. That is, $(V, E, \sigma, \mu)$ is a fuzzy graph.

Let $(V, E, \sigma, \mu)$ be a fuzzy graph. Define $\rho : V \to [0, m]$ by $\forall v \in V, \rho(v) = m\sigma(v)$. Define $w$ as before, i.e., $w(e) = m\mu(e)$. Then $(V, E, \rho, \omega)$ is a weighted graph with a weight on the vertices. Clearly, for all $x, y \in V, w(xy) \leq \rho(x) \wedge \rho(y)$.

## 8.3 $t$-Conorm Fuzzy Graphs

It is interesting to note the results that do not hold for an arbitrary $t$-conorm and those that do hold with similar proofs as those with the $t$-conorm maximum.

The **strength** of a path $P$ is defined to be $\vee_{i=1}^n \nu(x_{i-1}x_i)$, where $P : x_{1,\ldots,}x_n$. In other words, the strength of a path is defined to be the weight of the strongest arc of the path. Let $Q$ be another path. We say that the strength of $Q$ is less than the strength of $P$ if $\vee_{i=1}^{m} \nu(y_{i-1}y_i)$ is less than $\vee_{i=1}^{n} \nu(x_{i-1}x_i)$, where $Q : y_1, ..., y_m$.

**Definition 8.3.1** Let $G = (V, E)$ be a graph. Let $\tau$ be a fuzzy subset of $V$ and $\nu$ be a fuzzy subset of $E$. Then $(\tau, \nu)$ is called a **fuzzy $\oplus$-subgraph** of $G$ w.r.t. a $t$-conorm $\oplus$ if for all $uv \in E$, $\nu(uv) \geq \tau(u) \oplus \tau(v)$.

Let $k$ be a positive integer. Define $\nu^k(u, v) = \wedge\{\nu(uu_1) \oplus ... \oplus \nu(u_{n-1}v) | P : u = u_0, u_1, ..., u_{n-1}, u_n = v$ is a path of length $k$ from $u$ to $v\}$. Let $\nu_{\oplus}^{\infty}(u, v) = \wedge\{\nu^k(u, v) | k \in \mathbb{N}\}$, where $\mathbb{N}$ denotes the positive integers. A **weakest** path joining two nodes $u, v$ has strength $\nu_{\oplus}^{\infty}(u, v)$.

**Definition 8.3.2** Let $(\tau, \nu)$ be a fuzzy graph w.r.t. $\oplus$. Let $xy \in E$. Then $xy$ is called a **bridge** if $\nu_{\oplus}'^{\infty}(u, v) > \nu_{\oplus}^{\infty}(u, v)$ for some $u, v \in V$, where $\nu'(xy) = 1$ and $\nu' = \nu$ otherwise.

The following result for $t$-norms is known. The isomorphism involving fuzzy graphs and complementary fuzzy graphs allows the following result to be converted from the result involving $t$-norms. However, we provide the proof to demonstrate as an example how the conversion occurs.

**Theorem 8.3.3** *Let $(\tau, \nu)$ be a fuzzy graph w.r.t. $\oplus$. Let $xy \in E$. Let $\nu'$ be the fuzzy subset of $E$ such that $\nu'(xy) = 1$ and $\nu' = \nu$ otherwise. Then* (3) $\Rightarrow$ (2) $\Leftrightarrow$ (1).

*(1) $xy$ is a bridge with respect to $\oplus$.*
*(2) $\nu_{\oplus}'^{\infty}(x, y) > \nu(xy)$.*
*(3) $xy$ is not a strongest arc of any cycle.*

***Proof*** (2) $\Rightarrow$ (1) : Suppose $xy$ is not a bridge. Then $\nu_{\oplus}'^{\infty}(x, y) = \nu_{\oplus}^{\infty}(x, y) \leq \nu(xy)$.

(3) $\Rightarrow$ (2) : Suppose $\nu_{\oplus}'^{\infty}(x, y) \leq \nu(xy)$. Then there exists a path from $x$ to $y$ not involving $xy$ that has strength $\leq \nu(xy)$. This path together with $xy$ forms a cycle of which $xy$ is the strongest arc.

(1) $\Rightarrow$ (2) : Suppose $xy$ is a bridge w.r.t. $\oplus$. Then there exists $u, v \in V$ such that $\nu_{\oplus}'^{\infty}(u, v) > \nu_{\oplus}^{\infty}(u, v)$, where $\nu'(xy) = 1$ and $\nu' = \nu$ otherwise. Thus if $Q$ is a weakest path in $G$ from $u$ to $v$, then $Q$ must contain $xy$. Hence $\nu(xy) \leq \nu_{\oplus}^{\infty}(u, v) < \nu_{\oplus}'^{\infty}(u, v)$. Let $P$ be a weakest path from $x$ to $y$ in $G$. Then $\nu_{\oplus}^{\infty}(x, y) \leq \nu(xy)$ and so the strength of $(Q - xy) \cup P$ is $\leq$ the strength of $Q$. Hence the path $(Q - xy) \cup P$ must contain $xy$ and so $P$ must contain $xy$. Hence $\nu(xy) = \nu_{\oplus}^{\infty}(x, y)$. Thus $xy$ is a weakest path from $x$ to $y$. Hence $\nu(xy) < \nu_{\oplus}'^{\infty}(x, y)$, where strict inequality holds since $\nu_{\oplus}'^{\infty}(x, y)$ is the strength of a weakest path in $G - xy$.

**Example 8.3.4** Let $V = \{x, y, z\}$ and $E = \{xy, xz, yz\}$. Let $\oplus$ denote algebraic sum. Define the fuzzy subset $\nu$ of $E$ as follows: $\nu(xy) = 0.5$, $\nu(xz) = 0.4$, and $\nu(yz) = 0.4$. Define the fuzzy subset $\tau$ of $V$ by $\tau(x) = \tau(y) = \tau(z) = 0.2$. Then $(\tau, \nu)$ is a $\oplus$-fuzzy subgraph of $G = (V, E)$ and a $\vee$-fuzzy subgraph of $G$. Now $xy$ is a $\oplus$-bridge since $\nu_{\oplus}'^{\infty}(x, y) = 0.4 \oplus 0.4 = 0.66 > 0.5 = \nu(xy)$ and so (2) and therefore (1) holds by Theorem 8.3.3. However, $xy$ is a strongest edge of the cycle, $x, y, z, x$. Thus (3) does not hold.

Suppose that $\oplus = \vee$. We show that (2) implies (3) of Theorem 8.3.3. Suppose that (3) does not hold. Then $xy$ is a strongest edge of some cycle. Now $\nu'^k(u, v) = \wedge\{\nu'(uu_1) \vee ... \vee \nu'(u_{n-1}v) | P : u = u_0, u_1, ..., u_{n-1}, u_n = v$ is a path of length $k$ from $u$ to $v\}$ and $\nu'^{\infty}_{\oplus}(u, v) = \wedge\{\nu'^k(u, v) | k \in \mathbb{N}\}$. Thus $\nu'^{\infty}_{\vee}(x, y) \leq \nu(xy)$ by definition of $\nu'^{\infty}_{\vee}(x, y)$. Thus (2) does not hold.

**Example 8.3.5** Let $G$ be the graph in Example 8.3.4. Let $\tau$ and $\nu$ be defined as in Example 8.3.4. Then $(\tau, \nu)$ is a fuzzy $\vee$-tree, but not a fuzzy $\oplus$-tree.

A fuzzy graph $H = (\tau, \nu)$ is called a partial fuzzy supergraph of $G = (\sigma, \mu)$ if $\tau \supseteq \sigma$ and $\nu \supseteq \mu$. The fuzzy graph $H = (P, \tau, \nu)$ is called a **fuzzy super graph** of $G = (V, \sigma, \mu)$ if $P \supseteq V$, $\tau(x) = \sigma(x)$ for all $x \in V$, and $\nu(xy) = \mu(xy)$ for all $x, y \in V$.

We say that the partial fuzzy super graph $(\tau, \nu)$ spans the fuzzy graph $(\sigma, \mu)$ if $\tau = \sigma$. In this case, we call $(\tau, \nu)$ a spanning fuzzy supergraph of $(\sigma, \mu)$. For any fuzzy subset $\tau$ of $V$ such that $\rho \supseteq \sigma$, the partial fuzzy supergraph of $(\sigma, \mu)$ induced by $\tau$ is the minimal partial fuzzy supergraph of $(\sigma, \mu)$ that has vertex set $\tau$. This is the partial fuzzy graph $(\tau, \nu)$, where $\nu(xy) = \tau(x) \oplus \tau(y) \oplus \mu(xy)$.

A crisp graph that has no cycles is called **acyclic** or a **forest.** A connected forest is called a **tree**. A fuzzy graph is called a **forest** if the graph consisting of its edges with weight $\neq 1$ is a forest and a **tree** if this graph is also connected. We call the fuzzy graph $G = (\sigma, \mu)$ a **fuzzy forest** if it has a partial fuzzy spanning supergraph $F$ which is a forest, where for all the edges $xy$ with $\nu(xy) = 1$, we have $\mu(xy) > \nu^{\infty}(x, y)$. In other words, if $xy$ is in $G$, but has weight $= 1$ in $F$, there is a path in $F$ between $x$ and $y$ whose strength is less than $\mu(xy)$. It is clear that a forest is a fuzzy forest.

**Example 8.3.6** . Let $V = \{x, y, z\}$ and $E = \{xy, xz, yz\}$. Let $\oplus$ be the $t$-conorm algebraic sum. Define $\sigma : V \to [0, 1]$ by $\sigma(x) = \frac{7}{8}, \sigma(y) = \frac{1}{8}, \sigma(z) = \frac{1}{8}$ and $\mu(xy) = \frac{58}{64}, \mu(xz) = \frac{57}{64}, \mu(yz) = \frac{7}{8}$. Then $\mu((xy)) = \frac{58}{64} \geq \frac{7}{8} \oplus \frac{1}{8} = \sigma(x) \oplus \sigma(y)$. Then it follows that $(\sigma, \mu)$ is a $\oplus$-fuzzy supergraph of $(V, E)$. Let $t \in [0, 1]$. Let $\sigma^t = \{u \in V | \sigma(u) \leq t\}$ and $\mu^t = \{uv \in E | \mu(uv) \leq t \oplus t\}$. Let $t = \frac{8-\sqrt{6}}{8}$. Then $xy \in \mu^{\frac{8-\sqrt{6}}{8}}$, $y \in \sigma^{\frac{8-\sqrt{6}}{8}}$, but $x \notin \sigma^{\frac{8-\sqrt{6}}{8}}$. Thus $(\sigma^{\frac{8-\sqrt{6}}{8}}, \mu^{\frac{8-\sqrt{6}}{8}})$ is not a graph.

Let $(\sigma, \mu)$ be $\oplus$-fuzzy graph. Let $t \in [0, 1]$. Let $\sigma^t = \{u \in V | \sigma(u) \leq t\}$ and $\mu^t = \{uv \in E | \mu(uv) \leq t\}$. Let $xy \in \mu^t$. Then $t \geq \mu(xy) \geq \sigma(x) \otimes \sigma(y)$. Thus $\sigma(x) \leq t$ and $\sigma(y) \leq t$. Hence $x, y \in \sigma^t$. Thus $(\sigma^t, \mu^t)$ is a graph.

Conversely, suppose $(\sigma^t, \mu^t)$ is a graph for all $t$, where $\sigma^t = \{u \in V | \sigma(u) \leq t\}$ and $\mu^t = \{uv \in E | \mu(uv) \leq t \otimes t\}$. Suppose $\mu(xy) = t$. Then assuming $\oplus$ is continuous, there exists $t_1$ such that $t = t_1 \oplus t_1$ for some $t_1 \leq t$. $(f(t_1) = t_1 \oplus t_1, f(0) = 0, f(1) = 1$ and since $f$ is continuous, given $t$, there exists $t_1$ such that $f(t_1) = t$.) Thus $xy \in \mu^{t_1}$. Since $(\sigma^t, \mu^t)$ is a graph for all $t$, $(\sigma^{t_1}, \mu^{t_1})$ is a graph. Hence $x, y \in \sigma^{t_1}$. Thus $\mu(xy) = t = t_1 \oplus t_1 \geq \sigma(x) \otimes \sigma(y)$. Hence $(\sigma, \mu)$ is a $\otimes$- fuzzy graph.

Suppose $(\sigma^t, \mu^t)$ is a graph for all $t$, where $\sigma^t = \{x \in V | \sigma(x) \leq t\}$ and $\mu^t = \{xy \in E | \mu(xy) \leq t\}$. Suppose $\mu(x) = t$. Then by assumption $x, y \in \sigma^t$. Hence $\mu(xy) \geq t \vee t$, but not necessarily $t \oplus t$.

We next consider $t$-conorm weighted graphs. Let $(V, E, w)$ be a weighted graph and let $\rho$ be a weighted $t$-conorm. Recall that if $\oplus$ is a $t$-conorm, then $\rho(a, b) = m \oplus (a, b) = m(a \oplus b)$ is a weighted $t$-conorm. We have the following result as a corollary to Theorem 8.3.3.

Let $(V, E, w)$ be a weighted graph. The **strength** of a path $P$ is defined to be $\vee_{i=1}^{n} w(x_{i-1}x_i)$, where $P : x_{1,...,}x_n$. In other words, the strength of a path is defined to be the weight of the strongest arc of the path. Let $Q$ be another path. We say that the strength of $Q$ is less than the strength of $P$ if $\vee_{i=1}^{k} w(y_{i-1}y_i)$ is less than $\vee_{i=1}^{n} w(x_{i-1}x_i)$, where $Q : y_1, ..., y_k$.

**Definition 8.3.7** Let $G = (V, E, w)$ be a weighted graph. Then $G$ is called a **weighted $\rho$-subgraph** of $G$ w.r.t. a weighted $t$-conorm $\rho$ if for all $uv \in E$, $\nu(uv) \geq \tau(u) \oplus \tau(v)$.

Let $k$ be a positive integer. Define $\nu^k(u, v) = \wedge\{\nu(uu_1) \oplus ... \oplus \nu(u_{n-1}v) | P : u = u_0, u_1, ..., u_{n-1}, u_n = v$ is a path of length $k$ from $u$ to $v\}$. Let $\nu_{\oplus}^{\infty}(u, v) = \wedge\{\nu^k(u, v)| k \in \mathbb{N}\}$, where $\mathbb{N}$ denotes the positive integers. A **weakest** path joining two nodes $u, v$ has strength $\nu_{\oplus}^{\infty}(u, v)$.

**Definition 8.3.8** Let $(V, E, w)$ be a weighted graph w.r.t. $\rho$. Let $xy \in E$. Then $xy$ is called a **bridge** if $w_{\oplus}^{\prime\infty}(u, v) > w_{\oplus}^{\infty}(u, v)$ for some $u, v \in V$, where $w'(xy) = m$ and $w' = w$ otherwise.

**Theorem 8.3.9** *Let $(V, E, w)$ be a weighted fuzzy graph w.r.t. $\rho$. Let $xy \in E$. Let $w'$ be the weighted subset of $E$ such that $w'(xy) = m$ and $w' = w$ otherwise. Then* (3) $\Rightarrow$ (2) $\Leftrightarrow$ (1).

*(1) $xy$ is a bridge with respect to $\rho$.*
*(2) $w_{\oplus}^{\prime\infty}(x, y) > w(xy)$.*
*(3) $xy$ is not a strongest arc of any cycle.*

## 8.4 Nonstandard Analysis

In this section, we consider lattice homomorphisms of one lattice onto another. Our goal is to illustrate our results using the nonstandard interval $[0, 1]^*$ in nonstandard analysis. Consequently, we first review some basic properties of nonstandard analysis. We follow the approach in [14]. We do not provide a formal construction. A formal construction can be found in [12, 20]. We also consider lattice homomorphisms with an illustration involving nonstandard analysis. In particular, we consider a nonstandard weighted graph, i.e., a graph where the weights of the edges are from a nonstandard interval.

Let $F$ be a field and $<$ a relation on $F$. Suppose $<$ satisfies the following properties:

(1) $\forall x, y \in F$ such that $x \neq y$, either $x < y$ or $y < x$;
(2) $\forall x, y, z \in F, x < y$ and $y < z$ implies $x < z$;

(3) $\forall x, y, z \in F, x < y$ implies $x + z < y + z$;
(4) $\forall x, y, z \in F, x < y$ and $0 < z$ implies $xz < yz$.

Then $<$ is called an order on $F$ and $(F, <)$ is called an **ordered field**.

Let $\mathbb{R}$ denote the field of real numbers. Let $\mathbb{R}^*$ denote a nonstandard universe, [20], with the following properties:

(1) $(\mathbb{R}, +, \bullet, 0, 1, <)$ is an ordered subfield of $(\mathbb{R}^*, +, \bullet, 0, 1, <)$;

(2) $\mathbb{R}^*$ has a positive infinitesimal element $\varepsilon$, that is $\varepsilon \in \mathbb{R}^*$ is such that $\varepsilon > 0$ and $\varepsilon < r$ for all positive real numbers $r$.

(3) For all $n \in \mathbb{N}$ and every function $f : \mathbb{R}^n \to \mathbb{R}$, there is a natural extension $f^* : (\mathbb{R}^*)^n \to \mathbb{R}^*$. The natural extensions of the field operations $+, \bullet : \mathbb{R}^2 \to \mathbb{R}$ coincide with the operations in $\mathbb{R}^*$. Similarly, for every $A \subseteq \mathbb{R}^n$, then is a subset $A^* \subseteq (\mathbb{R}^*)^n$ such that $A^* \cap \mathbb{R}^n = A$.

(4) $\mathbb{R}^*$ equipped with the above assignments of extensions of functions and subsets behaves logically like $\mathbb{R}$.

**Definition 8.4.1** $\mathbb{R}^*$ is called the **ordered field of hyperreals**.

Now $\varepsilon$ has an additive inverse $-\varepsilon$. Clearly, $-\varepsilon$ is a negative infinitesimal. For every positive real number $r$, $\varepsilon^{-1} > r$. Thus $\varepsilon^{-1}$ is a positive infinite element and $-\varepsilon^{-1}$ is a negative infinite element.

**Definition 8.4.2** . (1) Let $\mathbb{R}_{fin} = \{x \in \mathbb{R}^* \mid |x| \leq n \text{ for some } n \in \mathbb{N}\}$. $\mathbb{R}_{fin}$ is called the set of **finite hyperreals**.

(2) Let $\mathbb{R}_{\text{inf}} = \mathbb{R}^* \backslash \mathbb{R}_{fin}$. $\mathbb{R}_{\text{inf}}$ is called the set of **infinite hyperreals**.

(3) Let $M = \{x \in \mathbb{R}^* \mid |x| \leq \frac{1}{n} \text{ for all } n \in \mathbb{N}, n > 0\}$. $M$ is called the set of **infinitesimal hyperreals**.

We see that $M \subseteq \mathbb{R}_{fin}$, $\mathbb{R} \subseteq \mathbb{R}_{fin}$, and $M \cap \mathbb{R} = \{0\}$. If $\delta \in M \backslash \{0\}$, then $\delta^{-1} \notin \mathbb{R}_{fin}$.

**Proposition 8.4.3** *$\mathbb{R}_{fin}$ is a subring of $\mathbb{R}^*$ and $M$ is and ideal of $\mathbb{R}_{fin}$.*

**Definition 8.4.4** Define the relation $\approx$ on $R^*$ by for all $x, y \in R^*$, $x \approx y$ if and only if $x - y \in M$. If $x \approx y$, we say that $x$ and $y$ are **infinitely close**.

It follows that not only is $\approx$ an equivalence relation on $\mathbb{R}^*$, but also a congruence relation.

**Theorem 8.4.5** . *(Existence of Standard Parts) Let $r \in \mathbb{R}_{fin}$. Then there exists a unique $s \in \mathbb{R}$ such that $r \approx s$. We call $s$ the **standard part** of $r$ and write $st(r) = s$.*

**Corollary 8.4.6** $\mathbb{R}_{fin} = \mathbb{R} + M$.

**Corollary 8.4.7** *Define $st : \mathbb{R}_{fin} \to \mathbb{R}$ by for all $r \in \mathbb{R}$, $st(r) = s$, where $s$ is the standard part of $r$. Then $st$ is a homomorphism of $\mathbb{R}_{fin}$ onto $\mathbb{R}$ such that $Ker(st) = M$.*

Let $h$ be a homomorphism of a lattice $(L_1, \leq_1)$ onto a lattice $(L_2, \leq_2)$. If the lattice $(L_1, \leq_1)$ is bounded with bottom element $0_1$ and top element $1_1$, then $(L_2, \leq_2)$ is bounded, and the bottom and top elements of $(L_2, \leq_2)$ is $0_2 = h(0_1)$ and $1_2 = h(1_1)$, respectively.

Let $(L_1, \vee_1, \wedge_1, 0_1, 1_1)$ and $(L_2, \vee_2, \wedge_2, 0_2, 1_2)$ be lattices. (Define $<_i$ on $L_i$ by for all $x, y \in L_i, x <_i y$ if and only if $x = x \wedge_i y$ and $x \neq y, i = 1, 2$.)

We assume in the following that $h$ is a homomorphism of $L_1$ onto $L_2$.

Define the relation $\approx$ on $L_1$ by for all $x, y \in L_1, x \approx y$ if and only if $h(x) = h(y)$. Then clearly $\approx$ is an equivalence relation on $L_1$. Now $\approx$ is also a congruence relation on $L_1$ since $h(x \vee_1 y) = h(x) \vee_2 h(y), h(x \wedge_1 y) = h(x) \wedge_2 h(y)$. (Suppose $x \approx y$ and $x' \approx y'$. Then $h(x) = h(y)$ and $h(x') = h(y')$. Thus $h(x \vee_1 x') = h(x) \vee_2 h(x') = h(y) \vee_2 h(y') = h(y \vee_1 y')$. Hence $x \vee_1 x' \approx y \vee_1 y'$. A similar result for $\wedge_1$and $\wedge_2$.

Let $x, y \in L_1$. Suppose $x <_1 y$. Then $x = x \wedge_1 y$ and $x \neq y$ and so $h(x) = h(x \wedge_1 y) = h(x) \wedge_2 h(y)$. It is not necessarily the case that $h(x) \neq h(y)$.

Now $x_1 \approx x_2$ and $y_1 \approx y_2 \Leftrightarrow h(x_1) = h(x_2)$ and $h(y_1) = h(y_2)$. Suppose $x_1 \approx x_2$ and $y_1 \approx y_2$. Then $h(x_1 \vee_1 y_1) = h(x_1) \vee_2 h(y_1) = h(x_2) \vee_2 h(y_2) = h(x_2 \vee_1 y_2)$. Thus $x_1 \vee_1 y_1 \approx x_2 \vee_1 y_2$. A similar result holds for $\wedge$. Hence $\approx$ is a congruence relation.

For all $x \in L_1$. Let $[x]$ denote the equivalence class of $x$ with respect to $\approx$ . Let $L_1^* = \{[x] | x \in L_1\}$. Define $\vee$ and $\wedge$ on $L_1^*$ as follows: For all $[x], [y] \in L_1^*, [x] \vee [y] = [x \vee_1 y]$ and $[x] \wedge [y] = [x \wedge_1 y]$. Define $f : L_1 \to L_1^*$ by for all $x \in L_1, f(x) = [x]$. Then $f$ is a function of $L_1$ onto $L_1^*$ such that $f(x \vee_1 y) = [x \vee_1 y] = [x] \vee [y] = f(x) \vee f(y)$ and similarly for $\wedge$. Define $\leq$ on $L_1^*$ by for all $[x], [y] \in L_1^*, [x] \leq [y]$ if and only if $h(x) \leq_2 h(y)$. Then $[x] = [y]$ if and only if $h(x) = h(y)$. Suppose $[x] < [y]$. Then $h(x) \leq h(y)$, but $h(x) \neq h(y)$, else $[x] = [y]$. We have $L_1^*$ is a lattice and $f$ is a homomorphism of $L_1$ onto $L_1^*$. Define $g : L_1^* \to L_2$ by for all $[x] \in L_1^*, g([x]) = h(x)$. Then $[x] = [y] \Leftrightarrow x \approx y \Leftrightarrow h(x) = h(y)$. Thus $g$ is a one-to-one function of $L_1^*$ into $L_2$. In fact, $g$ maps $L_1^*$ onto $L_2$ since $h$ is onto $L_2$. Now $g \circ f(x) = g(f(x)) = g([x]) = h(x)$ for all $x \in L_1$. Thus $g \circ f = h$. Clearly, $[x] \leq [y]$ if and only if $g(x) \leq_2 g(y)$. Hence $g$ is a lattice isomorphism of $L_1^*$ onto $L_2$. Let $L_1'$be a sublattice of $L_1$ such that $h|_{L_1'}$ is a an isomorphism of $L_1'$ onto $L_2$. Then for all $y \in L_1$, there exists unique $x \in L_1'$ such that $x \approx y$ : Let $x \in L_1'$ be such that $h(x) = h(y)$. ($x$ exists since $h$ maps $L_1'$ onto $L_2$.) Then $x$ is unique since $h$ is one-to-one on $L_1'$

Assume $L_2 \subseteq L_1$. Let $c_1$ be a complement on $L_1$. Define $c_2 : L_2 \to L_2$ by $c_2(y) = c_1(y)$, where $y \in L_2$. Assume $c_1(y) \in L_2$. Then, assuming $0_1 = 0_2$ and $1_1 = 1_2$, $c_2(0_2) = c_1(0_2) = c_1(0_1) = 1_1 = 1_2$.

Let $c_2$ be a complement of $L_2$. Define $c_1 : L_1 \to L_1$ by for all $x \in L_1, c_1(x) = c_2(h(x))$. Then, assuming $1_1 = 1_2$, we have that

$$c_1(0_1) = c_2(h(0_1)) = c_2(0_2) = 1_2 = 1_1.$$

These assumptions are the case for $st$, i.e., $L_1 = [0, 1]^*$ and $L_2 = [0, 1]$. In fact, $st$ is the identity map on $L_1$.

**Definition 8.4.8** Let $c$ be a complement on $L$. We say that $c$ is **infinitesimally involutive** if for all $x \in L, c(c(x) \approx x$.

**Proposition 8.4.9** *Let $L_2$ be a sublattice of $L_1$ and $h$ a lattice homomorphism of $L_1$ onto $L_2$. Let $c_2$ be a complement on $L_2$. Define $c_1 : L_1 \to L_1$ by for all $x \in L_1, c_1(x) = c_2(h(x))$. The $c_1$ is a complement on $L_1$. If $c_2$ is involutive and $h$ is the identity restricted to $L_2$, then $c_1$ is infinitesimally involutive.*

***Proof*** Let $x \in L_1$. Then since $c_2(h(x)) \in L_2$,

$$\begin{aligned} c_1(c_1(x)) &= c_1(c_2(h(x)) \\ &= c_2(h(c_2(h(x))) \\ &= c_2(c_2(h(x))) \\ &= h(x) \\ &\approx x. \end{aligned}$$

**Example 8.4.10** Let $L_1 = [0, 1]^*$ and $L_2 = [0, 1]$. Let $h = st$. Let $c_2(x) = 1 - x$ for all $x \in L_2$. Let $x \in L_1$. Then there exists $a \in [0, 1]$ and $m \in M$ such that $x = a + m$. Now $c_1(c_1(x)) = c_1(c_2(h(x))) = c_1(c_2(a)) = c_1(1 - a) = c_2(1 - a) = a \approx x$, where $c_1(x) = c_2(h(x))$.

Let $i_1$ be a $t$-norm on $L_1$. Define $i_2 : L_2 \times L_2 \to L_2$ by $\forall(x_2, y_2) \in L_2 \times L_2, i_2(x_2, y_2) = i_1(x_2, y_2)$. Assume $i_1(x_2, y_2) \in L_2$ and that $L_2 \subseteq L_1$. Then $i_2(y_2, 1_2) = i_1(y_2, 1_2) = i_1(y_2, 1_1) = y_2$. Assume $0_1 = 0_2$ and $1_1 = 1_2$. Now $i_2(0_2, y_2) = i_1(0_2, y_2) = i_1(0_1, y_2) = 0_1 = 0_2$.

Let $i_2$ be a $t$-norm on $L_2$. Define $i_1 : L_1 \times L_1 \to L_1$ by $\forall(x_1, y_1) \in L_1 \times L_1, i_1(x_1, y_1) = i_2(h(x_1), h(y_1))$. Then $i_1(x_1, 1_1) = i_2(h(x_1), h(1_1)) = i_2(h(x_1), 1_2) = h(x_1) \neq x_1$. Assume $h(h(x_1)) = h(x_1)$. Then $h(x_1) \approx x_1$. Now $i_1(0_1, y_1) = i_2(h(0_1), h(y_1)) = i_2(0_2, h(y_1)) = 0_2$.

**Definition 8.4.11** $i : L \times L \to L$. We say that $i$ satisfies the boundary conditions **infinitesimally** if $i(x, 1) \approx x$ and $i(0, x) \approx 0$. We say that $i$ is an **infinitesimal $t$-norm** if it satisfies the definition of a $t$-norm except for the boundary conditions which it satisfies infinitesimally.

An infinitesimal $t$-conorm is defined similarly.

**Proposition 8.4.12** *Suppose that $L_2$ is a sublattice of $L_1$ and that $h$ is a homomorphism $L_1$ onto $L_2$. Let $i_2$ be a $t$-norm on $L_2$. Define $i_1 : L_1 \times L_1 \to L_1$ by for all $(x, y) \in L_1 \times L_1, i_1(x, y) = i_2(h(x), h(y))$. If $h$ is the identity on $L_2$ and $h$ preserves $\leq$, then $i_1$ is an infinitesimal $t$-norm on $L_1$.*

***Proof*** Let $x, y \in L_1$. Then $i_1(x, y) = i_2(h(x), h(y)) = i_2(h(y), h(x)) = i_1(y, x)$. Let $x, y, z \in L_1$. Then

$$\begin{aligned} i_1(x, i_1(y, z)) &= i_2(h(x), h(i_1(y, z)) = i_2(h(x), h(i_2(h(y), h(z))) \\ &= i_2(h(x), i_2(h(y), h(z))) = i_2(i_2(h(x), h(y)), h(z)) \\ &= i_2(h(i_2(h(x), h(y)), h(z)) = i_1(i_2(h(x), h(y)), z) \\ &= i_1(i_1(x, y), z)) \end{aligned}$$

since $L_2 \subseteq L_1$ and $h$ is the identity on $L_2$. Let $x \in L_1$. Then $i_1(x, 1_1) = i_2(h(x), h(1_1)) = i_2(h(x), 1_2) = h(x)$. Now $h(x) \in L_2$ and $h(h(x) = h(x)$. Hence $h(x) \approx x$. Thus $i_1(x, 1_1) \approx x$. Suppose $x \leq_1 y$. Then $h(x) \leq_2 h(y)$.Thus $i_1(z, x) = i_2(h(z), h(x)) \leq_2 i_2(h(z), h(y)) = i_1(z, y)$.

**Example 8.4.13** Let $L_1 = [0, 1]^*$, $L_2 = [0, 1]$, and $h = st$. Let $x, y \in L_1$. Then there exists, $a, b \in \mathbb{R}$ and $m, m' \in M$ such that $x = a + m$ and $b = b + m'$. Now $x \leq y$ if and only if $a + m \leq b + m'$ if and only if ($a = b$ and $m \leq m'$ or $a < b$). Hence $st(x) \leq st(y)$ if $x \leq y$.

**Proposition 8.4.14** *Suppose that $L_2$ is a sublattice of $L_1$ and that $h$ is a homomorphism $L_1$ onto $L_2$. Let $i_1$ be a t-norm $L_1$ such that $i_1$ maps $L_2 \times L_2$ into $L_2$. Define $i_2 : L_2 \times L_2 \to L_2$ by for all $x, y \in L_2, i_2(x, y) = i_1(x, y)$. If $1_1 = 1_2$,then $i_2$ is a t-norm on $L_2$.*

***Proof*** . Let $x \in L_2$. Then $i_2(x, 1_2) = i_1(x, 1_2) = i_1(x, 1_1) = x$. Let $x, y, z \in L_2$. Suppose $x \leq_2 y$. Then $x \leq_1 y$. Thus $i_2(z, x) = i_1(z, x) \leq_1 i_1(z, y) = i_2(z, y)$. Hence $i_2(z, x) \leq_2 i_2(z, y)$. The commutative and associative properties hold for $i_2$ since they hold for $i_1$.

**Proposition 8.4.15** *Suppose that $L_2$ is a sublattice of $L_1$ and that $h$ is a homomorphism $L_1$ onto $L_2$. Let $u_2$ be a t-conorm on $L_2$. Define $u_1 : L_1 \times L_1 \to L_1$ by for all $(x, y) \in L_1 \times L_1, u_1(x, y) = u_2(h(x), h(y))$. If $h$ is the identity on $L_2$ and $h$ preserves $\leq$, then $u_1$ is an infinitesimal t-conorm on $L_1$.*

**Proposition 8.4.16** *Suppose that $L_2$ is a sublattice of $L_1$ and that $h$ is a homomorphism $L_1$ onto $L_2$. Let $u_1$ be a t-conorm $L_1$ such that $u_1$ maps $L_2 \times L_2$ into $L_2$. Define $u_2 : L_2 \times L_2 \to L_2$ by for all $x, y \in L_2, u_2(x, y) = u_1(x, y)$. If $0_1 = 0_2$, then $u_2$ is a t-conorm on $L_2$.*

## 8.5 Nonstandard Weighted Graphs

We next consider a nonstandard weighted graph $G = (V, E, w)$, where $w : E \to [0, m]^*$ and where $[0, m]^* = \{x \in \mathbb{R}^* | 0 \leq x \leq m\}$. Define $g : [0, m]^* \to [0, 1]^*$ by for all $a \in [0, m]^*, g(a) = \frac{a}{m}$. Then $g$ is a one-to-one function of $[0, m]^*$ onto $[0, 1]^*$. In fact, $g$ is an isomorphism of $([0, m]^*, \vee, \wedge, 0, m)$ onto $([0, 1]^*, \vee, \wedge, 0, 1) : g(a \wedge b) = \frac{a \wedge b}{m} = \frac{a}{m} \wedge \frac{b}{m} = g(a) \wedge g(b)$ and $g(a \vee b) = \frac{a \wedge \vee b}{m} = \frac{a}{m} \vee \frac{b}{m} = g(a) \vee g(b)$. Also, $g$ is continuous and preserves $<$.

Recall that for all $x \in [0, m]^*$, there exist $r \in \mathbb{R}$ and $y \in M$ such that $x = r + y$ and that $r$ is called the standard part of $x$, written $st(x) = r$. As a matter of fact, $r \in [0, m]$. The isomorphism $g$ also preserves $\approx$: Suppose $a, b \in [0, m]^*$ are such that $a \approx b$. Then $a - b \in M$ and so $a = b + x$ for some $x \in M$. Now $g(a) = \frac{a}{m} = \frac{b+x}{m} = g(b+x)$. Now $g(b) = \frac{b}{x}$ and $g(b+x) - g(b) = \frac{x}{m} \in M$. Thus $g(b+x) \approx g(b)$ and so $g(a) \approx g(b)$.

## References


1. Klement, E.P., Mesiar, R.: $L$-fuzzy sets and isomorphic lattices. are all the new results really new? Mathematics **6**, 146 (2018)
2. Trillas, E., De Soto, A.R.: On a new view of a fuzzy set. Trans. Fuzzy Sets Syst. **2**(1) (2023)
3. Global Slavery Index 2018, The Walk Free Foundation
4. Smarandach, F.: Neutrosophy: Neutrosophic Probability. Sets and Logic: Analytic Synthesis and Synthetic Analysis. American Research Press, Rehoboth (1998)
5. Yager, R.R.: Pythagorean membership grades in multi-criteria decision making. IEEE. Trans. Fuzzy Syst. **22**, 958–965 (2014)
6. Trillas, E., Garcia-Honrado, I.: A reflection on dialectic synthesis. New Math. Natural Comput. **15**(1), 31–46 (2019)
7. Atannasov, K.T.: Intuitionistic Fuzzy Sets, Theory and Applications, Studies in Fuzziness and Soft Computing, vol. 35. Springer, Physica-Verlag (1999)
8. Trafficking in Persons: Global Patterns, The United Nations Office for Drugs and Crime, Trafficking Persons Citation Index
9. Mordeson, J.N., Mathew, S., Malik, D.S.: Fuzzy Graph Theory: with Applications to Human Trafficking, Studies in Fuzziness and Soft Computing, vol. 365. Springer (2018)
10. Weinberger, A.: Reducing fuzzy algebra to classical algebra. New Math. Natural Comput. **1**(1), 27–64 (2005)
11. Head, T., A metatheorem for deriving fuzzy theorems from crisp versions. Fuzzy Sets Syst. **75** 349–358 (1995); 277–278 (1996)
12. Robinson, G. A.: Nonstandard Analysis. North Holland Publishing Co. (1966)
13. Davis, I.: An introduction to nonstandard analysis. https://mathematics.uchicago.edu/ VIGRE August 14, 2009
14. Goldberg, I.: Lecture notes on nonstandard analysis. UCLA Summer School in Logic (2014)
15. Rayo, D.A.B.: Introduction to nonstandard analysis. https://math.uchigago.edu REU Papers 2–15
16. Imamura, T.: Note on the definition of neutrosophic logic (2018). arXiv:1811.02961v1 [math.I.O]
17. Mordeson, J.N., Josy, A., Mathew, S.: Weighted graphs and fuzzy graphs. Trans. Fuzzy sets Syst. **2**(1) (2023)
18. Klir, G.J., Yuan, B.: Fuzzy Sets and Fuzzy Logic: Theory and Applications, p. 07458. Prentice Hall, Upper Saddle River (1995)
19. Rosenfeld, A.: Fuzzy Graphs. In: Zadeh, L.A., Fu, K.S., Shimura, M. (eds.) Fuzzy Sets and Their Applications, pp. 77–95. Academic, New York (1975)
20. Mordeson, J.N., Mathew, S., Gayathri, G.: Fuzzy Graph Theory: Applications to Global Problems. Studies in Fuzziness and Soft Computing, vol. 424. Springer (2023)

# Chapter 9
# Fuzzy Results from Crisp Results

In this chapter, we first prove that many results of fuzzy mathematics can be derived from crisp versions. We demonstrate this by the use of the work of Head and Weinberger, [1, 2]. We then show how these results can be used to derive fuzzy results from crisp on undirected power graphs, [3]. In Sect. 9.1, we outline the work in [1]. In Sects 9.2–9.5, we present the results of [2] leading up to and including the metatheorem which demonstrates how results from crisp sets can be used to derive results for fuzzy sets. In Sect. 9.6, we present some results on undirected power graphs of semigroups. In Sect. 9.7, we show how results concerning fuzzy graphs can be determined from the previous sections.

## 9.1 Derivation of Fuzzy Results from Crisp Results

The results in this section are from [1].

Let $X$ be a set and $C(X) = \{0, 1\}^X$ the set of all functions from $X$ into $\{0, 1\}$. Let $J = [0, 1)$. We let $\mathcal{P}(X)$ denote the power set of $X$ and $\mathcal{FP}(X)$ the fuzzy power set of $X$. Define Chi: $\mathcal{P}(X) \to C(X)$ by for all $Y \in \mathcal{P}(X)$, $\text{Chi}(Y)(x) = 1$ if $x \in Y$ and 0 otherwise.

**Definition 9.1.1** Let Rep: $\mathcal{FP}(X) \to C(X)^J$ be defined by for all $f \in \mathcal{FP}(X)$ and for all $j \in J$, $\text{Rep}(f)(j)(x) = 0$ if $f(x) \leq j$ and $f(x) = 1$ otherwise.

**Proposition 9.1.2** *Rep is an injective function.*

**Proposition 9.1.3** *The image of the function Rep consists of all those functions $G : J \to C(X)$ such that for all $j$ in $J$, $G(j) = \vee\{G(r) | r > j\}$.*



S. Mathew et al., *Weighted and Fuzzy Graph Theory*, Studies in Fuzziness and Soft Computing 429, https://doi.org/10.1007/978-3-031-39756-1_9

***Proof*** Suppose that $G = \text{Rep}(f)$ for some $f$ in $\mathcal{FP}(X)$. Let $j \in J$ and $x \in X$. Then $G(j)(x) = \text{Rep}(f)(j)(x) = 0$ if and only if $f(x) \leq j$. However, $f(x) \leq j$ if and only if $f(x) \leq r$ for all $r > j$. Consequently, $G(j)(x) = 0$ if and only if $G(r)(x) = 0$ for all $r > j$. Thus $G(j) = \vee\{G(r)|r > j\}$. Since this equation holds for all $x$ in $X$, $G(j) = \vee\{G(r)|r > j\}$.

Suppose now that $G : J \to C(X)$ has the property that for all $j$ in $J$, $G(j) = \vee\{G(r)|r > j\}$. Let $f : X \to [0, 1]$ be the function defined by for all $x$ in $X$ that $f(x) = \wedge\{s \in J|G(s)(x) = 0\}$. We show that $\text{Rep}(f) = G$ by observing that for all $j \in J$ and for all $x \in X$, the following five assertions are equivalent: (1) $\text{Rep}(f)(j)(x) = 0$, (2) $f(x) \leq j$, (3) for all $r > j$, $f(x) \leq r$, (4) for all $r > j$, $G(r)x) = 0$, (5) $G(j)(x) = 0$. ■

Note that for all $f$ in $\mathcal{FP}(X)$, $\text{Rep}(f)$ is an order reversing function from $J$ into $C(X)$, i.e., for all $x \in X$, $\text{Rep}(f)(r)(x) \leq \text{Rep}(f)(j)(x)$ whenever $r \geq j$. This is implicit in the statement of Proposition 9.1.3.

**Proposition 9.1.4** . *Rep commutes with infs of finite sets of fuzzy subsets and with sups of arbitrary sets of fuzzy subsets.*

**Definition 9.1.5** Let $X$ have an $n$-ary operation $* : X^n \to X$, where $n$ is a positive integer. We enlarge the domain of $*$ to provide an $n$-ary operation $* : \mathcal{FP}(X)^n \to \mathcal{FP}(X)$ on $\mathcal{FP}(X)$ by the standard convolution method:

$$*(f_1, ..., f_n)(x) = \vee\{\wedge\{\{f_1(x_1), ..., f_n(x_n)\}|x = *(x_1, ..., x_n)\}.$$

The operation $*$ on $X$ also provides an operation $*$ on $\mathcal{P}(X)$ defined, for subsets $X_1, ..., X_n$ of $X$ as follows:

$$*(X_1, ..., X_n) = \{*(x_1, ..., x_n)|x_i \in X_i, i = 1, ..., n\}.$$

**Proposition 9.1.6** *For each n-ary operation $*$ on $X$, with $n \geq 1$, $C(X)$ is closed with respect to the convolution extension $*$ to $\mathcal{FP}(X)$. The bijection Chi: $\mathcal{P}(X) \to C(X)$ commutes with the $*$ operations on $\mathcal{P}(X)$ and $C(X)$.*

**Proposition 9.1.7** *For each n-ary operation $*$ on $X$, with $n \geq 1$, the representation function $f : \mathcal{FP}(X) \to C(X)^J$ commutes with the convolution extension of $*$.*

Propositions 9.1.2, 9.1.4, and 9.1.7 say that Rep is an injective homomorphism of the algebra $\mathcal{FP}(X)$ into the product algebra $C(X)^J$ that establishes an algebraic and order-theoretic isomorphism of $\mathcal{FP}(X)$ with its image $I(X)$. Since Rep defines an isomorphism of $\mathcal{FP}(X)$ onto $I(X)$ and the projection of $I(X)$ into each coordinate space of $C(X)^J$ is a surjection, the following result holds.

**Subdirect Product Theorem 9.1.8** Let $X$ be an algebra having $n$-ary operations $*_1, ..., *_k$ for various values on $n \geq 1$. Then Rep:$\mathcal{FP}(X) \to C(X)^J$ is a representation of the $(\wedge, \vee, *_1, ..., *_k)$-algebra $\wedge, \vee, *_1, ..., *_k)$-algebra $\mathcal{FP}(X)$ as a subdirect product of copies of the $(\wedge, \vee, *_1, ..., *_k)$-algebra $C(X)$.

The interested reader should examine [1, pp. 355–357] to see that many algebraic results from the literature follow from Theorems 9.1.8 and 9.1.9. The papers [4–8] are pertinent here.

**Metatheorem 9.1.9** Let $X$ be an algebra having $n$-ary operations $*_1, ..., *_k$ for various values of $n \geq 1$. Let $\mathcal{FP}(X)$ be provided with operations $\wedge, \vee, *_1, ..., *_k$. Let $D(v_1, ..., v_m)$ and $E(v_1, ..., v_k)$ be expressions over the variable set $\{v_1, ..., v_m\}$ and the operation set $\{\wedge, \vee, *_1, ..., *_k\}$. Let $C_1, ..., C_m$ be classes of fuzzy subsets of $X$ that are closed under projections. The inequality

$$D(f_1, ..., f_m) \; REL \; E(f_1, ..., f_m)$$

holds for all fuzzy subsets $f_i$ in $C_i, i = 1, ..., m$ if and only if it holds for all crisp sets $f_i$ in $C_i, i = 1, ..., m$, where $REL$ is any one of the three relations $\leq, =$, or $\geq$ .

***Proof*** We consider only the case $\leq$ since the case $\geq$ follows by symmetry and the case $=$ follows by combining $\leq$ and $\geq$. If the inequality $\leq$ fails, then there exists $x \in X$ such that $D(f_1, ..., f_m) > E(f_1, ..., f_m)$. Let $j = E(f_1, ..., f_m)$. Then (recalling that $\text{Rep}(f_i) \in C(X)^J$ and so $\text{Rep}(f_i)(j)$ is a characteristic function), we have that

$$\begin{aligned} D(\text{Rep}(f_1)(j), ..., \text{Rep}(f_m)(j))(x) &= D(\text{Rep}(f_1), ..., \text{Rep}(f_m))(j)(x) \\ &= \text{Rep}(D(f_1, ..., f_m))(j)(x) = 1, \end{aligned}$$

but

$$\begin{aligned} E(\text{Rep}(f_1)(j), ..., \text{Rep}(f_m)(j))(x) &= E(\text{Rep}(f_1), ..., \text{Rep}(f_m))(j)(x) \\ &= \text{Rep}(E(f_1, ..., f_m))(j)(x) = 0. \end{aligned}$$

■

**Corollary 9.1.10** *For each algebra $X$, having $n$-ary operations $*_1, ..., *_k$ for various $n \geq 1$, the three $(\wedge, \vee, *_1, ..., *_k)$-algebras $\mathcal{P}(X)$, $C(X)$, and $\mathcal{FP}(X)$ all satisfy precisely the same $(\wedge, \vee, *_1, ..., *_k)$-identities.*

**Example 9.1.11** Let $x \in X$. Let $D(f_1, f_2)(x) = (f_1 \cap f_2)(x)$. Then $D(f_1, f_2)(x) = f_1(x) \wedge f_2(x)$. Let $j \in J$. Then

$$D(\text{Rep}(f_1(j), \text{Rep}(f_2)(j))(x) = \text{Rep}(f_1(j)) \cap \text{Rep}(f_2(j))(x)$$
$$\text{Rep}((D(f_1, f_2))(j)(x) = \text{Rep}(f_1 \cap f_2)(j)(x)$$

Suppose that $f_1(x) = \frac{1}{2}$ and $f_2(x) = \frac{1}{4}$. Let $j = \frac{1}{3}$. Then $\text{Rep}(f_1)(\frac{1}{3})(x) = 1$ since $f_1(x) = \frac{1}{2} > \frac{1}{3}$and $f_2(\frac{1}{3}) = 0$ since $f_2(x) = \frac{1}{4} < \frac{1}{3}$. Thus

$$\begin{aligned} \text{Rep}(f_1(j)) \cap \text{Rep}(f_2(j))(x) &= \text{Rep}(f_1(j))(x) \wedge \text{Rep}(f_2(j))(x) \\ &= 1 \wedge 0 = 0. \end{aligned}$$

**Example 9.1.12** Let $(X, *)$ be a group and $f$ be a fuzzy subgroup of $X$. (We write $xy$ for $x * y$.) Then $f$ is a normal fuzzy subgroup of $G$ if and only if $\forall x, y \in G$, $f(xyx^{-1}) = f(y)$. Now $f$ is a normal fuzzy subgroup of $G$ if and only if $f^j$ is a normal subgroup of $G$ for all $j \in J$.

Let $x \in X$. Let $D(f, *)(x) = f(x)$. Let $j \in J$ and $y \in X$. Then $\text{Rep}(f)(j)(xyx^{-1}) = 1$ iff $f(xyx^{-1}) > j$ iff $xyx^{-1} \in f^{j\cdot}$ Also, $\text{Rep}(f)(j)(y) = 1$ iff $f(y) > j$. Thus the Metatheorem holds.

## 9.2 Fuzzy Results from Crisp Results Continued

We now present some of the results of [2].

Let $X$ be a set and $J = [0, 1)$. Let $\mathcal{P}(X)$ denote the power set of $X$ and $\mathcal{P}(X)^J = \{\alpha | \alpha : J \to \mathcal{P}(X)\}$. Let $\mathcal{FP}(X)$ denote the fuzzy power set of $X$.

In the previous section, Rep:$\mathcal{FP}(X) \to C(X)^j$ while in Sects. 9.2–9.6, Rep: $\mathcal{FP}(X) \to \mathcal{P}(X)^J$.

**Definition 9.2.1** Let Rep:$\mathcal{FP}(X) \to \mathcal{P}(X)^J$ be defined by for all $f \in \mathcal{FP}(X)$, Rep$(f)$ is that function in $\mathcal{P}(X)^J$ such that for all $j \in J$,

$$\text{Rep}(f)(j) = \{x \in X | f(x) > j\}.$$

We see that Rep$(f)$ maps any $j$ in $J$ to the strong level set of $f$.

**Proposition 9.2.2** *Rep is injective.*

***Proof*** Let $f, g \in \mathcal{FP}(X)$. Suppose that $\text{Rep}(f) = \text{Rep}(g)$. Then for all $j \in J$, $\{x \in X | f(x) > j\} = \{x \in X | g(x) > j\}$. Suppose there exists $x \in X$ such that $f(x) \neq g(x)$, say $f(x) > g(x)$. Let $j \in J$ be such that $f(x) > j > g(x)$. Then $x \in \text{Rep}(f)(j)$, but $x \notin \text{Rep}(g)(j)$. Thus $\text{Rep}(f) \neq \text{Rep}(g)$. ∎

Proposition 9.2.2 shows that a fuzzy subset of $X$ is completely determined by its strong level sets.

**Proposition 9.2.3** *Rep*$(\mathcal{FP}(X)) = \{G \in \mathcal{P}(X)^J | G(j) = \cup_{r>j} G(r)$ *for all* $j \in J\}$.

***Proof*** Let $f \in \mathcal{FP}(X)$. Let $G = \text{Rep}(f)$. Let $j \in J$. We show $G(j) = \cup_{r>j} G(r)$. Now $G(j) = \{x \in X | \text{ not}(f(x) \leq j)\}$ and so $G(j)$ is the inverse image of the open interval $(j, 1)$ under $f$. Also, $\cup_{r>j}(r, 1) = (j, 1)$. Thus

$$\begin{aligned} G(j) &= f^{-1}((j, 1)) \\ &= f^{-1}(\cup_{r>j}(r, 1)) \\ &= \cup_{r>1} f^{-1}((r, 1)) \\ &= \cup_{r>j} G(r). \end{aligned}$$

Thus $\text{Rep}(\mathcal{FP}(X)) \subseteq \{G \in \mathcal{P}(X)^J | G(j) = \cup_{r>j} G(r)$ for all $j \in J\}$.

Let $G \in \mathcal{P}(X)^J$ be such that $G(j) = \cup_{r>j} G(r)$ for all $j \in J$. We must show that there exists $f \in \mathcal{FP}(X)$ such that $\text{Rep}(f) = G$. Define $f : X \to [0, 1]$ by for all $x \in X$,

$$f(x) = \vee\{j \in J | x \in G(j)\}.$$

Let $j \in J$. We need to show that $\text{Rep}(f)(j) = G(j)$. Suppose $x \in \text{Rep}(f)(j)$. Then $f(x) > j$. By the definition of $f$, there exists $s > j$ such that $x \in G(s)$. However, $G(s) \subseteq \cup_{r>j} G(r) = G(j)$. Thus $x \in G(j)$. Hence $\text{Rep}(f)(j) \subseteq G(j)$. Let $x \in G(j)$. Then $x \in \cup_{r>j} G(r)$. Thus $x \in G(r)$ for some $r > j$. Hence $f(x) \geq r$. Hence $f(x) > j$. Thus $x \in \text{Rep}(f)(j)$. Hence $\text{Rep}(f)(j) \supseteq G(j)$ and so $\text{Rep}(f)(j) = G(j)$. Therefore, $\text{Rep}(\mathcal{FP}(X)) \supseteq \{G \in P(X)^J | G(j) = \cup_{r>j} G(r) \text{ for all } j \in J\}$ and so wc havc cquality. ■

Proposition 9.2.3 shows that a collection of sets indexed by $J$, $G \in \mathcal{P}(X)^J$, is the collection of strong level subsets of some fuzzy subset, $f$, if and only if $G(j) = \cup_{r>j} G(r)$ for all $j \in J$.

**Corollary 9.2.4** *Let $f \in \mathcal{FP}(X)$ and $j \in J$. Then $f^j = \cup_{r>j} f^r$.*

***Proof*** From Proposition 9.2.3, we have

$$\begin{aligned} f^j &= \{x \in X | f(x) > j\} = \text{Rep}(f)(j) \\ &= \cup_{r>j} \text{Rep}(f)(r) = \cup_{r>j} \{x \in X | f(x) > r\} = \cup_{r>j} f^r. \end{aligned}$$

■

## 9.3 Fuzzifiable Operations

**Definition 9.3.1** An operation on $\mathcal{P}(X)$ is called **fuzzifiable** if it commutes with nested union. More explicitly, let $\omega$ be an operation on $\mathcal{P}(X)$. If $\omega$ is a nullary operation, then $\omega$ is fuzzifiable. Suppose $\omega$ is an operation with arity $n > 0$. The $\omega$ is called **fuzzifiable** if the following holds. Let $Z$ be a linearly ordered index set. For any $i \in \{1, ..., n\}$ and any $z \in Z$, let $S_{i,z} \subseteq X$ so that if $z_1 < z_2$, then $S_{,z_1} \subseteq S_{i,z_2}$. Then the following holds:

$$\cup_{z \in Z} \omega(S_{1,z}, ..., S_{n,z}) = \omega(\cup_{z \in Z} S_{1,z}, ..., \cup_{z \in Z} S_{n,z}).$$

**Proposition 9.3.2** *The union operation $\omega = \cup$ is fuzzifiable.*

***Proof*** Let $x \in \cup_{z \in Z} \omega(S_{1,z}, ..., S_{n,z})$. Then $x \in \omega(S_{1,z}, ..., S_{n,z})$ for some $z \in Z$. Since $w$ is union, we have $x \in S_{i,z}$ for some $i \in \{1, ..., n\}$. Hence $x \in \cup_{z \in Z} S_{i,z}$ and so $x \in \omega(\cup_{z \in Z} S_{1,z}, ..., \cup_{z \in Z} S_{n,z})$ since $\omega$ is union.

Suppose $x \in \omega(\cup_{z \in Z} S_{1,z}, ..., \cup_{z \in Z} S_{n,z})$. Then $x \in \cup_{z \in Z} S_{i,z}$ for some $i \in \{1, ..., n\}$. Thus $x \in S_{i,z}$ for some $z \in Z$ and so $x \in \omega(S_{1,z}, ..., S_{n,z})$. Thus $x \in \cup_{z \in Z} \omega(S_{1,z}, ..., S_{n,z})$. ■

**Proposition 9.3.3** . *The intersection operation* $\omega = \cap$ *is fuzzifiable.*

***Proof*** Let $x \in \cup_{z \in Z} \omega(S_{1,z}, ..., S_{n,z})$. Then $x \in \omega(S_{1,z}, ..., S_{n,z})$ for some $z \in Z$. Since $w$ is intersection, we have $x \in S_{i,z}$ for all $i \in \{1, ..., n\}$. Hence $x \in \cup_{z \in Z} S_{i,z})$ for all $i$ and so $x \in \omega(\cup_{z \in Z} S_{1,z}, ..., \cup_{z \in Z} S_{n,z})$.

Suppose $x \in \omega(\cup_{z \in Z} S_{1,z}, ..., \cup_{z \in Z} S_{n,z})$. Then $x \in \cup_{z \in Z} S_{i,z}$ for all $i \in \{1, ..., n\}$. Hence $x \in \omega(\cup_{z \in Z} S_{1,z}, ..., \cup_{z \in Z} S_{n,z})$. ■

If $S$ is a subset of $X$, we let $\overline{S}$ denote the subalgebra of $X$ generated by $S$.

**Proposition 9.3.4** *Let* $Z$ *be a linearly ordered set. For all* $z \in Z$, *let* $S_z$ *be a subset of* $X$ *such that* $z_1 < z_2$, *then* $S_{z_1} \subseteq S_{z_2}$. *Then* $\cup_{z \in Z} \overline{S_z} = \overline{\cup_{z \in Z} S_z}$.

***Proof*** Let $x \in \cup_{z \in Z} \overline{S_z}$. Then there exists $z \in Z$ such that $x \in \overline{S_z}$. Since $S_z \subseteq \cup_{z \in Z} S_z$, we have $x \in \overline{\cup_{z \in Z} S_z}$.

Let $x \in \overline{\cup_{z \in Z} S_z}$. Then there exists $x_1, ..., x_n \in \cup_{z \in Z} S_z$ such that $x \in \overline{\{x_1, ..., x_n\}}$. Hence there exists $z_1, ..., z_n \in Z$ such that $x_i \in S_{z_i}$ for each $i \in \{1, ..., n\}$. Let $z = z_1 \vee ., , , . \vee z_n$. Then $x \in \overline{S_z}$ and so $x \in \cup_{z \in Z} \overline{S_z}$. ■

Proposition 9.3.4 says that a unary operation on $\mathcal{P}(X)$ which takes a subset $S$ and returns $\overline{S}$, the subalgebra generated by $S$, is fuzzifiable.

**Proposition 9.3.5** . *Suppose* $\omega$ *has arity* $n > 0$. *Let* $Z$ *be any linearly ordered index set. For any* $i \in \{1, ..., n\}$ *and any* $z \in Z$, *let* $S_{i,z} \subseteq X$ *be such that if* $z_1 < z_2$, *then* $S_{i,z_1} \subseteq S_{i,z_2}$. *Then*

$$\cup_{z \in Z} \omega(S_{1,z}, ..., S_{n.z}) = \omega(\cup_{z \in Z} S_{1,z}, ..., \cup_{z \in Z} S_{n,z}).$$

***Proof*** Let $x \in \cup_{z \in Z} \omega(S_{1,z}, ..., S_{n.z})$. Then there exists $z \in Z$ such that $x \in \omega(S_{1,z}, ..., S_{n.z})$. Hence for each $i \in \{1, ..., n\}$, there exists $x_i \in S_{i,z}$ such that $x = \omega(x_1, ...x_n)$. Thus $x_i \in \cup_{z \in Z} S_{i,z}$. Hence $x \in \omega(\cup_{z \in Z} S_{1,z}, ..., \cup_{z \in Z} S_{n,z})$.

Let $x \in \omega(\cup_{z \in Z} S_{1,z}, ..., \cup_{z \in Z} S_{n,z})$. Then there exists $x_i \in \cup_{z \in Z} S_{i,z}$ such that $x = \omega(x_1, ..., x_n)$. Thus there exists $z_i \in Z$ such that $x_i \in S_{i,z_i}$. Let $z = z_1 \vee ... \vee z_n$. Then $x_i \in S_{i,z} >$ Thus $x \in \omega(S_{1,z}, ..., S_{n,z})$ and so $x \in \cup_{z \in Z} \omega(S_{1,z}, ..., S_{n.z})$. ■

Proposition 9.3.5 says that if $\omega$ is an operation on $\mathcal{P}(X)$ which is the extension of an operation on $X$, then $\omega$ is fuzzifiable.

**Proposition 9.3.6** *If* $\omega$ *is a fuzzifiable operation, then* $\omega$ *is order preserving.*

***Proof*** Suppose that $\omega$ has arity $n > 0$. Let $S_1, ..., S_n, T_1, ..., T_n \subseteq X$ be such that $S_i \subseteq T_i$ for $i = 1, ..., n$. We need to show that $\omega(S_1, ..., S_n) \subseteq \omega(T_1, ..., T_n)$. Since $\omega$ is fuzzifiable it commutes with nested union. Thus

$$\omega(S_1, ..., S_n) \cup \omega(T_1, ..., T_n) = \omega(S_1 \cup T_1, ..., S_n \cup T_n).$$

Since $S_i \subseteq T_i$, $S_i \cup T_i = 1, ..., n$. Hence

$$\omega(S_1, ..., S_n) \cup \omega(T_1, ..., T_n) = \omega(T_1, ..., T_n).$$

Therefore, $\omega(S_1, ..., S_n) \subseteq \omega(T_1, ..., T_n)$. ∎

**Corollary 9.3.7** *Complement is not fuzzifiable.*

***Proof*** Complement is not order preserving. ∎

## 9.4 Fuzzification of Fuzzifiable Operations

If $\mathcal{P}(A)$ is an algebra of type $\Omega'$, where all of the fundamental operations are fuzzifiable, then we can define an algebraic structure on $\mathcal{FP}(A)$ of the same type.

**Definition 9.4.1** $\mathcal{FP}(A)$ can be made into an algebra of type $\Omega'$ as follows. For any fuzzy subsets $f_1, ..., f_n \in \mathcal{FP}(A)$, define

$$\omega(f_1, ..., f_n)^j = \omega(f_1^j, ..., f_n^j).$$

If $\omega$ is a nullary operation on $\mathcal{P}(A)$, then we select the characteristic function of $\omega$, $\omega = \chi(\omega)$.

**Proposition 9.4.2** *Definition 9.4.1 properly defines the fuzzy set $\omega(f_1, ..., f_n)$.*

***Proof*** According to Proposition 9.2.3, it is sufficient to show that

$$\omega(f_1^j, ..., f_n^j) = \cup_{r>j}\omega(f_1^r, ..., f_n^r)$$

for each $\omega \in \Omega'(n)$ with $n > 0$.

Because $f_i$ is a fuzzy subset of $A$ for each $i \in \{1, ..., n\}$, we know that $f_i^j = \cup_{r>j} f_i^r$. Therefore

$$\omega(f_1^j, ..., f_n^j) = \omega(\cup_{r>j} f_1^r, ..., \cup_{r>j} f_n^r).$$

But strong level sets are nested, and $\omega$ is fuzzifiable, so

$$\omega(\cup_{r>j} f_1^r, ..., \cup_{r>j} f_n^r) = \cup_{r>j}\omega(f_1^r, ..., f_n^r).$$

By transitivity, we have the desired result. ∎

Some of the operations on $\mathcal{P}(A)$ were already extended to $\mathcal{FP}(A)$ before Definition 9.4.1. For example intersection and union were defined by meet, $\wedge$, and the join, $\vee$, on $\mathcal{FP}(A)$, respectively. It needs to be verified that this new, more general way of extending an operation from $\mathcal{P}(A)$ to $\mathcal{FP}(A)$ agrees with those prior individual definitions.

**Proposition 9.4.3** *The fuzzification of the union operator agrees with the prior definition of join,* $\vee$, *on* $\mathcal{FP}(A)$.

***Proof*** We need to prove that for any $f, h \in \mathcal{FP}(A)$, we have $(f \vee h)^j = f^j \cup h^j$ for all $j \in J$. Let $a \in (f \vee h)^j$. Then $(f \vee h)(a) > j$. Hence either $f(a) > j$ or $h(a) > j$. Thus $a \in f^j$ or $a \in h^j$. In either case, $a \in f^j \cup h^j$. The argument can be read in reverse to give the other inclusion. ■

**Proposition 9.4.4** *The fuzzification of the intersection operation agrees with the prior definition of meet,* $\wedge$, *on* $\mathcal{FP}(A)$.

***Proof*** We need to prove that for any $f, h \in F(A)$, we have $(f \wedge g)^j = f^j \cap h^j$ for all $j \in J$. Suppose $a \in (f \wedge h)^j$. Then $(f \wedge h)(a) > j$. Thus $f(a) \wedge h(a) > j$. Therefore, $f(a) > j$ and $h(a) > j$. Thus $a \in f^j$ and $a \in h^j$. Therefore, $a \in f^j \cap h^j$.

Again, this argument can be read in reverse to give the other inclusion. ■

## 9.5 $\mathcal{FP}(A)$ and $\mathcal{P}(A)$ Generate the Same Variety of Algebras

In the last section, we saw that if $\mathcal{P}(A)$ is an algebra of type $\Omega'$, where all of the fundamental operations are fuzzifiable, then $\mathcal{FP}(A)$ is an algebra of the same type where the corresponding operations are the fuzzifications. Now we create another algebra by raising $\mathcal{P}(A)$ to the power of $J$. The elements of $\mathcal{P}(A)^J$ are the functions from $J$ to $\mathcal{P}(A)$ and the operations are taken pointwise from $\mathcal{P}(A)$. Thus $\mathcal{FP}(A)$, $\mathcal{P}(A)$, and $\mathcal{P}(A)^J$ are all algebras of type $\Omega'$. In this section, we will show that they generate the same variety of algebra.

Earlier we introduced the $\chi : \mathcal{P}(A) \to \mathcal{FP}(A)$ function. It sends any subset of $A$ to its characteristic function. We already know that $\chi$ is injective. Now we show that is an $\Omega'$ homomorphism too. It extends Proposition 1.1.7 of [2] which says that $\chi$ is a lattice embedding.

**Proposition 9.5.1** *The function* $\chi : \mathcal{P}(A) \to \mathcal{FP}(A)$ *is a homomorphism of algebras.*

***Proof*** Suppose $\omega \in \Omega'$. If $\omega$ is a nullary operation then $\omega$ is some subset of $A$. Then $\chi(\omega)$ is the characteristic function of this subset. But this is just how we defined the fuzzification of a nullary operation. Thus $\chi$ preserves nullaries.

Suppose arity$(\omega) = n$ and $n > 0$. We need to show $\chi(\omega(S_1, ..., S_n) = \omega(\chi(S_1), ..., \chi(S_n))$. Let $j \in J$. It is enough to show that the $j$-th strong alpha cut of the left hand side is equal to the $j$-th strong alpha cut of the right hand side. Note that all of the strong alpha cuts of a characteristic function are equal. They are the original subset. Therefore,

$$\begin{aligned}\chi(\omega(S_1, ..., S_n))^j &= \omega(S_1, ..., S_n)\\ &= \omega(\chi(S_1)^j, ..., \chi(S_n)^j)\\ &= \omega(\chi(S_1), ..., \chi(S_n))^j.\end{aligned}$$

In the last step we pull the exponent $j$ through the operation $\omega$. This follows from the classical definition of the fuzzification of an operation on $\mathcal{P}(A)$. ■

**Proposition 9.5.2** *The function Rep:*$\mathcal{FP}(A) \to \mathcal{P}(A)^J$ *is a homomorphism of algebras.*

***Proof*** Suppose $\omega \in \Omega'$. If $\omega$ is a nullary operation symbol, then $\delta_{P(A)}(\omega)$ is a subset of $\mathcal{P}(A)$. The fuzzification of this $\delta_{\mathcal{FP}(A)}(\omega)$, is the character function of $\delta_{\mathcal{P}(A)}(\omega)$. Thus $\text{Rep}(\delta_{\mathcal{FP}(A)}(\omega))$ is the function which sends any $j \in J$ to the classical subset $\delta_{\mathcal{P}(A)}(\omega)$. This is exactly what the nullary operation $\delta_{\mathcal{P}(A)^j}(\omega)$ does. therefore, $\text{Rep}(\delta_{\mathcal{FP}(A)}(\omega)) = \delta_{\mathcal{P}(A)^j}(\omega)$.

Now suppose that $\text{arity}(\omega) = n$ and $n > 0$. Let $j \in J$. Let $f_1 ..., f_n \in \mathcal{FP}(A)$. Then

$$\begin{aligned}\text{Rep}(\omega(f_1, ..., f_n))(j) &= \omega(f_1, ..., f_n)^j\\ &= \omega(f_1^j, ..., f_n^j)\\ &= \omega(\text{Rep}(f_1)(j), ..., \text{Rep}(f_n)(j))\\ &= \omega(\text{Rep}(f_1), ..., \text{Rep}(f_n))(j).\end{aligned}$$

But $j$ was arbitrary. Therefore, $\text{Rep}(\omega(f_1, ..., f_n)) = \omega(\text{Rep}(f_1), ...,\text{Rep}(f_n))$. ■

The last two propositions combine to give the following theorem.

**Theorem 9.5.3** *Given any type of algebra $\Omega'$, if $\mathcal{P}(A)$ is an algebra of type $\Omega'$ and the fundamental operations, $\delta_{\mathcal{P}(A)(\Omega')}$, are fuzzifiable, then $\mathcal{FP}(A)$ and $\mathcal{P}(A)$ generate the same variety of $\Omega'$ algebras.*

***Proof*** We just saw that $\chi$ embeds $\mathcal{P}(A)$ into $\mathcal{FP}(A)$. Thus $\mathcal{P}(A)$ is isomorphic to a subalgebra of $\mathcal{FP}(A)$. Next we saw that Rep embeds $\mathcal{FP}(A)$ into $\mathcal{P}(A)^J$. Thus $\mathcal{FP}(A)$ is isomorphic to a subalgebra of a product of $J$ copies of $\mathcal{P}(A)$. Therefore, $\mathcal{FP}(A)$ and $\mathcal{P}(A)$ generate the same variety of algebras. ■

**Corollary 9.5.4** *$\mathcal{FP}(A)$ and $\mathcal{P}(A)$ satisfy the same $\Omega'$ identities.*

**Corollary 9.5.5** *$\mathcal{FP}(A)$ and $\mathcal{P}(A)$ satisfy the same lattice identities.*

***Proof*** We saw that the lattice operations $\wedge$ and $\vee$ are fuzzifiable. ■

**Corollary 9.5.6** *$\mathcal{FP}(A)$ is a distributive lattice under $\wedge$ and $\vee$.*

Theorem 9.5.3 is a generalization of Theorem 9.1.8. As noted in that paper, many results in the literature on fuzzy sets can be unified and proved more simply using this theorem. Also using this theorem, many results from the literature can be generalized by removing restrictions that are not really necessary.

## 9.6 Undirected Power Graphs of Semigroups

The results from this section are taken from [3].

Let $(S, *)$ be a semigroup. The **undirected power graph** of $S$, $\mathcal{G}(S)$, is an undirected graph whose vertex set is $S$ and two vertices $a, b \in S$ are said to be **adjacent** if $a \neq b$ and there exists $m \in \mathbb{N}$ such that $a^m = b$ or $b^m = a$. We write $ab$ if $a$ and $b$ are adjacent, where we consider $ab$ to be an undirected edge in $\mathcal{G}(S)$. Let $E = \{ab | a, b \in V \text{ are adjacent}\}$.

Let $S$ be a finite semigroup. An element $e \in S$ is called an **idempotent** if $e^2 = e$. We denote the set of all idempotents by $E(S)$. We show that $E(S)$ is not empty. Let $x \in S$. Consider $x, x^2, x^4, \ldots$. Since $S$ is finite, there exist positive integers $s$ and $t$ such that $x^{2^t} = x^{2^s}$ where $t > s \geq 1$. Then $x^{2^t} = (x^{2^t})^{2^{t-s}}$. Let $a = x^{2^{s-s}}$ and $k = 2^{t-s}$. Then $a^k = a$. If $k = 2$, then $a$ is an idempotent. Suppose $k > 2$. Now $a^k a^{k-2} = aa^{k-2}$. Thus $(a^{k-1})^2 = a^{k-1}$ and so $e = a^{k-1}$ is an idempotent. We have shown that for each $a \in S$, there exists $m \in \mathbb{N}$ such that $a^m$ is an idempotent. Also, it is easy to verify that if $a^m = e$ and $a^n = f$ for some $m, n \in \mathbb{N}$, then $e = a^{mn} = f$.

Define the relation $\rho$ on $S$ by for all $a, b \in S$, $a\rho b$ if and only if there exists $m \in \mathbb{N}$ such that $a^m = b^m$. Then $\rho$ is an equivalence relation on $S$. In the following, we show for all $a, b \in S$, $a\rho b$ if and only if $a$ and $b$ are connected by a path in $\mathcal{G}(S)$. Thus the equivalence classes of $\rho$ are precisely the components of the graph $\mathcal{G}(S)$. It also follows that if $S$ is commutative, then $\rho$ is a congruence relation on $S$ and $S/\rho$ is isomorphic to $E(S)$.

**Lemma 9.6.1** *Let $S$ be a finite semigroup and let $\rho$ be the equivalence relation on $S$ defined above. Then for all $a, b \in S$, $a\rho b$ if and only if there exists $m_1, m_2 \in \mathbb{N}$, $e \in E(S)$ such that $a^{m_1} = b^{m_2} = e$.*

***Proof*** Suppose $a\rho b$ for $a, b \in S$. Then there exists $m \in \mathbb{N}$ such that $a^m = b^m$. Since $S$ is finite, there exist $n_1, n_2 \in \mathbb{N}$ such that $a^{n_1} = e$ and $b^{n_2} = f$, where $e, f \in E(S)$. Then $e = a^{mn_1n_2} = b^{mn_1n_2} = f$. Hence $m_1 = m_2 = mn_1n_2$.

Conversely, let $a, b \in S$ be such that $a^{m_1} = b^{m_2} = e$ for some $m_1, m_2 \in \mathbb{N}$, $e \in E(S)$. Then $a^m = b^m$, where $m = m_1m_2$. Hence $a\rho b$. ■

**Theorem 9.6.2** *Let $S$ be a finite semigroup. Let $a, b \in S$. Then $a$ and $b$ are connected by a path in $\mathcal{G}(S)$ if and only if $a\rho b$.*

***Proof*** Suppose $a, b \in S$ are connected by a path, say $a, c_1, c_2, \ldots, c_k, b$ in $\mathcal{G}(S)$. Since $a$ and $c_1$ are adjacent, either $a^m = c_1$ or $c_1^m = a$ for some $m \in \mathbb{N}$. Then by Lemma 9.6.1, we have $a\rho c_1$. Similarly, $c_i \rho c_{i+1}$ for $i = 1, 2, , , k-1$, and $c_k \rho b$. By the transitivity of $\rho$, we have that $a\rho b$.

Conversely, suppose that $a\rho b$. Then $a^m = b^m$ for some $m \in \mathbb{N}$. By the Lemma, $a^n = e = b^n$ for some $n \in \mathbb{N}$. Hence both $a$ and $b$ are adjacent to $e$ Thus $a$ and $b$ are connected by a path. ■

Let $f, g$ be idempotents. If $f, g$ are connected by a path, then $f\rho g$ and so $f^{m_1} = g^{m_2} = e$ for some $m_1, m_2 \in \mathbb{N}$ and $e \in E(S)$. However, $f^{m_1} = f$ and $g^{m_2} = g$. Thus

$f = g = e$. Hence no two idempotents are connected by a path. Also, every vertex in $\mathcal{G}(S)$ is adjacent to only one idempotent. We can thus see that every component of $\mathcal{G}(S)$ contains a unique idempotent to which every other vertex of that component is adjacent. This is summarized in the following corollary. The equivalence class of $e \in E(S)$ is $C_e$, where

$$C_e = \{a \in S | a\rho e\} = \{a \in S | \exists m \in \mathbb{N} \text{ such that } a^m = e\}.$$

(Recall that $a, b$ in $S$ are adjacent if there exists $m \in \mathbb{N}$ such that $a^m = b$ or $b^m = a$. Hence $a \in C_e$ implies that $a$ and $e$ are adjacent.)

**Corollary 9.6.3** *The components of $\mathcal{G}(S)$ are precisely $\{C_e | e \in E(S)\}$. Each component $C_e$ contains the unique idempotent $e$. If $S$ is commutative, then $C_e$ is a subsemigroup of $S$ for each $e \in E(S)$.*

(Let $e$ be an idempotent and let $a \in C_e$. Suppose $b \in \mathcal{G}(S)$ is such that $a$ and $b$ are adjacent. Then there exists $n \in \mathbb{N}$ such that $a^n = b$ or $b^n = a$. Now there exists $m \in \mathbb{N}$ such that $a^m = e$. Suppose that $a^n = b$. Then $a^{nm} = e^n = e$ and $a^{nm} = b^n$ and so $b^n = e$. In this case, $b$ and $e$ are adjacent. Suppose that $a = b^n$. Then $b^{mn} = e$ and so $b$ and $e$ are adjacent. Hence we have that $b \in C_e$.)

Suppose $\mathcal{G}(S)$ is connected. Then $S = C_e$. Thus if for all $a, b \in S \backslash \{e\}$, $a$ and $b$ are not adjacent, then $\mathcal{G}(S)$ is a tree with root $e$.

**Example 9.6.4** (1) Suppose $S = \{e, a, b, c\}$ is the Klein-group. Then $a^2 = b^2 = c^2 = e$. Here $\mathcal{G}(S)$ is a tree.

(2) Suppose $S = \{e, a, a^2\}$ is a cycle group of order 3. Then $a$ and $a^2$ are adjacent. Thus $\mathcal{G}(S)$ is a cycle.

## 9.7 Fuzzy Subgraphs of Undirected Power Graphs

Let $f$ be a fuzzy semigroup of $S$. Then for all $a, b \in S$, $f(a * b) \geq f(a) \wedge f(b)$. Consequently, every level set and strong level set of $f$ is a subsemigroup of $S$ : Let $j \in [0, 1)$ and let $f^j = \{a \in S | f(a) > j\}$. Then for $a, b \in f^j$, $f(a * b) \geq f(a) \wedge f(b) > j$ and so $a * b \in f^j$.

Define the fuzzy subset $\mu$ of $E$ by for all $ab \in E$, $\mu(ab) = f(a) \wedge f(b)$. Then $(S, f, \mu)$ is a fuzzy subgraph of $\mathcal{G}(S)$.

Let $j \in [0, 1)$. Let $a, b \in f^j$. Suppose there exist $m \in \mathbb{N}$ such that either $a^m = b$ or $b^m = a$. Since $f(a^m) \geq f(a) \wedge ... \wedge f(a) > j$, we have that $a^m \in f^j$ and similarly $b^m \in f^j$. Thus $\mathcal{G}(f^j)$ is an undirected power graph. Now $\mu(ab) = f(a) \wedge f(b) > j$ and so $ab \in \mu^j$. Hence $(f, \mu)$ is a fuzzy subgraph of $\mathcal{G}(f^j)$.

Now suppose $a, b \in f^j$ are such that $a \neq b$ and $a^m = b^m$ for some $m \in \mathbb{N}$. Then $a$ and $b$ are connected by a path in $\mathcal{G}(f^j)$, say $a = a_0, a_1, ..., a_k = b$. We have $f(a_i) > j, i = 0, 1, ..., k$, and $\mu(a_{i-1}a_i) = f(a_{i-1}) \wedge f(a_i) > j, i = 1, ..., k$.

In [1, p. 355], it is discussed how many results concerning semigroups can be carried over to fuzzy semigroups. We now can see that many of these results concerning undirected power graphs can be carried over to fuzzy undirected power graphs. For example, suppose $S$ is a finite group. Then $\mathcal{G}(S)$ is connected [3, Corollary 2.6]. Let $f$ be a fuzzy subgroup of $S$. Let $j \in [0, 1)$. Then the strong level set, $f^j = \{a \in S | f(a) > j\}$ is a subgroup of $S$ (assuming closure under projections). Since $\mu(ab) = f(a) \wedge f(b)$, any $a, b$ in $f^j$, $a \neq b$, is connected by a path and this path is of strength $> j$. Hence the graph $(f^j, \mu^j)$ is connected and so is $\mathcal{G}(f^j)$.

We can now see that results obtained here for fuzzy sets can be immediately converted to results concerning weighted graphs.

## References


1. Head, T., A metatheorem for deriving fuzzy theorems from crisp versions. Fuzzy Sets Syst. **75**, 349–358 (1995); 277–278 (1996)
2. Weinberger, A.: Reducing fuzzy algebra to classical algebra. New Math. Nat. Comput. **1**(1), 27–64 (2005)
3. Chakrabarty, I., Ghosh, S., Sen, M.K.: Undirected power graphs of semigroups. Semigroup Forum (2009)
4. Ajmal, N.: The lattice of fuzzy normal subgroups is modular. Inf. Sci. **83**, 199–209 (1995)
5. Ajmal, N., Thomas, K.V.: The lattice of fuzzy subgroups and fuzzynormal subgroups. Inf. Sci. **76**, 1–11 (1994)
6. Golan, J.S.: Norms, semirings, and power algebras. In: Ramamujan Conference on Algebra and Its Applications, Madras
7. Kuroki, N.: On fuzzy ideals and fuzzy biideals in semigroups. Fuzzy Sets Syst. **5**, 203–215 (1981)
8. Kuroki, N.: On fuzzy semigroups. Inf. Sci. **53**, 201–236 (1991)

# Bibliography


1. Balci, E.A., Dundar, P.: Average edge-distance in graphs. Selcuk J. Appl. Math. **11**(2), 63–70 (2010)
2. Bandelt, H.J., Chepoi, V.: Metric graph theory and geometry: a survey. Contemp, Math. (1991)
3. Bollobas, B., Scott, A.D.: A proof of a conjecture of Bondy concerning paths in weighted graphs. J. Combin. Theory **66**, 283–292 (1996)
4. Bondy, J.A., Murty, U.S.R.: Graph theory and Applications. McMillan Press Ltd, London (1976)
5. Caceres, J., Puertas, H.J.: Searching for geodetic boundary vertex sets. Electron. Notes Discret. Math. **19**, 25–31 (2005)
6. Caceres, J., Marquez, A., Oellerman, O.R., Puertas, M.L.: Rebuilding convex sets in graphs **297**, 26–37 (2005)
7. Changat, M., Mulder, H.M., Sierksma, G.: Convexities related to path properties on graphs. Disc. Math. **290**, 117–131 (2005)
8. Chartrand, G., Erwin, D., Johns, G.L., Zhang, P.: Boundary vertices in graphs. Discret. Math. **263**, 25–34 (2003)
9. Chartrand, G., Erwin, D., Johns, G.L., Zhang, P.: On boundary vertices in graphs. J. Combin. Math. Combin. Comput. **48**, 39–53 (2004)
10. Chartrand, G., Zhang, P.: Distance in Graphs- Taking a Long View. AKCE J. Graphs. Combin. **1**(1), 1–13 (2004)
11. Chartrand, G., Erwin, D., Johns, G.L., Zhang, P.: Boundary vertices in graphs. Discret. Math. **263**(1–3), 25–34 (2003)
12. Diestel, R.: Graph Theory. Graduate Texts in Mathematics, vol. 173, 2nd edn. Springer (2000)
13. Dirac, G.A.: Some theorems on abstract graphs. Proc. Lond. Math. Soc. (2–3), 69–81 (1952)
14. Dobrynin, A.A., Gutman, I., Klavzar, S., Zigert, P.: Wiener index of hexagonal systems. Acta Appl. Math. **72**, 247–294 (2002)
15. Estrada., E., Patlewicz, G.: From molecular graphs to drugs. A review on the use of topological indices in drug design and discovery. Indian J. Chem. **42**, 1315–1329 (2003)
16. Farber, M., Jamison, R.E.: On local convexity in graphs. Discr. Math **66**, 231–247 (1987)
17. Farber, M., Jamison, R.E.: Convexity in graphs and hypergraphs. SIAM J. Alg. Disc. Math. **7**(3), 433–444 (1986)
18. Fujisawa, J.: Heavy cycles in weighted graphs. Electron. Notes Discret. Math. **22**, 195–199 (2005)




S. Mathew et al., *Weighted and Fuzzy Graph Theory*, Studies in Fuzziness and Soft Computing 429, https://doi.org/10.1007/978-3-031-39756-1

19. Fujisawa, J., Yohimotto, K., Zhang, S.: Heavy cycles passing through some specified vertices in weighted graphs. J. Graph Theory **49**(2), 93–103 (2005)
20. Gavril, F.: The intersection graphs of subtrees in trees are exactly the chordal graphs. J. Combin. Theory, Ser. B **16**, 47–56 (1974)
21. Harary, F., Loukasis, E., Tsouros, C.: The geodetic number of a graph, math. Comput. Model. **17**(11), 89–95 (1993)
22. Harary, F., Norman, R.: Graph Theory as Mathematical Model in Social Science. University of Mishigan, Institute of Social Research (1953)
23. Harris, J.M., Hirst, J.L., Mossinghoff, M.: Combinatorics and Graph Theory. Springer (2008)
24. Hasegawa, Y., Saito, A.: Graphs with small boundary. Discret. Math. **307**, 1801–1807 (2007)
25. Hernando, C., Mora, M., Pelayo, I.M., Seara, C.: Some structural, metric and convex properties of the boundary of a graph. Electron. Notes Discret. Math. **24**, 203–209 (2006)
26. Hernando, C., Jiang, T., Mora, M., Pelayo, I.M., Seara, C.: On the Steiner, geodetic and hull numbers of graphs. Discret. Math. **293**(1–3), 139–154 (2005)
27. Hernando, C., Mora, M., Pelayo, I.M., Seara, C., Caceres, J., Puertas, M.L.: On Geodetic Sets Formed by Boundary Vertices, vol. 21. Elsevier Science (2006)
28. Hernando, C., Mora, M., Pelayo, I.M., Seara, C.: On monophonic sets in graphs, submitted to J. Combin. Theory, Ser. B (Accepted)
29. Hosoya, H.: Topological index; a newly proposed quantity characterizing the topological nature of structural isomers of saturated hydrocarbons. Bull. Chem. Soc. Jpn. **44**, 2332–2339 (1971)
30. Howorka, E.: A characterization of distance-hereditary graphs. Quart. J. Math. Oxford Ser. **2**(28), 417–420 (1977)
31. Ilic, A., Klavzar, S., Stevanovic, D.: Calculating the degree distance of partial Hamming graphs. MATCH Commun. Math. Comput. Chem. **63**, 411–426 (2010)
32. Jicy, N., Mathew, S.: Connectivity analysis of a network by levels. Annals of. Pure Appl. Math. **8**, 57–63 (2014)
33. Klavzar, S., Gutman, I.: Wiener number of vertex weighted graphs and a chemical application. Discret. Appl. Math. **80**, 73–81 (1997)
34. Kys, P., Bratislava.: Graphs with the same peripheral and center eccentric vertices. Math. Bohem. **125**(3), 331–339 (2000)
35. Lada, A., Ringel, G., Serra, O.: Another tree conjecture. Bull. Inst. Combin. Appl. **18**, 83–85 (1996)
36. Mordeson, J.N., Nair, P.S.: Fuzzy Graphs and Fuzzy Hypergraphs. Physica-Verlag, Heidelberg (2000)
37. Mordeson, J.N., Mathew, S.: Nonstandard fuzzy mathematics. Adv. Fuzzy Sets Syst. **27**(1), 35–52 (2022)
38. Mordeson, J.N., Mathew, S.: Fuzzy mathematics, and nonstandard analysis; application to the theory of relativity. Trans. Fuzzy Sets Syst. **1**(1), 143–154 (2022)
39. Muller, T., Por, A., Sereni, J-S.: Lower bounding the boundary of a graph in terms of its maximum or minimum degree. Discret. Math. **308**(24), 6581–6583 (2008)
40. Randic, M.: On generalization of Wiener index for cyclic structures. Acta Chim. Slov. **49**, 483–496 (2002)
41. Raut, N.K.: Degree based topological indices of isomers of organic compounds. Int. J. Sci. Res. Publ. **4**, 1–4 (2014)
42. Rodriguez-Velazquez, J.A., Kamisalic, A., Domingo-Ferrer, J.: On reliability indices of communication networks. Comput. Math. Appl. **58**, 1433–1440 (2009)
43. Sampathkumar, E.: Convex Sets in a graph. Indian J. Pure Appl. Math. **15**(10), 1065–1071 (1984)
44. Sibi, S., Swarna Priya, R.M.: Using graph theory concepts to determine the boiling point formula for Choloalkanes. Int. J. Eng. Trends Technol. **5**, 125–129 (2013)
45. Soltan, V.P.: Metric Convexity in graphs. Studia Univ. Babes- Bolyai Math. **36**(4), 3–43 (1991)
46. Trillas, E., De Soto, A.R.: On the search for speculations. New Math. Nat. Comput. **18**(1), 9–18 (2022)

47. Vijayabarathi, A., Anjaneyulu, G.S.G.N.: Wiener index of a graph and chemical applications. Int. J. Chem. Tech. Res. **5**, 1847–1853 (2013)
48. Wang, Y., Hu, Y.: The wiener index for weighted trees. WSEAS Trans. Math. **11**(12), 55–788 (2012)
49. Wiener, H.: Structural determination of Praffin boiling points. J. Amer. Chem. Soc. **69**, 17–20 (1947)
50. Zhang, S., Li, X., Broersma, H.: Heavy paths and cycles in weighted graphs. Discret. Math. **223**, 327–336 (2000)
51. Zhang, S., Broersma, H.: A fan type condition for heavy cycles in weighted graphs. J. Graphs Combin. **18**, 193–200 (2002)
52. Zhao, K.: A simple proof of Whitney's theorem on connectivity in graphs. Math. Bohem. **136**, 25–26 (2011)

# Index

**Symbols**
$I$-gate free, 171
$I_s$-convex, 168
$\delta$-distance between two vertices, 156
$d$-gate free, 171
$k$-partite, 4
$r$-regular, 4
$w$-distance, 155
$w$-length, 155
$\alpha$-edge count, 142
$\alpha$-saturated, 146
$\alpha$-sequence, 133
$\alpha$-strong, 57
$\alpha$-strong path, 58
$\alpha$-vertex count, 142
$\beta$ -sequence, 133
$\beta$-edge count, 142
$\beta$-saturated, 146
$\beta$-strong, 57
$\beta$-strong path, 58
$\beta$-vertex count, 142
$\delta$-edge, 57
$\delta^*$-edge, 57
$\theta$—evaluation, 91
$\theta$—weighted graph, 91
$\theta$-evaluation, 91
$p$-bridge, 55
$p$-cut vertex, 54
$t$-conorm fuzzy graph, 179
$t$-torm fuzzy graph, 179
$u - v$ strength reducing set, 81
$u - v$ strength reducing set of edges, 81
2-sum, 41

**A**
Acyclic graph, 5
Adjacency matrix, 5
Adjacent edges, 2
Adjacent vertices, 2
Arity, 202

**B**
Binary sequence, 134
Bipartite graphs, 5
Block, 7
Bridge, 7, 187

**C**
Chord, 125
Circuit, 5
Complement, 184
Complement of a weighted graph, 116
Complete bipartite graph, 5
Complete k-partite graph, 5
Component, 5
Connected, 5
Connectivity of a graph, 8
Convex, 168
Cut-vertex, 6
Cycle, 5
Cycle connectivity, 94
Cycle connectivity $C^G_{u,v}$, 91
Cycle connectivity between vertices, 91
Cycle-extremal, 22
Cyclically balanced, 96
Cyclically balanced graphs, 98
Cyclic cut vertex, 95
Cyclic edge connectivity, 99



S. Mathew et al., *Weighted and Fuzzy Graph Theory*, Studies in Fuzziness and Soft Computing 429, https://doi.org/10.1007/978-3-031-39756-1

Cyclic edge cut, 99
Cyclic end vertex, 100
Cyclic strong weight, 98
Cyclic vertex connectivity, 99
Cyclic vertex cut, 98

**D**
Degree of a vertex, 2
Diameter, 6
Disconnected, 5
Distance matrix, 113

**E**
Eccentricity, 6
Edge-disjoint path, 7
Edge-weighting, 25
Embedding, 183
End-vertex, 2
Epimorphism, 183

**F**
Finite hyperreals, 191
Forest, 7
Fully strong gated, 171
Fuzzifiable operations, 201
Fuzzification, 203
Fuzzy algebra, 182
Fuzzy graph, 179
Fuzzy subgraph, 188
Fuzzy super graph, 189

**G**
Gate, 169
Gated, 169
Gated set, 169
Gated with respect to an interval function, 169
Generalized Menger's theorem, 80
Geodetic convex sets, 117
Graph, 1, 2

**H**
Hamilton cycle, 17
Homomorphism of algebras, 204

**I**
Independence number, 9
Induced subgraph, 3
Infinite hyperreals, 191
Infinitesimal hyperreals, 191
Infinitesimally involutive, 193
Internal vertex, 7
Interval function, 161
Isolated vertex, 2
Isomorphic lattices, 181
Isomorphisms, 181, 183

**L**
Lattice embedding, 182
Lattice homomorphism, 183
Line graph, 9
Linguistic variables, 181
Locamin cycle, 90
Longest heaviest path, 28
Loop, 4

**M**
Maximum bandwidth path, 54
Maximum spanning tree, 11
Maximum strong degree, 76
Max-max composition, 113
Membership function, 180
Metatheorem, 199
Minimal $u - v$ strength reducing set, 81
Minimum degree, 3
Minimum strong degree, 76
Monomorphism, 183
Monophonic block, 127
Monophonic convex set, 125
Mulder's interval, 161
Multigraph, 4

**N**
Neutrosophic fuzzy sets, 181
Node strength sequence, 78
Nonnegative-induced, 26
Non-separable graph, 7
Nonstandard analysis, 182
Nonstandard weighted graphs, 194

**O**
Optimal, 11
Optimal cycle, 16
Order, 2
Ordered field, 191
Ordered field of hyperreals, 191
Order homomorphism, 183
Order-theoretic isomorphism, 198

**P**
Parallel edges, 3
Partial block, 61
Partial bridge, 55
Partial cut vertex, 54
Partially strong gated, 171
Partial forest, 66
Partial trees, 66
Path extremal, 46
Pendant edge, 7
Pendant triangle, 13
Positive-induced, 25
Power graph, 206
Precisely weighted graphs, 71
Prefiber, 10
Pythagorean fuzzy sets, 181

**R**
Radius, 6
Regular, 4

**S**
Saturated graph, 146
Saturation, 146
Self-centered, 6
Semigroup, 206
Simple graph, 4
Spanning subgraph, 3
Standard part, 191
Star graph, 5
Strength, 187
Strength of a path, 53
Strength of connectedness, 54
Strength reducing set of vertices, 80, 81
Strong $u - v$ path, 58
Strong components, 174
Strong cycle, 58
Strong degree of a vertex, 75
Strong edge count, 141
Strong edges, 57
Strongest path, 54, 59
Strongest strong cycles, 89
Strongest strong distance, 156
Strong gate, 169
Strong geodesic distance, 156
Strongly connected, 174
Strongly disconnected, 174
Strongly gated, 169
Strongly restricted, 171
Strong paths, 57, 59
Strong sequence, 133
Strong size, 82
Strong strength reducing set of edges, 85
Strong vertex count, 141
Strong weight, 85, 99
Strong weighted interval, 162
Subdirect product theorem, 198
Subgraph, 3
Supergraph, 3

**T**
Totally weighted, 11
Trail, 4
Transfer principle, 182
Tree, 7
Triangle-free, 28
Trigraph, 37
Tritree, 22, 37

**U**
Undirected power graph, 197
Unsaturated graph, 146
Unweighted graph, 11

**V**
Vertex-connectivity, 8
Vertex-cut, 8
Vertex-strength sequence, 78
Vertex weighted graph, 19
Vertex-weighting of a graph, 25

**W**
Walk, 4
Weight of a path, 161
Weight of a subgraph, 11
Weighted $u - v$ geodesic, 105, 117
Weighted center, 110
Weighted complement, 116
Weighted degree, 11
Weighted diameter, 109
Weighted diametral vertices, 109
Weighted digraphs, 47
Weighted distance, 105
Weighted distance between two vertices, 105
Weighted distance matrix, 113
Weighted eccentricity, 108
Weighted eccentric vertex, 109
Weighted edge connectivity, 86
Weighted geodetic block, 119
Weighted geodetic boundary vertex, 123
Weighted geodetic closed interval, 117

Weighted geodetic closure, 118
Weighted geodetic convex, 118
Weighted geodetic interior vertex, 123
Weighted graph, 1, 11, 179
Weighted monophonic, 125
Weighted monophonic $u - v$ path, 125
Weighted monophonic boundary vertex, 130
Weighted monophonic closed interval, 125
Weighted monophonic closure, 125
Weighted monophonic convex, 125
Weighted monophonic interior vertex, 130
Weighted periphery, 110
Weighted radius, 109
Weighted self centered graph, 111
Weighted union, 22
Weighted vertex connectivity, 85
Widest path, 54

**Z**
Zero sequence, 139

The manufacturer's authorised representative in the EU is Springer
Nature Customer Service Centre GmbH, Europaplatz 3, 69115 Heidelberg,
Germany. If you have any concerns regarding our products, please
contact ProductSafety@springernature.com

Printed and bound by CPI Group (UK) Ltd, Croydon, CR0 4YY
01/10/2026
02225853-0002